ENVIRONMENTAL IMPACT STATEMENT GLOSSARY

A Reference Source for
EIS Writers, Reviewers, and Citizens

Edited by

Marc Landy

IFI/PLENUM • NEW YORK-WASHINGTON-LONDON

Library of Congress Cataloging in Publication Data

Main entry under title:

Environmental impact statement glossary.

Includes index.
1. Environmental impact statements — Dictionaries. 2. English language — Glossaries, vocabularies, etc. I. Landy, Marc.
TD194.6.E63   333.7   79-19586
ISBN 0-306-65185-8

©1979 IFI/Plenum Data Company
A Division of Plenum Publishing Corporation
227 West 17th Street, New York, N.Y. 10011

Printed in the United States of America

£47·25

ENVIRONMENTAL IMPACT STATEMENT GLOSSARY

A Reference Source for
EIS Writers, Reviewers, and Citizens

To HAROLD and JANICE

PREFACE

This reference book sets out to provide a useful glossary
to writers, reviewers and citizens interested in the EIS process.
Over the last decade, environmental impact statements have developed a rich, but sometimes confusing vocabulary. The purpose of
this book is to help people understand and communicate more effectively by presenting, analyzing and comparing terminology used by
various EIS organizations. This book is not a technical glossary
or the standard glossary, but rather the first complete effort to
assemble and examine the translation of complex technical EIS language into a vocabulary aimed directly at the lay reader.

Two major factors are responsible for the confusion over
EIS terminology: the diversity of professional vocabularies and
multiple interpretations of federal regulations. Due to the diversity and breadth of professional jargons needed to describe the
wide range of EIS subject matter, both quantitatively and qualita-
tively, EIS terminology has become a complex amalgam of professional languages. The second factor of multiple interpretations of
federal regulations contributes to the confusion over EIS procedural terminology--providing a disjointed vocabulary charged with individual interpretation. As a result of this complexity and confusion, the need for a uniform or standard terminology has been
advocated by many organizations, ranging from the Council on Environmental Quality on the federal level, down to city planning
departments on the local level. This book, the <u>Environmental Impact Statement Glossary</u>, is the first step in the standardization
plan proposed for EIS terminology. A more detailed discussion
follows in the section of this book entitled, "Four Year Standardization Plan: 1978-1982".

All terms and definitions were prepared by members of the
EIS community. A written survey of four hundred organizations,
agencies and subagencies was made in the summer of 1978. The response to the survey was excellent, providing a wide range of terms
from organizational circulars, environmental impact statements, executive orders, handbooks, laws and regulations. This book contains approximately 4,000 entries developed from over 85 documents.

This glossary was written in the hope that it will help people understand and communicate more effectively--ultimately improving the quality of EIS decision making. I realize the limitations of this initial edition and accept the challenge of providing a more useful and effective reference tool in future editions.

Many people have contributed their time and energy to this project. My thanks go to Oscar Graham and Rick Thacker for reviewing portions of the manuscript. The publisher of this book, Plenum Publishing Corporation, have been most helpful with particular appreciation to Frank Columbus, Assistant Vice President; Stephen Dyer, Assistant Managing Editor; and Georgia Prince, Director of Special Projects. Without the cooperation of EIS organizations this book would not be possible. Special thanks goes to the following individuals and organizations for their contributions. American Medical Association, Ward Duel, Assistant Director, Department of Environmental, Public and Occupational Health; American Society of Civil Engineers, Irving Amron, Editor of Information Servies; Connecticut Lung Association, Karen J. Matczak, Program Manager; Four Corners Regional Commission, Stuart H. Huntington, Special Assistant to the Federal Cochairman; Great Lakes Basin Commission, Nancy W. Huang, Information Specialist; Hawaii Office of Environmental Quality Control, Office of the Govenor; Indiana State Planning Services Agency; Massachusetts Executive Office of Environmental Affairs, Raymond E. Gherlardi, Associate Planner; Michigan Department of Agriculture, Emmanuel T. Nierop, Environmental Advisor; Mississippi Marine Resources Council, Philip L. Lewis, Marine Program Administrator; National Institute of Environmental Health Sciences; New Jersey Department of Environmental Protection, Division of Marine Services, Office of Coastal Zone Management, David N. Kinsey, Chief; Oregon Department of Environmental Quality, Air Quality Division; Pacific Northwest River Basins Commission, David L. Ricks, Resource Planner; Texas State Soil and Water Conservation Board, A.C. Spencer, Executive Director; U.S. Department of the Interior, Bureau of Reclamation; U.S. General Services Administration; U.S. Water Resources Council; Virginia Council on the Environment; and the Virginia Department of Conservation and Economic Development.

CONTENTS

EXPLANATORY NOTES

AUDIENCE

   The environmental impact statement is an informational document used by different kinds of people with different kinds of informational needs. Below is a brief discussion of how the writer,
reviewer, citizen and student might benefit from the use of this
glossary.

   Writers. A basic sourcebook for translating complex scientific analysis into a comprehensible text for the lay reader.
Offers easy access to specific themes as well as random terms. A
reference tool for clarification, comparison and translation of
critical terms. Provides a source for the construction of glossaries tailored to the specific needs of a particular environmental
impact statement.

   Reviewers. Easy access to both thematic and random terms.
Offers convenient grouping of pertinent themes to reviewers with
specific reviewing responsibilities. An opportunity to probe the
limits of important terms and issues that may apply to a specific
environmental impact statement.

   Citizens. Offers an introduction to the basic concepts and
terminology associated with the EIS process. Provides an opportunity for the review of an EIS thematically as well as step-by-step.

   Students. Provides a supplement to standard textbooks in
the areas of conservation, ecology, environmental impact statements, environmental studies, geography and planning. Presents
the basic concepts and terminology currently used in the EIS process. Presents the student with a reference tool for unfamiliar
terms found in professional journals.

GLOSSARY ORGANIZATION AND STRUCTURE

The glossary is divided into sixteen environmental impact
statement themes. The thematic arrangement has been chosen because: (1) it provides the user with an opportunity to search
for terms in two ways: thematically by using the thematic glossaries and alphabetically by using the alphabetical index; (2)
the thematic presentation more accurately presents the way in
which EIS analysis and review is conducted--theme by theme; (3)
by placing terms together, the strengths and weaknesses of individual themes are exposed; and (4) facilitates the starting point
for a standardization plan. The organization and structure of the
thematic glossaries are portrayed in Table 1.

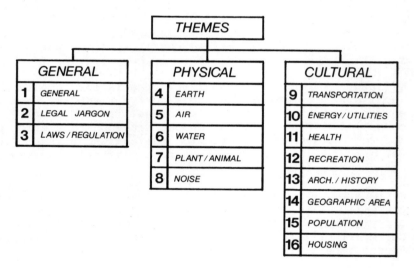

Table 1.   Thematic Glossaries

Themes 1-13 were developed from over eighty-five documents, while themes 14-16 are a reprint of the U.S. Department
of Commerce, Bureau of the Census, "Census Users' Dictionary."
Much of the data concerning geographic, population and housing
themes are taken from the Bureau of the Census data and it seems
appropriate to include the "Census Users' Dictionary" in its original form.

ENTRY ORGANIZATION AND STRUCTURE

These general rules have been followed in the preparation
of the glossary: (1) alphabetization is letter-by-letter; (2) compound terms have been listed to the first word, i.e., Environmental
Impact--not Impact, Environmental; and (3) an effort has been made
to maintain the original entry in its complete form - however,
in some cases an entry has been shortened or technical passages
omitted.

Explanatory Chart

The organization and structure of entries are portrayed in
Table 2.

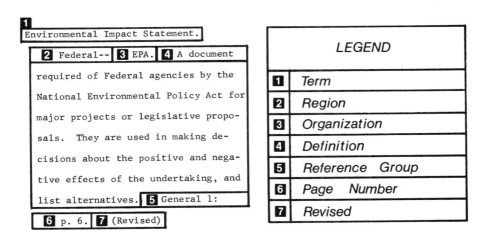

Table 2. Explanatory Chart

Key to Explanatory Chart

(1) Term. All words capitalized followed by a period.

(2) Region. The region of influence covered by the organization.
     In the case of multiple definitions, the ordering of entries
     is from the largest region to the smallest region. The order-
     ing of regions is as follows: Federal, National, Regional,
     State and City.

(3)   <u>Organization</u>. A brief description or abbreviation of the
      type of organization defining the entry. In the case of mul-
      tiple definitions, the ordering of entries is alphabetical.
      For example, State--Board is followed by State--Commission.

(4)   <u>Definition</u>. The definition of a term which may be followed
      by multiple definitions.

(5)   <u>Reference Group</u>. Ties the entry to the document found in the
      bibliography. For example, Air 2, refers the user to the
      "Air" reference group in the bibliography--the second entry.

(6)   <u>Page Number</u>. The page number where the entry can be found
      in the document.

(7)   <u>(Revised)</u>. A notation that informs the user that the entry
      has been shortened or altered.

ENVIRONMENTAL IMPACT STATEMENT INSTITUTE

        The EIS Institute, a small research organization, is ded-
icated to the study of the EIS process; coordination of information between EIS practitioners; and the publication of information
to both the EIS community and the interested public. Founded in
1979, major areas of interest include the standardization of EIS
terminology and content. Mailing address: 8928 19th Street West,
Tacoma, Washington 98466.

FOUR YEAR STANDARDIZATION PLAN FOR EIS TERMINOLOGY: 1978-1982

        The purpose of this plan is to: (1) develop a uniform EIS
terminology based on input from EIS practitioners and edited by an
editorial board, and (2) compile, edit and publish the Standard
Environmental Impact Statement Glossary by 1982.   The procedure
involved in standardizing EIS terminology is a complex problem
concerning both procedural and analytical terms.   Procedural terminology, which describes governmental regulations and processes,
will be the most difficult type, due to legal and regulatory constraints imposed by governmental agencies.   Analytical terminology,
which consists mainly of professional vocabularies, are more easily
standardized, with many professional vocabularies currently using
standard technical glossaries.   It is important to note that the
standardization plan is designed to compile EIS terminology that
translates scientific and procedural terminology into a vocabulary
aimed at the lay reader.

        The standardization plan is built around six major goals,
portrayed in Table 3.

a	b	c	d				
			Y e a r				
GOAL	OBJECTIVE	STRATEGY	78	79	80	81	+
1	ESTABLISH CURRENT USEAGE	SURVEY EIS COMMUNITY	█				
2	PUBLISH INTERIM GLOSSARY	EIS GLOSSARY		█			
3	RESPONSE TO EIS GLOSSARY	COMMENT FORMS		█	█		
4	EDITORIAL BOARD	EIS ACADEMY			█	█	
5	STANDARD TERMINOLOGY	STANDARD EIS GLOSSARY				█	
6	CONTINUOUS REVIEW	COMMENT FORMS					█

Table 3.   Summary of Standardization Plan

Goal #1
Establish Current Useage
(1978)

To develop current useage used by EIS
practitioners, a survey of 400 EIS organizations, agencies and subagencies
was made in 1978. The information
gathered in this survey forms the bulk
of this book and represents a reasonable cross section of current useage.
This goal has been accomplished by the
publication of this book.

Goal #2
Publish Interim Glossary
(1979)

The purpose of the interim glossary is
to: (1) provide a useful glossary to
the EIS community until a standard
glossary can be compiled, edited and
published, (2) point out the need for
a uniform terminology by illustrating
multiple definitions of some terms, and
(3) present EIS practitioners an opportunity to comment on terms found in the
interim glossary. This goal has been
accomplished with the publication of
this book.

Goal #3
EIS Practitioners Re-
sponse to the Interim
Glossary (1979-1980)

A one year reviewing period of the interim glossary will allow practitioners
an opportunity to comment on terms presented in the glossary. A format for
responses to the glossary are contained
at the end of this section entitled,
"Comment Form".

Goal #4
Establish EIS Academy
(1979-1980)

Each of the sixteen EIS themes will be
edited by an editor with expertise in
that field. Final responsibility and
selection of terms will be made by each
editor.

Goal #5
Publish Standard EIS
Glossary (1982)

The publication of the Standard EIS
Glossary will present a uniform terminology based on contributions of EIS
practitioners and edited by the EIS
Academy.

Goal #6
Continuous Review
(1981 and beyond)

To accommodate the evolutionary changes
in EIS analysis and review, practitioners will be encourage to update terms.

COMMENT FORM

The purpose of the Four Year Standardization Plan is to develop an EIS terminology based on information provided by EIS practitioners. Please respond to the following: (1) clarification, addition or deletion of terms found in the EIS Glossary, (2) reorganization of any or all of the EIS Glossary, (3) addition or deletion of any of the sixteen EIS themes, (4) submit formal or informal glossaries, laws or regulations used by EIS practitioners, (5) interest in the EIS Academy, and (6) other. Please mail information to: EIS Institute, 8928 19th Street West, Tacoma, Washington 98466.

Term   Page Number

Comment

Name

Organization

Address

General Glossaries

1. GENERAL

Index

Cartographic Feature
Census
Census Of Population And
      Housing
City Planning
Classification
Coffin
Comensurable Values
Community Boundaries
Community Development
Community Profile
Compatibility Values
Compliance Schedule
Comprehensive Plan
Comprehensive Planning Program
Computer Graphics
Conceptual Framework
Conceptual Plan
Concern
Condemnation
Conservation
Conservation District
Conservation Easements
Conservation Plan Map
Conservation Standards
Consumptive Use
Contingency Plan
Contour
Contour Map
Control Equipment
Correlation
Cost
Cost Benefit Analysis
Cost Effective Analysis
Cost Revenue Analysis
Criteria
Critical
Critical Resources
Culture
Cutie-Pie
Data
Data Base
Datum
DDT
Decision Theory
Demography
Density-Dependent Factor
Density-Independent Factor
Design Life
Design Solution

Development
Development Plan
Development Potentials
Diagrammatic Representation
Direct Impact
Disturbance
Diurnal
Draft Environmental Impact
      Statement (DEIS)
Duration Curve
Easement
Economic Growth
Effectiveness Index
Efficiency
Eminent Domain
Emission Standard
Engineering Solution
Environment
Environmental Analysis Report
      (EAR)
Environmental Components
Environmental Criteria
Environmental Design Arts
Environmental Effect
Environmental Fragility
Environmental Impact Analysis
Environmental Impact Assessment
Environmental Impact Report
      (EIR)
Environmental Impact Statement
      (EIS)
Environmental Inventory
Environmental Monitoring
Environmental Parameters
Environmental Protection Agency
      (EPA)
Environmental Quality
Environmental Stress
Environmental Surveillance
Environs
Exceedence Frequency
Exogenous Independent Variable
Fair Market Value
Feasibility Study
Feedback
Final Environmental Impact
      Statement (FEIS)
Fixed Assests
Fixed Capital
Fluorides

Forecast
Fragile Area
Framework
Framework Plan
Frequency Curve
Functional Plans
Functional Planning
Functional Relationships
Functional Relationships Dia-
    gram
Game Theory
Geographical Setting
Goal
Gradient
Graphic Scale
Gravity Model
Guideline
Half-Life
Hazzardous Waste
Heavy Metals
High Density Polyethylene
Highest And Best Use
Holistic
Housing Market
Human Resources Development
    Systems
Hypothesis
Implementation Plan
Improvement
Independent Measure
Index Number
Indirect Effect
Infiltration
Infrastructure
Input-Output Analysis
Inputs
Interdisciplinary Approach
Interdisciplinary Team
Interpretation
Intuitive Design
Inventory
Issue
Joint Costs
Joint Use Corridor
Labor Market Area
Labor Shed
Land And Water System
Land Development
Landscape
Landscape Treatment

Land Use Allocation
Land Use Constraints
Land Use Survey
Land Use Type
Line Plot Survey
Local Economic Effects
Location Theory
Long-Range Planning
Long-Term Impacts
Lot
Major Land Use
Major Service Area
Management Plan
Management Unit
Man-Made Environment
Man-Made Resources
Market Appeal
Market Capture Or Penetration
    Rate
Market/Supply Orientation
Market Value
Masking
Master Plan
Maximum Practicability
Mesoscale Analysis
Mid-Range Planning
Mitigate
Mixed Use
Model
Monitoring
Monitoring Program
Multifunctional
Multiple Use Planning
Municipal Bond
Municipality
Natural Pollution
Natural Resources
Need
Neighborhood
Neoteric Area
Net Acre
Nonconforming Use
Nonconsumptive Use
Nonpoint Pollution
Non-Reactive Pollutant
Nutrients
Off-Road Vehicles (ORV)
On-Site Construction
Open Space
Operational Definition

Operational Map
Operational Plans
Opportunity Cost
Ordinal Value
Pandemic
Paper Plan
Parameter
Parcel
Parent Plan
Perceptual Variables
Performance Standard Zoning
Permit
Permeable Material
Photo Map
Physical Barrier
Physical Planning
Plan
Plan Controls
Plan Element
Planned-Unit Development (PUD)
Planning
Planning Horizon
Planning Level
Planning Period
Planning Process
Planning-Programming-Budgeting
    System (PPBS)
Planning Region
Plan Selection
Plat
Platting
Plot
Point Source
Policy Plan
Political Jurisdiction
Pollutant
Pollution
Polychlorinated Biphenyls
    (PCB's)
Polyvinyl Chloride
Pre-Existing Use
Preservation
Pressure Group
Primary Impacts
Pristine State
Probabilistic Model
Process
Process Planning
Process Weight
Profile View

Program
Programming
Property Improvement
Property Tax
Prototype
Psychological Barrier
Public Domain
Quadrat
Rating Curve
Real Property Tax Base
Reasonably Available Control
    Technology (RACT)
Receptor
Redevelopment
Reevaluation
Regional Plan
Regional Planning Agency (RPA)
Regression Analysis
Regulatory Agencies
Rehabilitation
Related Land
Relational Study
Remote Sensing
Renewable Resource
Resource
Resource Allocation
Resource Base Analysis
Resource Inventory
Revenue-Cost Analysis
Right-Of-Way Acquisition
Scale
Scenario
Scenic Area
Secondary Impacts
Secondary Pollutants
Sensitivity Analysis
Separable Cost
Service Area
Short-Range Planning
Significant
Significant Deterioration
Simulation
Site
Site Analysis
Site Plan
Site Specific
Slag
Slurry
Social Analysis
Social Equity

Sorption
Special Interest Group
Special Management Zone
Specific Cost
Sprawl
Stagnation
Standard Deviation
Standard Metropolitan Statistical Area (SMSA)
Stationary Source
Statistics
Strategic Planning
Strategy
Stratification
Study Area
Subdivision
Subjective
Substantial
Substantive Comment
Surrounding Region
Surveillance System
Survey
Symbolic Resources Sites

Synergism
Synoptic Model
Systematic Sample
Systems Evaluation
Systems Planning Or Systems Analysis
Theory
Time-Contour Map
Trade Area
Trade-Off Analysis
Transect
Turbidity
Urban Renewal
Use
Variance
Visual Character
Visual Landscape Character
Visual Sensitivity Level
Visual Variety Class
Zonation
Zone
Zoning Ordinance

Terms

Abatement. Federal--Forest Service. The reduction in degree or
     intensity of pollution. General 1: p. 1.

Abiotic. Federal--Forest Service. The nonliving, material (as
     opposed to conceptual) components of the environment such
     as air, rocks, soil (in general), water, coal, peat, plant
     litter, etc. General 15: p. 23.

Abiotic Community. Federal--Forest Service. A broad class of
     resource products which includes all material yields of
     commercial value whose initial condition was that of a non-
     living substance--such as water, steam, minerals, sand,
     gravel, petroleum, and natural gas. General 15: p. 23.

Absorption. Federal--EPA. The penetration of one substance into
     or through another. General 1: p. 1.

     Federal--Interior. The taking in of fluids or other sub-
     stances by cells or tissues. General 3: p. 1.

Accessibility. Federal--Forest Service. The relative ease or
     difficulty of getting to or from someplace. For example,
     a potential land use evaluation procedure may locate areas
     with characteristics highly favorable for use but which are
     separated from the nearest road by some impediment to travel
     such as bluffs or a river. Such an area would be character-
     ized as having low accessibility. General 15: p. 23.

Acquisition Of Land. Federal--Forest Service. Purchase of full
     land ownership rights. General 15: p. 23.

Activity Planning. Federal--Forest Service. U.S. Forest Service
     useage. Under the "multiple use planning" system "activ-
     ity planning" referred to the functional plans for such
     items as timber management, recreation, transportation and
     fire protection. General 15: p. 24.

Address Coding Guide (ACG). Federal--DOT. A geocoding technique
     developed by the U.S. Census Bureau for recording geographic
     information. A computerized street directory containing one
     record for each block face. Transportation 3: p. 11.

Adhesion. Federal--EPA. Molecular attraction which holds the
     surfaces of two substances in contact, such as water and
     rock particles. General 1: p. 1.

Administrative Jurisdiction. Federal--DOT. A geographic area
     used as the basis for performing one or more administrative
     functions; e.g., county, town, sewer district, water dis-
     trict, school district, police precinct. Transportation 3:
     p. 12.

Adsorption. Federal--Army. The adherence of substances to the
     surfaces of bodies with which they are in contact, but not
     in chemical combination. General 9: p. 1.

     Federal--EPA. The attachment of the molecules of a liquid
     or gaseous substance to the surface of a solid. General 1:
     p. 1.

Adulterants. Federal--EPA. Chemical impurities or substances
     that by law do not belong in a food, plant, animal, or
     pesticide formulation. General 1: p. 1.

Adversary Procedure. Federal--Forest Service. This procedure is
     sometimes advocated as the best method for resolving (in
     or out of court) land use planning issues having or invol-
     ving readily identified, opposing, antagonistic parties
     or interests. Its theoretical strength lies in the strong
     presentation and contrasting of both pro and con interpre-
     tations of what the real facts are for the judging parties.
     Its theoretical weakness lies in the distortions of the rel-
     ative merits of the opposing cases that may result from
     unequal (1) ability to finance presentations and defenses,
     (2) access to information, (3) theatrical and intellectual
     abilities in making presentations and cross examinations,
     and, (4) limits the ability of judging parties to be truly
     objective. General 15: p. 25.

Adverse Land Use. Federal--Forest Service. Not in conformity with
     the planning or social requirements of the community.
     General 15: p. 25.

Advocacy Planning. Federal--Forest Service. More generally used
     in the sense of planners working directly with socio-econ-
     omically defined special interest groups to translate their
     aspirations for a living environment into formal planning
     goals and a set of procedures for attaining those goals.
     Presumably conflicts between different socioeconomic
     groups values are to be resolved by adversary procedures
     during which the arguments for and against all proposals
     are presented. General 15: p. 25.

Aerial Photograph. National--Soil Conservation. A photograph of
     the earth's surface taken from airborne equipment, sometimes

called aerial photo or air photograph. Earth 4: p. 5g.

Aerobic. <u>Federal--EPA</u>. Life or processes that depend on the
       presence of oxygen. General 1: p. 1.

       <u>Federal--GSA</u>. Condition characterized by the presence of
       dissolved oxygen in the aquatic environment; also the life
       or processes occuring only in the presence of oxygen. With
       few exceptions, all life requires aerobic conditions.
       General 2: p. IX-17.

       <u>State--Water Resources</u>. Requiring, or not destroyed by,
       the presence of free elemental oxygen. Energy/Utility 2:
       Appendix E.

Aerosol. <u>Federal--EPA</u>. A suspension of liquid or solid particles
       in a gas. General 1: p. 1.

Aesthetics. <u>Federal--Army</u>. The measurement of natural and man-
       made elements relative to a certain class of community
       values representative of the notion of beauty. Recreation
       2: p. A-29.

       <u>Federal--Forest Service</u>. Evaluations and considerations
       concerned with the sensory quality of resources (sight,
       sound, smell, taste and touch) and especially with respect
       to judgement about their pleasurable qualities. General
       15: p. 25.

       <u>Federal--GSA</u>. That which people find beautiful or attrac-
       tive. The quality of being aesthetic is not the opposite
       of the qualities of "practicality" or "reality", but rather
       another aspect or way of experiencing the same real world
       phenomena. Thus, blue skies, uncontaminated water, and
       uncluttered urban landscapes all have aesthetic impact,
       because they imply health, pleasure and security. Legal
       Jargon 18: Appendix B-1.

       <u>National--Soil Conservati on</u>. The appeal or beauty of ob-
       jects, animals, plants, scenes, natural or improved areas
       to the viewer and his appreciation of such items. Earth 4:
       p. 5g.

Agricultural Pollution. <u>Federal--EPA</u>. The liquid or solid wastes
       from farming, including: runoff from pesticides, fert-
       ilizers, and feedlots; erosion and dust from plowing; animal
       manuare and carcasses, crop residues, and debris. General
       1: p. 1.

Air Rights. <u>Federal--DOT</u>. The right to the use of air space
    over property owned by another. Through the erection of
    a platform over the existing use, additional "land" can,
    in effect, be created. Transportation 3: p. 12.

Algicide. <u>State--Water Resources</u>. Any substance or chemical
    applied to kill or control algal growth. Energy/Utility
    2: Appendix E.

Allopatric. <u>Federal--Interior</u>. Having separate and mutually ex-
    clusive areas of geographical distribution. General 3:
    p. 4.

Alternate Costs. <u>National--Soil Conservation</u>. The cost of pro-
    viding the same or equivalent benefits from the most likely
    economically feasible alternative source available in the
    area to be served. Earth 2: p. 16g.

Alternative. <u>Federal--Forest Service</u>. The different means by
    which objectives or goals can be attained. They need not
    be obvious substitutes for one another or perform the same
    specific function. General 15: p. 27.

Ambient. <u>Federal--Forest Service</u>. Referring to the quality of
    some specific environmental factor--such as the "ambient"
    temperature or "ambient" air pollution levels. General 15:
    p. 27.

    <u>Federal--Interior</u>. The natural conditions (or environment)
    at a given place and time. General 4: Glossary.

Ambient Standards. <u>Federal--Interior</u>. Maximum allowable levels
    of specific polluting materials permitted by State, Federal,
    or local laws. General 3: p. 5.

Ameliorate. <u>State--Transportation</u>. To make better or more toler-
    able; to improve. Transportation 1: p. vi.

Anaerobic. <u>Federal--EPA</u>. Life or processes that can occur with-
    out free oxygen. General 1: p. 1.

    <u>Federal--GSA</u>. Condition characterized by the absence of
    dissolved oxygen; also the life or processes occuring in
    the absence of oxygen. Several bacteria live in anaerobic
    conditions. General 2: p. IX-17.

Analysis Zone. <u>Federal--DOT</u>. A discrete geographical subdivision
    into which a larger given study area is divided for purposes
    of finegrain analysis of phenomena occuring within and be-
    tween zones. A traffic zone, for example, has been defined

as a portion of the study area, delineated as such for
particular land use and traffic analysis purposes. Transportation 3: p. 12.

Annexation. Federal--DOT. The process by which a municipality
absorbs surrounding land and brings it under its jurisdiction. Usually of unincorporated territory. Transportation
3: p. 13.

Anticipatory Planning. Federal--Forest Service. Any planning
approach which operates by attempting to forsee potential
issues and develop solutions to them before they become
real, overt problems. General 15: p. 155.

Anti-Degradation Clause. Federal--EPA. Part of air quality and
water quality laws that prohibits deterioration where
pollution levels are within the legal limit. General 1:
p. 2.

Federal--GSA. A provision in air quality and water quality
laws that prohibits deterioration of air and water quality
in areas where the pollution levels are presently below
those allowed. Legal Jargon 18: Appendix B-1.

Appropriation. Federal--Interior. An amount set aside by the
Congress or other legislative body to be expended for a
particular object or objects. Recreation 3: p. 4.

Appropriative Right. State--Water Resources. The right of beneficial use. Energy/Utility 2: Appendix E.

Areal Analysis. Federal--DOT. The study of specific geographic
areas or zones and their related characteristics. Involves
the grouping of data within a specified geographic area for
the purpose of analyses pertaining to events, facilities,
and/or people. Transportation 3: p. 14.

Area Reconnaissance. Federal--Forest Service. Engineering and
surveying usage. The projection of feasible road routes
on an area topographic map and their delineation on aerial
photographs. General 15: p. 30.

Area Source. Federal--GSA. Pollutants generated from multiple
sources over a broad geographical area, ranging from one
block to an entire city. General 2: p. IV-21.

Assemblage. Federal--DOT. The merging of separate properties
into a contiguous unit generally for new buildings or
projects. Accomplished by purchase of the individual

properties and their transfer into single ownership.
Transportation 3: p. 13.

Assimilative Capacity. Federal--Commerce. The amount of adverse
impacts (pollutants) that a water body or land area can
absorb and neutralize before it begins to display a significant reduction in biological diversity, chemical, and/
or physical quality. General 6: p. 310.

Attrition. Federal--EPA. Wearing away or grinding down a substance by friction. A contributing factor in air pollution, as with dust. General 1: p. 2.

Azimuth. National--Soil Conservation. The direction of a line
given as an angle, measured clockwise from a reference
direction, usually true north. Earth 4: p. 8g.

BACT. State--Committee. Best available control technology.
Air 3: no page number.

Bacteria. Federal--EPA. Single-celled microorganisms that lack
chlorophyll. Some cause disease, others aid in pollution
control by breaking down organic matter in air and water.
General 1: p. 2.

Baffle. Federal--EPA. A deflector that changes the direction of
flow or velocity or water, sewage, or particulate matter.
Also used to deaden sound. General 1: p. 2.

Base Case. Federal--ERDA. The activities as presently planned;
in environmental impact considerations and cost/benefit
analyses, the detailed plan of actions for which impact
is assessed and to which reasonable alternatives will be
compared. General 7: p. g-1.

Baseline Profile. Federal--Army. Used for a complete survey of
the environmental conditions and organisms existing in a
region prior to unnatural disturbances. General 9: p. 5.

Base Map. Federal--Forest Service. A map showing certain fundamental information, on which is compiled additional data
of specialized nature. General 15: p. 32.

National--Soil Conservation. A map showing certain basic
data to which other information may be added. Earth 4:
p. 8g.

Base Period. National--Soil Conservation. A period of time from
which comparisons of other time periods are made, normally

used with reference to price, population, production, or
other statistics. Earth 4: p. 8g.

Basic Data. State--Water Resources. Records of observations
    and measurements of physical facts, occurrences, and con-
    ditions, as they have occurred, excluding any material
    information developed by means of computation or estimate.
    In the strictest sense, basic data include only the record-
    ed notes of observations and measurements, although in
    general use it is taken to include computations or esti-
    mates necessary to present a clear statement of facts,
    occurrences, and conditions. Energy/Utility 2: Appendix
    E.

Bearing Weight. National--Soil Conservation. The maximum load
    that a material can support before failing. Earth 4:
    p. 9g.

Benefit. Federal--DOT. Result of an action that is seen as
    being useful, profitable, positive or promotive of human
    welfare. Transportation 3: p. 16.

Benefit-Cost Ratio. Federal--Interior. An economic indicator of
    efficiency, computed by dividing benefits by costs. Rec-
    reation 3: p. 4.

Bioaccumulation. Federal--Interior. The uptake of substances
    from the environment, other than food. Generally, the up-
    take of environmental pollutants. General 3: p. 12.

Bioassy. Federal--EPA. Using living organisms to measure the
    effect of a substance, factor, or condition. General 1:
    p. 2.

Biodegradable. Federal--Army. Can be broken down to simple in-
    organic substances by the action of decay organisms.
    General 9: p. 6.

    Federal--EPA. Any substance that decomposes quickly through
    the action of microorganisms. General 1: p. 2.

Biological Control. Federal--EPA. Using means other than chemical
    to control pests, such as predatory organisms, sterilization
    or inhibiting hormones. General 1: p. 2.

Biosphere. Federal--EPA. The portion of Earth and its atmosphere
    that can support life. General 1: p. 3.

Block Diagram. Federal--Forest Service. A drawn illustration of
    a piece of land (or water) in three dimensions which

displays the land's surface appearance bounded around the exposed sides by a cutaway representation of the underground (underwater) conditions at those edges. Commonly the land surface is tilted toward the viewer and image size is not reduced with distance from the viewer--as would truly occur in accordance with the rules of perspective. The end result is a pictoral representation which appears as if a block of land (or water) had been cut out of its true surroundings and lifted up so as to reveal both its surface and underground characterisitcs. General 15: p. 37.

Bond Issue. Federal--Interior. A borrowing technique, often used by governmental units to finance capital improvements. Recreation 3: p. 5.

Botanical Pesticide. Federal--EPA. A plant produced chemical used to control pests; for example nicotine or strychnine. General 1: p. 3.

Cadastral Survey. Federal--DOT. A survey relating to land boundaries and subdivisions, made to create units suitable for transfer or to define the limitations of title. Transportation 3: p. 18.

Cadmium. Federal--EPA. A heavy metal element that accumulates in the environment. General 1: p. 3.

Calcareous. Federal--Interior. Having sufficient accumulation of calcium carbonate to produce a pH of over 7.0. General 3: p. 16.

Capital Expenditure. Federal--DOT. An expenditure intended to benefit future periods, in contrast to a revenue expenditure, which benefits a current period; an addition to a capital asset. Transportation 3: p. 18.

Capital Improvement. Federal--DOT. Any substantial physical facility having a potentially long period of usefulness. For example, streets, traffic control facilities, water and sewage treatment facilities, and land acquisition and improvement. Also, any major nonrecurring expenditure of government. Transportation 3: p. 19.

Cartographic Feature. National--Soil Conservation. The natural or cultural objects shown on a map or chart. Earth 4: p. 11g.

Census. <u>National--Soil Conservation</u>. A complete counting, with
classification of a population or group at a point in time,
as regards to some well defined characteristic(s), e.g.,
of traffic on particular roads; usually has a governmental
and economic social connotation; often used in wildlife
surveys. Earth 4: p. 11g.

Census Of Population And Housing. <u>National--Soil Conservation</u>.
A census taken by the Bureau of Census every 10 years.
It includes number of people and housing units and various
population and housing characteristics. Earth 4: p. 11g.

City Planning. <u>Federal--DOT</u>. As traditionally conceived, a means
for systematically anticipating and achieving adjustment in
the physical environment of a city consistent with social
and economic goals and sound principles of civic design.
More recent concepts of planning extend to include complete
consideration of social values, optimization of the human
landscape, and the development of advocacy planning wherein the people to be affected are consulted. City planning
is usually effectuated by the lawful and reasonable manipulation of the regulatory, eminent-domain, and tax powers.
Transportation 3: p. 20.

Classification. <u>Federal--Forest Service</u>. The forming, sorting,
apportioning, grouping, or dividing of objects into classes
to form an ordered arrangement of items having a defined
range of characteristics. Classification systems may be
taxonomic, mathematical, or other types, depending upon
the purpose to be served. General 15: p. 43.

Coffin. <u>Federal--EPA</u>. A thick-walled container (usually lead)
used for transporting radioactive materials. General 1:
p. 4.

Commensurable Values. <u>Federal--Forest Service</u>. Resource yields
which can be objectively compared because their values can
be expressed quantitatively on the same measurement scale
(such as in dollars, pounds, or cubic feet). General 15:
p. 46.

Community Boundaries. <u>Federal--DOT</u>. The edges of the spatial
area that define a community unit. They may be physical
(i.e., a river or highway) or be the limits of activity or
shared behavior and perceptions. Transportation 3: p. 21.

Community Development. Federal--DOT. The physical implementation
of plans for the use of land in a jurisdiction; i.e., construction of residential, commercial and industrial buildings on land set aside for those purposes and the supporting

urban infrastructure, e.g., streets, utilities, and other
community facilities. Transportation 3: p. 21.

Community Profile. _Federal--DOT_. An outline description of a
community unit using certain key descriptors of that unit.
This might include demographic, income, employment, land
use distribution information and similar data. Transportation 3: p. 21.

Compatibility Values. _Federal--GSA_. Public interest considerations of health, safety, convenience, economy, and amen-
ity. General 2: p. X-12.

Compliance Schedule. _State--Committee_. A legally enforceable
detailed timetable of actions to be taken by a pollution
source to bring it into accord with the regulations.
Air 3: no page number.

Comprehensive Plan. _Federal--DOT_. An official public document
adopted by a local government as a policy guide to decisions about the physical, social and economic development of the community. Transportation 3: p. 22.

   Federal--Interior. A plan for development of a geographical region or area including policies, goals and interrelated plans for private and public land use, transportation systems, community facilities and all other elements
   and features that represent decisions of the people affected. Recreation 3: p. 7.

   National--Soil Conservation. A report from a government
   planning agency that describes how its area of jurisdiction
   should be developed, expressing both policies and a coordinated plan for public and private land use, a transportation
   system, and public services, and facilities. Also called
   a comprehensive development plan, general plan, master
   plan. Earth 4: p. 14g.

Comprehensive Planning Program. _National--Soil Conservation_.
A continuing process which includes research on the conditions and trends in physical, social, and economic development; preparation and adoption of a comprehensive plan;
programming of capital improvements; and initiation of the
regulatory and administrative measures for implementation
and maintenance of the plan. Earth 4: p. 14g.

Computer Graphics. _Federal--Forest Service_. Visual displays of
information produced by an electronic computer. This includes both 'hard copy' (paper, film) and cathode-ray tube
(CRT) displays. Among the common computer graphics are

maps, graphs, still drawings, and motion pictures. General 15: p. 49.

Conceptual Framework. Federal--DOT. The structure of ideas used to guide the development and evaluate the desirability of a plan, program or set of proposals. Transportation 3: p. 22.

Conceptual Plan. Federal--Army. A single line diagrammatic layout of a proposed area/project which shows the functional relationships of the various facilities on a topographic background without regard to detailed design. Recreation 2: p. A-32.

Concern. Federal--Forest Service. (1) Public apprehension about whether some land use action might produce some more or less vaguely defined undesirable result, (2) A point of dispute which can only be vaguely defined or in which the potential undesirable results of a land use action are more speculative than documentable. When the relationship between the action and undesirable result can be sharply defined, a point of dispute is referred to as an issue rather than a concern. General 15: p. 49.

Condemnation. Federal--Interior. The due process of law by which property of a private owner is taken for public use; payment of just compensation is required. Recreation 3: p. 7.

Conservation. Federal--Army. Supervision, management and maintenance of natural resources. General 9: p. 11.

Federal--Army. The planned management and use of a natural resource in such a manner to prevent exploitation, destruction, or neglect of that resource by natural or man-made causes. Recreation 2: p. A-32.

Federal--DOT. (1) The protection of the resources (minerals, water, forests, fisheries, wildlife, etc.) of man's environment against depletion of waste, and the safeguarding of its beauty, (2) An urban renewal strategy that emphasizes the protection of an existing, viable neighborhood against the encroachment of blight. It usually includes improvement of the area's amenities, organization of the residents, and strict code enforcement. Transportation 3: p. 22.

Federal--EPA. The protection, improvement, and use of natural resources according to principles that will assure

their highest economic and social benefits. General 1:
p. 4.

Conservation District. National--Soil Conservaticn. A public
organization created under state law enabling as a special-
purpose district to develop and carry out a program of soil,
water, and related resource conservation, use, and development within its boundaries; usually a subdivision of state
government with a local governing body. Often called a
soil conservation district or a soil and water conservation
district. Earth 4: p. 15g.

Conservation Easements. Federal--Interior. Rights acquired over
private property for the purpose of preserving certain identified resources or features. Recreation 3: p. 7.

Conservation Plan Map. National--Soil Conservation. A portion of
an aerial photograph or a composite mosaic of two or more
aerial photos covering a farm or ranch with planned land
use, field boundaries, fences, etc., portrayed thereon.
Earth 4: p. 15g.

Conservation Standards. National--Soil Conservation. Standards
for various types of soils and land uses, including criteria, techniques, and methods for the control of erosion
and sediment resulting from land disturbing activities.
Earth 4: p. 15g.

Consumptive Use. Federal--Forest Service. Those uses of resources
that reduce the supply--such as logging and mining. General 15: p. 52.

Contingency Plan. Federal--Army. A predetermined course of action
to modify a plan in the event that recognized uncertainties
or infrequent occurrences come to pass. Recreation 2:
p. A-32.

Contour. National--Soil Conservation. A line drawn on a map
connecting points of the same elevation. Earth 4: p.
15g. (Revised).

Contour Map. Federal--Forest Service. A map which portrays the
elevational features of an area by joining points of equal
elevation with continuous lines, called contour lines. General 15: p. 53.

Control Equipment. State--Environmental Protection. Any device
or contrivance which prevents or reduces emissions. Legal
Jargon 8: p. 6.

Correlation. <u>National--Soil Conservation</u>. An expression indicating the degree of association or mutual relationship between the value of two attributes, not necessarily a casual
or dependent relationship. Earth 4: p. 15g.

Regional--River Basin. The process of establishing a relationship between two or more related variables. It is a
simple correlation if there is only one independent variable; multiple correlation if there is more than one independent variable. Water 5: p. 1012.

Cost. <u>Federal--DOT</u>. The outlay or expenditure made to achieve
an objective or the loss or penalty incurred in gaining
something. Transportation 3: p. 23.

<u>Federal--Forest Service</u>. The negative (adverse) effects.
Costs may be monetary, social, physical, or environmental
in nature. General 15: p. 53.

Cost Benefit Analysis. <u>Federal--DOT</u>. An analytic method designed
to evaluate alternative programs in terms of their potential
benefits and likely costs, and to aid decisionmakers in
choosing them. When applied in the environmental sciences
it ideally weighs the social, ecological, and aesthetic as
well as the economic factors, and takes into account the
indirect consequences of different courses of action.
Transportation 3: p. 23.

<u>Federal--GSA</u>. Analysis comparing quantity and quality of
alternative means for accomplishing a specific objective.
The alternatives are ranked according to the degree of
economic efficiency with which they achieve the specific
goal. General 2: p. VII-20.

Cost Effective Analysis. <u>Federal--Forest Service</u>. Cost effectiveness analysis compares alternatives in terms of their contribution to a goal by using costs and other effectiveness
criteria. General 15: p. 53.

Cost Revenue Analysis. <u>Federal--DOT</u>. An analysis designed to provide information on the governmental costs of supplying
public improvements and services to urban land areas as
these costs related to available monies to finance such improvements and services and the revenues that will be generated by the urban land area provided with these improvements and services. Transportation 3: p. 23.

Criteria. <u>Federal--EPA</u>. The standards EPA has established for
certain pollutants, which not only limit the concentration,

but also set a limit to the number of violations per year.
General 1: p. 4.

Federal--Forest Service. (1) Predetermined rules for ranking alternatives in order of desirability to facilitate
and expedite the decision-making process, (2) A rule, or
test, by which something can be judged. General 15: p. 55.

Critical. Federal--Commerce. A condition, measurement, or point
at which some quality, property, or phenomena suffers a
definite change. An essential component. General 6:
p. 311.

Critical Resources. Federal--GSA. Resources which are in extremely short supply in the nation as a whole, or in a
particular area where a proposed Federal action or program is located. Legal Jargon 18: Appendix B-2.

Culture. Federal--DOT. A collective name for all behavior patterns socially acquired and socially transmitted by means
of symbols; hence the term is a name for all the distinctive achievements of human groups, including, not only such
items as language, tool-making, industry, art, science, law,
government, morals and religion, but also the material instruments or artifacts in which cultural achievements are
embodied and by which intellectual cultural features are
given practical effect, such as buildings, tools, machines,
communication devices, art objects, etc. Transportation 3:
p. 24.

Cutie-Pie. Federal--EPA. An instrument used to measure radiation
levels. General 1: p. 4.

Data. State--Water Resources. Records of observations and measurements of physical facts, occurrences, and conditions,
reduced to written, graphical, or tabular form. Energy/
Utility 2: Appendix E.

Data Base. Federal--DOT. Information organized for analysis or
used as the basis for a decision. Transportation 3: p. 24.

Datum. Federal--DOT. (1) An agreed standard point or plane of
stated elevation, noted by permanent bench marks on some
solid immovable structure, from which elevations are measured or to which they are referred, (2) A point, line or
surface used as a reference, as in surveying, mapping, or
geology, (3) Any numerical or geometrical quantity or set
of such quantities which may serve as a reference or base
for other such quantities. Transportation 3: p. 24.

State--Environmental Protection. A reference point; line
or plane used as a basis of measurement. General 6: p. 311.

DDT. Federal--EPA. The first chlorinated hydrocarbon insecticide.
It has a half-life of 15 years and can collect in fatty
tissues of certain animals. EPA banned registration and
interstate sale of DDT for virtually all but emergency uses
in the U.S. in 1972 because of its persistence in the environment and accumulation in the food chain. General 1:
p. 5.

Decision Theory. Federal--DOT. A body of general propositions
dealing with the process of making rational choices (particularly by complex public and private organizations) and
defining the limits or rationality in decisions governed
by uncertainties. Transportation 3: p. 25.

Demography. National--Soil Conservation. The statistical study
of human vital statistics and population dynamics. Earth
4: p. 17g.

Density-Dependent Factor. Federal--Interior. Any environmental
factor that is dependent upon population density to be
fully effective. General 3: p. 24.

Density-Independent Factor. Federal--Interior. Any environmental factor that operates without regard to the population
density. General 3: p. 24.

Design Life. State--Soil And Water Commission. The period of
time for which a facility is expected to perform its intended function. Earth 3: p. G-6.

Design Solution. Federal--Forest Service. A class of land use
planning analytic approaches. The planner/designer generates unique solutions to potential land use conflicts by
manipulating the appearance or other physical characteristics of any changes and/or their occurrence in space and
time. "Design solution" potentials are typically evaluated
at the individual "project level" or site specific level of
any land use planning process. However, this technique can
also be used at higher planning levels by examination of the
quality of solutions obtained by superimposing design pro-
totypes or current design practices upon the typical site
characteristics in a planning area. General 15: p. 59.

Development. Federal--Forest Service. In planning, the term "development" is typically used when referring to intensive
build-up types of land uses--i.e., uses predominantly consisting of buildings, paved areas and other types of site

alterations. "Development" in that sense utilizes the
land only as a location, not for its natural resource properties. "Development" is sometimes broken into broad use
categories--such as residential, commercial, industrial,
etc. General 15: p. 60.

Development Plan. National--Soil Conservation. A detailed construction plan including drawings and specifications.
Earth 4: p. 18g.

Development Potentials. Federal--DOT. The "highest and best"
or most productive use to which property may be put.
Transportation 3: p. 26.

Diagrammatic Representation. Federal--Forest Service. A land
use planning analytical technique in which the physical components (such as parking lots, lavatories, garbage cans,
picnic tables, beaches, etc.) in a plan are represented
symbolically, usually just by their names written inside
individual circles. The relationship between components are
then represented by lines connecting those which have necessary or desirable support functions. The individual components are then shifted so as to minimize the extent to which
different interconnecting lines cross each other. The resulting solution is assumed to represent the most effective
spatial relationship between project components, and thus
represents the idealized goal of the actual physical plan.
General 15: p. 61. (Revised).

Direct Impact. Federal--Commerce. A change in the built or natural environment that is either the immediate result of an
impacting activity or is linked to the impacting activity
through an identified chain of cause and effect without
further human intervention. General 6: p. 311.

Disturbance. Federal--Commerce. A disruption or perturbation;
significant changes in the equalibrium of natural or social
processes and resources from artificial or natural causes.
General 6: p. 311.

Diurnal. Federal--Army. Pertaining to phenomena of daily occurrence; of/that portion of the day in which light occurs.
General 9: p. 12.

Federal--Interior. Occuring every day, generally in daylight. Diurnal animals are generally active only during
the daylight hours. General 3: p. 26.

Draft Environmental Impact Statement (DEIS). State--Transporta-
tion. A written statement as required by the National

Environmental Policy Act containing a detailed assessment
of the anticipated significant beneficial and detrimental
effects which projects may have upon the quality of the environment. Transportation 1: p. viii.

Duration Curve. National--Soil Conservation. A graphical representation of the number of times a given flows are equalled
or exceeded during a certain period of record. Earth 4:
p. 20g.

State--Natural Resources. A curve that shows the percentage of time that specified discharges, lasting a certain
length of time, are equalled or exceeded. General 12:
p. 336.

Easement. Federal--DOT. An acquired right of use, interest, or
privilege (short of ownership) in lands owned by another,
such as easement of light, of building support, or right-
of-way. It may be permanent or limited in time. Transportation 3: p. 27.

Federal--Interior. A nonpossessing interest held by one
party in land of another party whereby the first person is
accorded partial use of such land for a specific purpose.
It restricts but does not abridge the rights of the fee
owner to the use and enjoyment of his land, subject to the
enjoyment of the easement holder's rights. General 3:
p. 28.

Economic Growth. Federal--Forest Service. Increased economic
activity in real terms over time. This does not necessarily correspond to increased production of physical goods.
General 15: p. 66.

Effectiveness Index. Federal--Forest Service. The index used in
cost-effectiveness analysis. General 15: p. 68.

Efficiency. Federal--DOT. A condition wherein the least amount
of input technologically necessary is utilized to produce
a given quantity and quality of product, stated another way:
achievement of the greatest end result per unit of resource
input. Transportation 3: p. 28.

Eminent Domain. Federal--Interior. The right of government to
take private property for public use. Recreation 3: p. 9.

Emission Standard. National--Soil Conservation. The maximum
amount of pollutant permitted to be discharged from a single
polluting source. Earth 4: p. 21g.

Engineering Solution. <u>Federal--Forest Service</u>. The actual application of technical solutions to solve (or at least minimize) problems associated with particular land uses or land
use practices. General 15: p. 70.

Environment. <u>Federal--Army</u>. The sum total or the resultant of
all external conditions which act upon organisms. General 9: p. 17.

  <u>Federal--EPA</u>. The sum of all external conditions affecting
the life, development and survival of an organism. General
1: p. 6.

  <u>Federal--Interior</u>. The sum total of all biological, chemical, and physical factors to which organisms are exposed.
General 3: p. 30.

  <u>National--Resources</u>. The air, water and earth, sometimes
called the biosphere. Energy/Utility 5: p. 5.

  <u>State--Board</u>. The aggregate of all the external conditions
and influences, affecting the life, development, and ultimately the survival of an organism. Air 2: no page number.

Environmental Analysis Report (EAR). <u>Federal--Forest Service</u>.
A report on environmental effects of proposed Federal actions which may require an Environmental Impact Statement
(EIS) under section 102 of the National Environmental Policy
Act (NEPA). The EAR is an "in-house" document of varying
degrees of formality which becomes the final document on
environmental impacts for those projects which, because
their effects are minor, do not require a formal EIS. Although not formally prescribed under NEPA, the EAR is the
document normally used to determine whether section 102 of
NEPA applies to the project in question, and as such is
subject to court challenge if no EIS is filed. General
15: p. 70.

Environmental Components. <u>Federal--DOT</u>. Those information processes concerned with the collection and maintenance of
data which describe the community to be served or affected
by an action. Transportation 3: p. 29.

Environmental Criteria. <u>Federal--Army</u>. Standards of physical,
chemical and biological (but sometimes including social,
aesthetic, etc.) components that define a given quality of
an environment. General 9: p. 17.

Environmental Design Arts. <u>Federal--DOT</u>. Those disciplines--in-
cluding architecture, landscape architecture, city planning,

urban design and graphic design--that are concerned with
the relationship between human values and activities and
the quality of the spatial environment. Transportation 3:
p. 29.

Environmental Effect. <u>Federal--Army</u>. Resultant of natural or man-
made perturbations of the physical, chemical or biological
components making up the environment. General 9: p. 17.

Environmental Fragility. <u>Federal--Forest Service</u>. In a land use
planning context the relative ability of resources to tolerate sustained use without degradation of the resource base
(i.e., lowering productivity or desirable resource quality)
is sometimes rated on a qualitative scale extending from
"fragile" to "durable". "Fragility" in this usage applies
to those resource types or locations which are more or less
susceptible to degradation under the presence of whatever
use categories have been considered in the evaluation process. For example, areas of vegetation-covered sand dunes
may be characterized as being "environmentally fragile" for
off-road vehicle uses, whereas beaches or unvegetated dunes
may be said to be "environmentally durable" for the same
uses. General 15: p. 84.

Environmental Impact Analysis. <u>Federal--Forest Service</u>. An activity that involves the consideration of the interaction of
physical, natural, social and economic factors and a determination of probable effects of the plan or proposal upon
these operating systems. General 15: p. 71.

Environmental Impact Assessment. <u>Federal--Interior</u>. A report
that analyzes environmental effects of a proposed Federal
action, and serves as a basis for determining whether the
action is major and significant. Recreation 3: p. 9.

Environmental Impact Report (EIR). <u>State--Board</u>. A detailed report setting forth the potential effects/impacts on the environment by a proposed project. Prepared pursuant to requirements outlined in CEQA and Resources Agency Guidelines.
Air 2: no page number.

Environmental Impact Statement (EIS). <u>Federal--EPA</u>. A document
required of Federal agencies by the National Environmental
Policy Act for major projects or legislative proposals.
They are used in making decisions about the positive and
negative effects of the undertaking, and list alternatives.
General 1: p. 6.

<u>Federal--Interior</u>. A Federal report which analyzes the po-
tential environmental impact of a proposed project on the

environment when the impact is determined to be major and
significant. Recreation 3: p. 10.

State--Board. An EIR prepared pursuant to NEPA; the federal term for EIR. Air 2: no page number.

State--Committee. A written assessment of the effect a
project will have on air quality, water quality, noise
levels and other environmental factors. Air 3: no page
number.

State--Transportation. A document required by the National
Environmental Policy Act of 1969 which requires inclusion
in every recommendation or report on proposals for legislation and other major Federal actions significantly affecting the quality of the human environment, a detailed statement by the responsible officials on: (1) the environmental
impact of the proposed action, (2) any adverse environmental effects which cannot be avoided should the proposal be
implemented, (3) alternatives to the proposed action, (4)
the relationship between local short-term uses of man's
environment and the maintenance and enhancement of long-
term productivity, and (5) any irreversible and irretrievable commitments of resources which would be involved in the
proposed action should it be implemented. Transportation
1: pp. viii-ix.

Environmental Inventory. Federal--Army. A listing of the components making up an environment--or a listing of types of
environments. General 9: p. 17.

Environmental Monitoring. Federal--Army. The systematic (simultaneous or sequential) measuring of various components
constituting the environment. General 9: p. 18.

Environmental Parameters. Federal--Army. Physical, chemical, or
biological components and their interactions which can be
stated in quantitative terms; a parameter is what is measured by a statistic. General 9: p. 18.

Environmental Protection Agency (EPA). National--Resources. An
agency of the Federal government responsible for: the establishment and enforcement of environmental protection
standards consistent with national environmental goals; the
conduct of research on the adverse effects of pollution and
on methods of equipment for controlling it; the accumulation
and dissemination of information to develop and strengthen
environmental protection programs and formulate related
policy. Energy/Utility 5: p. 6.

State--Board. The Federal agency established in 1970 by
Presidential Reorganization Plan #3 for the purpose of directing and coordinating federal programs to protect the environment to include air and water pollution, pesticides,
noise, radiation, and solid waste. Air 2: no page number.

Environmental Quality. Federal--Army. Human (individual or social) considerations of desirable ecological situations.
General 9: p. 18.

Federal--GSA. Environmental quality refers to the properties and characteristics of the environment, either generalized or local, as they impinge on human beings and other organisms. Environmental quality is a general term which can
refer to: (1) varied characteristics such as air and water
purity or pollution, noise, access to open space, and the
visual effects of buildings, and (2) the potential effects
which such characteristics may have on physical and mental
health. Legal Jargon 18: Appendix B-4.

Environmental Stress. Federal--Army. Perturbations likely to
cause observable changes in ecosystems; usually departures
from normal or optimum. General 9: p. 18.

Environmental Surveillance. Federal--ERDA. A program to monitor
the impact on the surrounding region of the discharges from
industrial operations. General 7: p. g-4.

Environs. Federal--Army. The surrounding geographic zones and
environmental systems that exert influence on the ecology,
economy and social functions of the area under study. Recreation 2: p. A-34.

Exceedence Frequency. Regional--River Basin. Percent of values
that exceed a specified magnitude. Water 5: p. 1014.

Exogenous Independent Variable. Federal--DOT. Facets of a situation outside the direct control of a decisionmaker. For
example, the characteristics of a neighborhood through which
a highway passes are a set of such variables. Transportation 3: p. 30.

Fair Market Value. National--Soil Conservation. That value that
would induce a willing seller to sell and a willing buyer
to buy, usually applied to real estate cases where the right
of eminent domain is being exercised. Earth 2: p. 22g.

Feasibility Study. Federal--DOT. A survey to determine the prac-
ticality of an enterprise before a financial or other major
commitment is made. Transportation 3: p. 31.

Feedback. Federal--DOT. (1) Broadly, any information about the
    results of a process, (2) The return of all or a portion of
    the output of any process or system to the input, especially
    when used to maintain the output within predetermined limits
    or for corrective purposes. Transportation 3: p. 31.

Final Environmental Impact Statement (FEIS). Federal--Forest
    Service. The final version of the statement of environ-
    mental effects required for major Federal actions under
    section 102 of the National Environmental Policy Act (NEPA).
    It is a revision of the draft environmental impact state-
    ment to include public and agency responses to the draft.
    It is a formal document which must meet legal requirements
    and is the document used as a basis for judicial decisions
    concerning compliance with NEPA. This term is also used
    for similar statements prepared to comply with state and
    local laws patterned after NEPA. General 15: p. 72.

Fixed Assets. Federal--DOT. (1) Assets of a long-term character
    which are intended to continue to be held or used, such as
    land, buildings, machinery, furniture or other equipment.
    (2) Those expenditures over $25 for items with more than
    one year of useful life. Transportation 3: p. 31.

Fixed Capital. Federal--DOT. The investment represented by fixed
    assets. Transportation 3: p. 31.

Fluorides. Federal--EPA. Gaseous, solid, or dissolved compounds
    containing fluorine that results from industrial processes.
    General 1: p. 7.

Forecast. Federal--Forest Service. A prediction of future con-
    ditions and occurrences based on the perceived functioning
    of the system in question. The accuracy of a forecast is
    highly dependent on a proper understanding of the system.
    A "forecast" differs in a strict sense from a projection,
    which is also a prediction of future conditions and occur-
    rences, but is based on extrapolation of past trends. While
    the distinction may exist, the dividing line between the two
    terms is imprecise. General 15: p. 81.

Fragile Area. Federal--Interior. An identifiable area where the
    natural ecosystem is unusual, sensitive and vulnerable. It
    could be destroyed, severely altered or irreversibly changed
    by man's instrusion. Recreation 3: p. 10.

Framework. State--Transportation. This term is used collectively
    to represent organizational structure, assignments of re-
    sponsibility and decisionmaking process. Transportation 1:
    p. ix.

Framework Plan. National--Soil Conservation. A statement describing the objectives or goals of a program and the general methods to be followed in working toward these goals. Frequently issued as one of the first reports in a comprehensive planning program; also called policies plan. Earth 4: p. 24g.

Frequency Curve. National--Soil Conservation. A graphical representation of the frequency of occurrence of specific events. Earth 4: p. 24g.

Functional Plans. Federal--Forest Service. Documents prepared for separate resources or activities that identify those separate resource or activity management actions that are necessary to implement the coordinated land management direction established by the National Forest land use plans. Examples include timber management plans, transportation plans, land ownership adjustment plans. General 15: p. 85.

National--Soil Conservation. A plan for one element or closely related elements of a comprehensive plan, for example, transportation, recreation, and open spaces. Such plans, of necessity should be closely related to the land use plan. Plans that fall short of considering all elements of a comprehensive plan may be considered functional plans. Thus, resource conservation and development plans and watershed plans are considered as functional plans. Earth 4: pp. 24-25g.

Functional Planning. Federal--Forest Service. Establishing the specific management actions to be taken in the management of a single resource or activity. Preferably this is undertaken only to detail the implementation of a more general plan involving multiple resources and activities. General 15: p. 85.

Functional Relationships. Federal--DOT. The core of a model is the statement of relationships between input and output variables. For example, the degree of noise nuisance might be predicted as a function of vehicular traffic and neighborhood characteristics. Transportation 3: p. 33.

Functional Relationships Diagram. Federal--Army. A schematic diagram which arranges various uses or facilities according to the interrelations of the individual parts to each other with emphasis on eliminating conflicts, stressing efficient operation of the entire system and establishing circulation routes. Recreation 2: p. A-37.

Game Theory. <u>Federal--Forest Service</u>. The study of decision
    problems in competitive situations. Game theory is con-
    cerned with the derivation of rules for making decisions
    when two or more persons or organizations are competing for
    some objective. General 15: p. 86.

Geographical Setting. <u>Federal--Army</u>. Description of a site's
    location with regard to the natural features within a large
    region. Recreation 2: p. A-37.

Goal. <u>Federal--DOT</u>. An end toward which effort is directed. A
    statement expressing some desired future state toward which
    an individual, group, or community wishes to move. Trans-
    portation 3: p. 33.

    <u>Federal--Forest Service</u>. (1) Characteristically, "goals"
    are enduring statements of purpose, often not attainable in
    the short term, and frequently incapable of expression in
    quantifiable terms, (2) The broad end toward which effort
    is directed. In the context of land use planning, goals
    are normally stated in terms of the fulfillment of broad
    public needs, the preservation of fundamental constitutional
    principles, the achievement of targeted levels of excel-
    lence, the alleviation of major problems, or other justi-
    fiable missions or purposes to be served by government
    effort. General 15: p. 87.

Gradient. <u>Federal--Army</u>. A more or less continuous change of
    some property in space. Gradients of environmental prop-
    erties are ordinarily reflected in gradients of biota.
    General 9: p. 22.

    <u>National--Engineering</u>. Change in elevation, velocity,
    pressure or other characteristics per unit length. Water
    1: no page number.

    <u>City--Planning</u>. Applied to a stream, it is the slope mea-
    sured along the course of the stream. Change in value of
    one variable with respect to another variable, i.e., geo-
    thermal gradient, change in temperature with depth. Gen-
    eral 5: p. 56.

Graphic Scale. <u>Federal--Forest Service</u>. A bar, or other such
    graphic device, on a scale drawing, divided into properly
    scaled-down measurement units--such as 100, 500, 1000,
    5000 yards or 0.25, 0.5, 1.0, 5.0 miles. In subsequent
    photographic reduction or enlargement of a drawing, the
    "graphic scale" becomes a device from which distances can
    be redily and accurately measured. General 15: p. 87.

Gravity Model. Federal--DOT. A representation of human, economic
        or functional interaction postulating that the force of
        attraction between two areas of activity is a function of
        the size or strength of some pertinent variable in one
        (population, number of jobs, square footage of retail
        space or trip ends, for example), and that is is inversely
        related to the intervening distance over which the inter-
        action must take place. Transportation 3: p. 34.

Guideline. Federal--Forest Service. U.S. Forest Service usage.
        A steering or usual course of action, but not mandatory nor
        requiring a superior's approval for deviation from it.
        General 15: p. 91.

Half-Life. Federal--EPA. The time taken by certain materials to
        lose half their strength. For example the half-life of
        DDT is 15 years; of radium 1,580 years. General 1: p. 8.

        Federal--Interior. Time required for one-half of a given
        substance to disintegrate. General 3: p. 39.

Hazzardous Waste. Federal--EPA. Waste materials which by their
        nature are inherently dangerous to handle or dispose of,
        such as old explosives, radioactive materials, some chemi-
        cals, and some biological wastes; usually produced in indus-
        trial operations. General 1: p. 8.

Heavy Metals. Federal--EPA. Metallic elements like mercury,
        chromium, cadmium, arsenic, and lead, with high molecular
        weights. They can damage living things at low concentra-
        tions and tend to accumulate in the food chain. General 1:
        p. 8.

High Density Polyethylene. Federal--EPA. A material used to make
        plastic bottles that produces toxic fumes when burned. Gen-
        eral 1: p. 8.

Highest And Best Use. Federal--DOT. The use of land to its max-
        imum permissible development, one that will bring maximum
        profit to the owner. It is the use that would justify the
        highest payment for the land if offered for sale. Trans-
        portation 3: p. 35.

Holistic. Federal--DOT. A method of planning which attempts to
        see its subject (be it a neighborhood, city, or region) as
        a single integrated system of mutually interdependent parts.
        Transportation 3: p. 35.

Housing Market. Federal--DOT. In simple terms the housing market
        is the composite of negotiations between buyers and sellers

(including lessees and lessors) for the acquisition or disposition of individual dwelling units which are in some degree of competition with each other. The units may be regarded as linked in a chain of substitutability in varying
degrees of closeness depending on qualitative considerations and buyer preferences. The market may be considered as consisting of clusters of substitutes cross-linked
in complex patterns. Transportation 3: p. 36.

Human Resources Development Systems. Federal--DOT. The complex
of people and organizations that provide the services required by individuals in satisfying their basic growth
needs. It normally consists of the education, library,
welfare, health, recreation, voter registration, vital
statistics and manpower functions. Transportation 3:
p. 36.

Hypothesis. Federal--Interior. A proposition, condition, or prin-
ciple which is assumed in order to explain certain facts
and guide in the investigation of others. General 3: p. 43.

Implementation Plan. Federal--EPA. An outline of steps needed to
meet environmental quality standards by a set time. General 1: p. 8.

Federal--GSA. A document of the steps to be taken to ensure attainment of environmental quality standards within
a specified time period. Implementation plans are required
by various laws. Legal Jargon 18: Appendix B-4.

Improvement. Federal--Army. A measure which directly increases
the value or usefulness of a particular facility or service
but which does enable that element to progress further.
Pertains mostly to inanimate objects. Recreation 2:
p. A-38.

Federal--DOT. Any physical addition to land that increases
its utility, beauty, income or value. Most improvements
increase only one or two of these. The rare improvement increases all of them. Transportation 3: p. 37.

Independent Measure. Federal--Army. A management action which
addresses only one planning objective and has no deliberate
beneficial contribution to other planning measures. Recreation 2: p. A-38.

Index Number. National--Soil Conservation. Expression of the relationship of a given situation with that of a base-period
value of 100; used to express the rate or degree change,
especially in prices or production. Earth 4: p. 28g.

Indirect Effect. Federal--Forest Service. A condition caused by
    an action or inaction through intermediary causal agents.
    An effect for which the causal linkages to the action or
    inaction are not readily apparent. Contrasts with direct
    effect. This is not a measure of importance, but merely a
    classification by causal linkages. Direct effects are usu-
    ally easier to detect and measure with certainty, but they
    may be either more or less important than indirect effects.
    General 15: p. 100.

Infiltration. Federal--GSA. Flow of a fluid into a substance
    through pores or small openings. General 2: p. VIII-29.

Infrastructure. Federal--DOT. The basic network or foundation of
    capital facilities or community investments which are nec-
    essary to support economic and community activities. Exam-
    ples of infrastructure elements include roads, sewer and
    water supply systems, school systems, etc. Transportation
    3: p. 38.

    Federal--Forest Service. The foundation underlying a
    nation's, region's, or community's economy (transportation
    and communication systems, power facilities, schools, hos-
    pitals, etc.) General 15: p. 100.

Input-Output Analysis. Federal--Forest Service. A quantitative
    study of the interdependence of a group of activities based
    on the relationship between inputs and outputs of the ac-
    tivities. The basic tool of analysis is a square input-
    output table, interaction model, for a given period that
    shows simultaneously for each activity the value of inputs
    and outputs, as well as the value of transactions within
    each activity itself. It has especially been applied to
    the economy and the industries into which the economy can
    be divided. General 15: p. 101.

Inputs. Federal--Forest Service. The basic resources of land,
    labor, and capital required in carrying out an activity.
    General 15: p. 101.

Interdisciplinary Approach. Federal--DOT. A method of planning
    that incorporates the expertise of two or more professional
    disciplines and requires the interaction of individuals in
    these disciplines to resolve complex planning issues.
    Transportation 3: p. 39.

Interdisciplinary Team. Federal--Army. A group of professionals
    such as recreation resource planners, landscape architects,
    engineers, biologists, architects, sociologists and ecolo-
    gists who lend their particular expertise to the analysis

of a common problem for the purpose of developing a balanced and sound solution. Recreation 2: p. A-39.

Interpretation. Federal--Army. The analysis of a particular
natural or man-made resource, event or force which seeks
to express in basic terms the structure, process, background and/or relationships of those elements to the general public. Recreation 2: p. A-39.

Intuitive Design. Federal--Army. The design of an area or facility in which the professional relies on personal opinion
or past experience without the aid of basic data, resource
analysis or other studies. Recreation 2: p. A-39.

Inventory. Federal--Forest Service. (1) The gathering of data
for future use, (2) The quantity or count of physical entities (such as trees, lakes, etc.) in an area. There are
two basic types of inventories in land use planning: (a)
Existing--measurement of what is actually "on the ground"
(in terms of number), and (b) Potential--total identified
undeveloped capability. General 15: p. 103.

Issue. Federal--Forest Service. A well defined, discrete subject
of controversy--such as whether the Supersonic Transport
(SST) development program should or should not proceed because of the possibility that their exhaust gases might significantly reduce the ozone concentration in the stratosphere, or whether phosphates in detergents should be
banned because of their alleged controlling role in the
eutrophication of many water bodies. When the point of
controversy can only be vaguely defined or when the causal-
ity relationships between use actions and undesirable results are more speculative than documentable the dispute is
referred to as a concern rather than an issue. General 15:
p. 104.

Joint Costs. National--Soil Conservation. The difference between
the cost of a multiple-purpose development as a whole and
the total of the separable costs for all project purposes.
Earth 2: p. 16g.

Joint Use Corridor. Federal--Interior. A corridor occupied by
more than one type of utility or transportation system
(i.e., pipelines, transmission lines, communication lines,
highways, railroads, etc.). Many joint use corridors develop as a result of new and different utilities or transportation systems being constructed alongside an existing
corridor. General 4: p. 3.

Labor Market Area. <u>Federal--DOT</u>. The area within which workers
    compete for jobs and employers compete for workers and for
    which labor market data are periodically published by the
    U.S. Department of Labor. Transportation 3: p. 40.

Labor Shed. <u>Federal--DOT</u>. A geographic area from which employees
    are drawn. It is not necessarily defined by the boundaries
    of political jurisdictions but rather the ease of commun-
    ication and travel times from residential areas to employ-
    ment centers. Transportation 3: p. 40.

Land And Water System. <u>Federal--Forest Service</u>. U.S. Forest
    Service usage. One of the six "systems" established by the
    U.S. Forest Service to have a systematic, orderly way to
    view and evaluate its many diverse but interrelated activ-
    ities. The Forest Service has developed this approach to
    better respond to the mandates of the Forest and Rangeland
    Renewable Resources Planning Act of 1974. It has grouped
    its various programs into these six "systems", each of
    which incorporates all the activities concerned with de-
    veloping and managing a specific resource. General 15:
    p. 107.

    <u>Federal--Forest Service</u>. The role of this system is to pro-
    tect, conserve, and enhance the basic resources of air,
    soil, and water on forest and range land. It provides the
    base on which other resources grow. Activities on Nation-
    al Forest land include land-use planning, administration
    of special land uses, easements, and land adjustments, and
    management of mineral areas in addition to the protection,
    conservation, and enhancement role. The six "systems" are:
    Land and Water, Timber Resources, Outdoor Recreation and
    Wilderness, Rangeland Grazing, Wildlife and Fish Habitat,
    and Human Community Development. General 15: p. 107.

Land Development. <u>Federal--DOT</u>. The improvement of land with
    utilities and services, making it more suitable for resale
    as developable plots for housing or other purposes. Trans-
    portation 3: p. 40.

Landscape. <u>Federal--Forest Service</u>. The sum total of the charac-
    teristics that distinguish a certain area on the earth's
    surface from other areas. These characteristics are a re-
    sult not only of natural forces but of human occupancy and
    use of the land. General 15: p. 110.

Landscape Treatment. <u>Federal--Army</u>. The improvement or enhance-
    ment of a natural or man-made landscape by the addition, de-
    letion, and/or rearrangement of the physical resources,

i.e., soil, vegetation, topography, water, etc. Recreation 2: p. A-40.

Land Use Allocation. <u>Federal--Forest Service</u>. The committing of a given area of land or a resource to one or more specific uses--e.g., to campgrounds, wilderness areas. General 15: p. 114.

Land Use Constraints. <u>Federal--Forest Service</u>. Any factor that may act to reduce management options or discourage various uses--e.g., erosive soils, unstable slopes, steep topography, presence of archeological or historic remains. General 15: p. 114.

Land Use Survey. <u>Federal--DOT</u>. A survey of the uses to which land is put in a particular area, usually summarized both in map form and statistically, that show developed and vacant land, streets, parkland, public buildings, etc. Transportation 3: p. 41.

Land Use Type. <u>Federal--Forest Service</u>. U.S. Forest Service usage. The primary use of a tract of land--e.g., crops, pasture, forest, urban and other. General 15: p. 116.

Line Plot Survey. <u>National--Soil Conservation</u>. A survey employing plots as sampling units. Plots of specific size are laid out, usually at regular intervals along parallel survey lines. Earth 4: p. 32g.

Local Economic Effects. <u>Federal--Forest Service</u>. Effects of an action or inaction on dollar market transactions in the immediate area of the action or inaction. The "local" area would include an area encompassing the primary sources of employment and retail outlets for basic commodities such as food utilized by the people of that area. General 15: p. 119.

Location Theory. <u>Federal--Forest Service</u>. The theory which attempts to explain and predict the location of individual, commerical, and industrial development. It is important to the planner for predicting population and industrial distributions which would result from alternative plans. General 15: p. 119.

Long-Range Planning. <u>Federal--Forest Service</u>. (1) U.S. Forest Service usage. Planning for the period covered by basic resource management plans, usually ten or more years, (2) Planning for a future more than 5 years distant. General 15: p. 120.

Long-Term Impacts. Federal--GSA. Those which will last long
    after the project is completed, such as increased demand
    for public services, unavoidable pollution impacts, and
    irreversible commitments of resources. Legal Jargon 18:
    Appendix B-5.

Lot. Federal--DOT. The smallest subdivision of land or of a
    block into which cities are normally sectioned for purposes
    of building development, its size varying with the local-
    ity. Transportation 3: p. 43.

Major Land Use. Federal--Forest Service. U.S. Forest Service
    usage. A grouping of primary uses together into classes
    with similar characteristics--e.g., cropland, pasture, and
    forest. General 15: p. 121.

Major Service Area. Regional--River Basin. Arbitrarily selected
    service areas containing significant portions of the reg-
    ion's population and industry. Energy/Utility 7: p. 256.
    (Revised).

Management Plan. National--Soil Conservation. A program of ac-
    tion designed to reach a given set of objectives. Earth
    4: p. 33g.

Management Unit. Federal--Forest Service. Under the "Multiple
    Use Planning" system used until 1972 they were subdivisions
    of the "multiple use management zone" in the ranger district
    multiple use plan. It was the finest delineation of land
    areas. Under the newly adopted U.S. Forest Service "Land
    Use Planning" system they are now subdivisions of planning
    units which identify the area of land or water to which
    specific management decisions apply. All lands within a
    planning unit are included in one of the identified manage-
    ment units. General 15: pp. 121-122.

Man-Made Environment. Federal--DOT. Those elements of the spatial
    environment that are products of man's design. Transpor-
    tation 3: p. 43.

Man-Made Resources. Federal--Army. Those elements and forces of
    the environment which are manufactured, introduced and/or
    developed by mankind, i.e., buildings, pollution, roads,
    synthetic materials, etc. Recreation 2: p. A-40.

Market Appeal. Federal--DOT. The extent to which a real estate
    project--or other commodity in the marketplace--has favor-
    able attributes which appeal to the consumers for which it
    is intended. Transportation 3: p. 43.

Market Capture Or Penetration Rate. Federal--DOT. That share, or
    percentage, of the consumer spending potential for a specif-
    ic class of goods and services, within a defined market area
    that is likely to be spent at a specific location, project,
    or group of establishments. Transportation 3: p. 43.

Market/Supply Orientation. Federal--DOT. The pattern of geograph-
    ic locations in which products are sold and from which sup-
    plies are purchased. Transportation 3: p. 44.

Market Value. Federal--DOT. The highest price estimated in terms
    of money which a property will bring if exposed for sale in
    the open market, allowing reasonable time to find a pur-
    chaser who buys with knowlege of all the uses for which it
    is capable of being used. Frequently referred to as the
    price at which a willing seller would sell and a willing
    buyer to buy, neither being under abnormal pressure. Used
    in appraisal, condemnation proceedings and assessments for
    taxes. Transportation 3: p. 44.

Masking. Federal--EPA. Blocking out one sight, sound, or smell
    with another. General 1: p. 9.

Master Plan. Federal--DOT. A comprehensive, long-range plan in-
    tended to guide the growth and development of a city, town,
    or region, expressing official contemplations on the course
    its transportation, housing, and community facilities should
    take, and making proposals for industrial settlement, com-
    merce, population, distribution, and other aspects of growth
    and development. Transportation 3: p. 44.

    National--Soil Conservation. In early state legislation on
    planning, this term was often defined and considered syn-
    onymous with comprehensive plan; more recently, used to mod-
    ify a functional element of a comprehensive plan such as
    a master highway plan or master recreation plan. Earth 4:
    p. 33g.

Maximum Practicability. Federal--Commerce. Best available tech-
    nology; all alternative mitigation measures have been con-
    sidered resulting in selection of measure, technique, or
    level which produces most environmentally desirable effect.
    General 6: p. 312.

Mesoscale Analysis. Federal--GSA. Total pollutant burden of a
    given area, determined from cumulative sum of emissions
    from all sources. General 2: p. IV-22.

Mid-Range Planning. Federal--Forest Service. Planning that com-
    bines elements of both long-range planning and short-range

planning. Generally, mid-range planning is designed for
that period beyond the immediate actions of short-range
planning, thus allowing a more general identification of
long-term objectives and goals, and yet within the time
period where reasonably accurate predictions of the future
with corresponding specific actions to be taken are prac-
tical--typically between three and ten years in the future.
Often, however, the planning process is merely divided into
the short-range and long-range planning with a single dividing point, say five years in the future, and the mid-
range classification is not used. General 15: p. 126.

Mitigate. Federal--Interior. In environmental usage, the reduction or control of adverse environmental impact through
various measures which seek to make the impact less severe,
less obvious, more acceptable, etc. General 4: p. 7.

Mixed Use. Federal--DOT. A variety of land uses in an area as
distinguished from the isolated uses and planned separat-
ism prescribed by many zoning ordinances. Transportation
3: p. 45.

Model. Federal--Forest Service. An idealized representation of
reality for purposes of describing, analyzing or understanding the behavior of some aspects of it. The term model is
applicable to a broad class of representations, ranging
from a relatively simple qualitative description of a system or organization to a highly abstract set of mathematical
equations. General 15: p. 131.

Monitoring. Federal--Army. The sampling and measuring of one or
more resource elements during the planning process required
to accurately establish environmental baseline conditions.
This is distinguished from monitoring of environmental parameters and processes during the operation phase of a completed project, or for the purpose of research. Legal Jargon 31: p. 10782.

Federal--EPA. Periodic or continuous sampling to determine
the level of pollution or radioactivity. General 1: p. 9.

Federal--GSA. Periodic or continuous determination of the
amount of pollutants or radioactive contamination present
in the environment. Legal Jargon 18: Appendix B-5.

Monitoring Program. Federal--Army. An orderly, recorded observation of a continuing or reoccuring phenomena for the purposes of analysis and problem solving. Recreation 2:
p. A-41.

Multifunctional. <u>Federal--Forest Service</u>. A joint effort by two
or more people, each person having a different functional
or resource responsibility such as timber management, engineering, etc., but not necessarily different training
or backgrounds. General 15: p. 133.

Multiple Use Planning. <u>Federal--Forest Service</u>. U.S. Forest
Service usage. The official planning system for the National Forest until replaced in 1972 by the "land use planning" system. Based primarily on subdivisions within National Forests, the principal document was the ranger district multiple use plan. Each National Forest was divided
into Ranger Districts, Management Zones, and Management
Units. General 15: pp. 134-135.

Municipal Bond. <u>Federal--DOT</u>. A bond issued by a state or local
governmental unit for a public purpose. They may be gen-
eral-obligation bonds to which the full faith and credit
of the issuing agency is pledged, or revenue bonds payable
from specified revenues. Transportation 3: p. 46.

Municipality. <u>Federal--DOT</u>. A town, city, or other district having powers of local self-government. Its powers are generally provided either by specific charter or by state
statute. Transportation 3: p. 46.

Natural Pollution. <u>Federal--Army</u>. The production and emission by
geological or non-human biological processes of substances
commonly associated with human activities (e.g., natural
oil seeps, hydrocarbons or toxins released by plants or
animals). General 9: p. 30.

Natural Resources. <u>Federal--Army</u>. Those natural elements of the
environment which are regarded as useful for consumption
and/or enjoyment by mankind such as minerals, vegetation,
water, wildlife, etc. Recreation 2: p. A-41.

Need. <u>Regional--Commission</u>. The quantity of a service, commodity,
or resource required to satisfy a projected essential requirement and objective, a goal or even a desire. General
13: p. 590.

Neighborhood. <u>Federal--DOT</u>. A local area whose residents are generally conscious of its existence as an entity and have informal face-to-face contacts and some social institutions
they recognize as their own. The term is often used to mean
nothing more than the geographic area within which residents
conveniently share the common services and facilities in
the vicinity of their dwellings. While sociologists have

generally destinguished between a neighborhood and a community, the two terms are used interchangeably in these Notebooks. Transportation 3: p. 47.

National--Soil Conservation. A primary informal group consisting of all persons who live in local proximity; often considered to be the locality served by an elementary school or neighborhood convenience shopping center. Earth 4: p. 35g.

Neoteric Area. Federal--Forest Service. U.S. Forest Service usage. Sites and areas which have been designated by the Forest Service as containing outstanding examples of man's modern culture which will obviously become historic properties in the future. An example would be a present-day building designed by a very popular architect. General 15: p. 140.

Net Acre. Federal--GSA. All developable land exclusive of road or utility rights-of-way, steep slopes, wetlands, or other legally protected lands. General 2: p. X-13.

Nonconforming Use. Federal--DOT. A building or use that is inconsistent with a district's zoning regulations. If erected after the enactment of the ordinance it may be ordered removed. Transportation 3: p. 49.

Nonconsumptive Use. Federal--Forest Service. Those uses of resources that do not reduce the supply, such as many types of recreation. General 15: p. 141.

Nonpoint Pollution. National--Soil Conservation. Pollutants whose sources cannot be pinpointed; can best be controlled by proper soil, water, and land management practices. Earth 4: p. 35g.

State--Natural Resources. Pollution from a widespread area, as opposed to pollution that occurs from an identifiable site. General 12: p. 337.

Non-Reactive Pollutant. Federal--DOT. One which is essentially inert and does not change through reactions with other atmospheric components. Transportation 3: p. 49.

Nutrients. Federal--EPA. Elements or compounds essential to growth and development of living things; carbon, oxygen, potassium, and phosphorus. General 1: p. 10.

Off-Road Vehicles (ORV). Federal--EPA. Forms of motorized transportation that do not require prepared surfaces--they can be used to reach remote areas. General 1: p. 10.

Federal--Interior. Abbreviation for off-road vehicles
such as motorcycles, four-wheel drive jeeps, trucks, snow-
mobiles, etc. General 3: p. 57.

On-Site Construction. State--Environmental Protection. All
physical activity necessary to begin and complete a particular facility and the total development of its site. Legal
Jargon 14: p. 3.

Open Space. Federal--DOT. Land or water that is not predominantly occupied by buildings or structures. Transportation 3:
p. 49.

Operational Definition. Federal--DOT. A definition of an abstract
concept in terms of simple observable procedures. Transportation 3: p. 50.

Operational Map. Federal--Forest Service. A precise, detailed,
large-scale map, usually with small contour intervals but
sometimes only planimetric (i.e., representing only the
horizontal position of features), constructed for specific
uses--such as administrative site development, road location, bridge sites, dam sites, campground development, ski
areas, and geological area maps. General 15: p. 122.

Operational Plans. National--Soil Conservation. The administrative and legislative actions taken by an agency to carry
out the decisions that were made in a comprehensive plan.
Such actions might include a capital improvements program,
intergovernmental coordinating committees, land development regulations, and construction codes. Earth 4:
p. 36g.

Opportunity Cost. National--Soil Conservation. The return to the
best alternative use by employing a unit of resource in a
given manner. Earth 2: p. 36g.

Ordinal Value. Federal--Forest Service. Numerical values assigned
to a variable which represents a ranking only. The numerical values have meaning only when compared to one another.
Because of this, mathematical operations performed on the
values (mean, modes, differences, etc.) will not necessarily be valid, and the results of such operations must be
carefully interpreted. General 15: p. 145.

Pandemic. Federal--EPA. Widespread throughout an area. General
1: p. 11.

Paper Plan. <u>Federal--Forest Service</u>. A term sometimes used to
describe formal plan documents which consist mainly of a
map on which land use regulation districts are indicated.
While a "paper plan" is accompanied by such explanatory
written information as is necessary to convey its meaning,
the basic regulatory document is the map. In contrast,
plans whose basic regulatory document consists of a set of
policies are sometimes called policy plans, whether accompanied or not by maps differentiating the planning area into districts where some specific policies apply while others
do not. General 15: p. 148.

Parameter. <u>Federal--Army</u>. A measurable, variable quantity as
distinct from a statistic or estimate. General 9: p. 32.

Parcel. <u>Federal--DOT</u>. A lot or group of lots considered as a
unit and generally in single ownership. A tract or plot
of land. Transportation 3: p. 50.

Parent Plan. <u>Federal--Forest Service</u>. U.S. Forest Service usage.
A plan establishing the decision framework in which other
plans are made. For example, the planning area guide is
the parent plan for individual National Forest land use
plans, and the National Forest land use plans are parent
plans for unit plans. General 15: p. 148.

Perceptual Variables. <u>Federal--DOT</u>. Factors which describe the
degree to which a shared identity exists among the residents of an area, establishing a sense of belonging, or
commonly held outlook, and influencing the probability of
shared activity patterns. Transportation 3: p. 51.

Performance Standard Zoning. <u>Federal--DOT</u>. Regulations providing general criteria for determining the acceptability of
certain industries, land uses and buildings as distinguished from specification standards or detailed requirements.
Technological measurements are provided for noise and vibration, smoke, odor, dust and dirt, glare and heat, fire
hazzards, noxious gases, industrial wastes, transportation
and traffic, aesthetics and psychological impacts. Transportation 3: p. 51.

Permit. <u>Federal--Commerce</u>. A writing, issued by a person in
authority, empowering the grantee to do some act not for-
bidden by law, but not allowed without such authorization.
General 6: p. 313.

Permeable Material. <u>Federal--Interior</u>. That which allows water
to pass through easily. General 3: p. 61.

Photo Map. <u>National--Soil Conservation</u>. A mosaic to which place
names, marginal data, and other map information, usually
including a grid of coordinate system, have been added.
Earth 2: p. 39g.

Physical Barrier. <u>Federal--DOT</u>. A natural or man-made obstruction or impediment to access that has three-dimensional or
spatial characteristics. Transportation 3: p. 52.

Physical Planning. <u>Federal--Forest Service</u>. A form of urban land
use planning which attempts to achieve a physically attractive urban open space environment by regulating the physical
relationships of structures to their sites and surroundings,
with such devices as set back requirements, building height
controls, maximum lot coverage by structures and unnatural
surfaces, etc. General 15: p. 152.

Plan. <u>Federal--Forest Service</u>. (1) A formalized statement of
goals, objectives, and policies, (2) An assemblage of management directions, (3) A predetermined course of action.
General 15: p. 153.

<u>State--Environmental Protection</u>. Any proposed course of
action expressed in writing or depicted graphically, or
both. Legal Jargon 14: p. 4.

Plan Controls. <u>Federal--Forest Service</u>. Any of the various legal
and administrative methods which are available for implementing a land use plan's goals, objectives, or policies.
"Plan controls" include such devices as ordinances, zoning,
use permit and hearing procedures, design criteria, performance standards or any other formal means or procedure
for attaining a land use plan's goals, objectives or policies. General 15: p. 154.

Plan Element. <u>Federal--Forest Service</u>. U.S. Forest Service usage.
The major land use planning element of the U.S. Forest Service are: legal requirements and authorities, national and
regional objectives and targets, management situation, basic
assumptions, data collection, land capability, alternative
considerations, plan selection, functional planning and documentation. General 15: p. 154.

<u>Federal--Forest Service</u>. A discrete topic area or issue of
concern in an overall plan--e.g., housing, open space, recreation, transportation. General 15: p. 154.

Planned-Unit Development (PUD). <u>Federal--DOT</u>. A residential development in which the subdivision and zoning regulations

apply to the project as a whole rather than to its individual lots (as in most tract housing). Densities are calculated on a project wide basis, permitting among other
things the clustering of houses and provision of common
open space. Potential advantages may include: improved
site design free of standard lot pattern limitations;
lower street and utility costs made possible by reduced
frontages; more useful open space; and greater flexibility
in the mixing of residential building types. Transportation 3: p. 52.

National--Soil Conservation. A special zone in some zoning
ordinances which permits a unit of land under control of a
single developer to be used for a variety of uses and densities, subject to review and approval by the local governing body. The location of the zone is usually decided on a
case-by-case basis. Earth 4: p. 39g.

Planning. Federal--Forest Service. The act of deciding, in ad-
vance, what to do. A dynamic effort to use decisions to
guide future actions and decisions. A means of solving
future problems by intent. One step in the process of
guiding the future. General 15: p. 155.

Planning Horizon. Federal--Forest Service. U.S. Forest Service
usage. The furthest point in time for which management
systems are specified in a management program for a planning unit. For example, in the 1970 timber review, the
planning horizon was 2020. General 15: p. 157.

Federal--Forest Service. The time period (fixed by the
planner) which will be considered in the planning process.
It is assumed to span all activities covered in the plan
and all future conditions and effects of alternative actions
which would influence the planning decisions. General 15:
p. 157.

Planning Level. Federal--Forest Service. Referring to the scale
of a planning effort--usually denoted by the size of the
project or the governmental level at which the planning is
being done, e.g., site or project planning, municipal planning, metropolitan regional planning, state planning, interstate and regional planning, national planning, etc. General 15: p. 157.

Planning Period. Federal--DOT. The period of time for which active planning is being done. Short term is usually less
than five years and long term any period over five years.
Transportation 3: p. 53.

Planning Process. Federal--DOT. Planning is a series of related
    actions and decisions that are organized around and moving
    toward the accomplishment of defined objectives. The goals
    and objectives themselves are viewed as the cornerstone of
    the planning process, since, in theory, they form the frame-
    work for public and private decisionmaking. Transportation
    3: p. 53.

Planning-Programming-Budgeting System (PPBS). Federal--DOT. A
    planning and decision-making tool that attempts to organ-
    ize information and analysis so that the consequences of
    alternative policies are clearly revealed and fully com-
    parable. It endevors to display the allocation of resources
    in terms of the function; to gather hard, quantitative
    information on the actual results of programs; and to ana-
    lyze the cost of the various efforts to achieve similar
    ends and rank them in terms of their effectiveness per
    dollar. Transportation 3: p. 53.

Planning Region. Federal--Interior. An area having relatively
    homogeneous physiographic, economic, social or cultural
    characteristics. Recreation 3: p. 18.

Plan Selection. Federal--Forest Service. U.S. Forest Service
    usage. Subjected to the planning process, it is essen-
    tially a choice governed by a reasonable and rational per-
    ception of how the available resources can be most effi-
    ciently managed to achieve optimum multibenefits for the
    American people. The process includes full consideration
    of the priorities and preferences expressed by the public
    at all levels to be affected by the plan. General 15:
    p. 158.

Plat. Federal--DOT. A diagram drawn to scale showing land bound-
    aries and subdivisions, together with all data essential
    to the legal description and identification of the several
    parcels shown thereon, and including one or more certifi-
    cates indicating due approval. It differs from a map in
    that it does not necessarily show additional cultural,
    drainage, or relief features. Transportation 3: p. 53.

    Federal--Forest Service. A diagram drawn to scale and
    showing essential data pertaining to boundaries and sub-
    divisions of a tract of land, as determined by survey or
    drawing to scale. The primary purpose of a plat is to
    show boundary survey and ownership information rather than
    map information. General 15: p. 158.

Platting. Federal--Army. The legal division of land, by public

record, usually preliminary to sale for development. General 9: p. 34.

Plot. Federal--Interior. An area of land that is studied or used
for an experimental purpose, in which sample areas are
often located. General 3: p. 63.

Point Source. Federal--EPA. A stationary location where pollutants are discharged, usually from an industry. General 1: p. 11.

State--Environmental Control. Any discernible confined
and discrete conveyance, including but not limited to any
pipe, ditch, channel, tunnel, conduit, well, discrete fis-
sure, container, rolling stock, or vessel or other floating craft, from which pollutants are or may be discharged.
Legal Jargon 25: p. 3.

Policy Plan. Federal--Forest Service. This term is sometimes
used to describe a formal plan document consisting entirely of a set of policies unaccompanied by maps differentiating the planning area into districts where some policies
apply while others do not. In contrast, a plan whose formal plan documents consist mainly of a map on which land
use regulation districts are indicated is sometimes called
a paper plan or "map plan". While "paper plans" are accompanied by such explanatory written information as is
necessary to convey their meaning, the basic regulatory
document is still the map. General 15: p. 159.

Political Jurisdiction. Federal--DOT. A governmental administrative jurisdiction such as a county, town, township
or metropolitan district which serves as the territorial
basis for political representation. Transportation 3:
p. 53.

Pollutant. Federal--EPA. Any introduced substance that adversely affects the usefulness of a resource. General 1: p. 12.

Federal--GSA. Any introduced gas, liquid or solid that
makes a resource unfit for a specific purpose. Legal
Jargon 18: Appendix B-5.

Federal--Interior. A residue (usually of human activity)
which has an undesirable effect upon the environment (particularly of concern when in excess of the natural capacity
of the environment to render it innocuous). General 4:
p. 9.

Pollution. <u>Federal--Army</u>. An undesirable change in atmospheric,
    land or water conditions harmfully affecting the material
    or aesthetic attributes of the environment. General 9:
    p. 35.

    <u>Federal--EPA</u>. The presence of matter or energy whose
    nature, location, or quantity produces undesired environ-
    mental effects. General 1: p. 12.

    <u>Federal--Forest Service</u>. Any substance or energy form
    (heat, light, noise, etc.) which alters the state of the
    environment from what would naturally occur. Especially
    associated with those altered states which human value
    judgements have decreed as bad. General 15: p. 160.

    <u>Federal--GSA</u>. The presence of matter or energy whose
    nature, location or quantity produces undesired environ-
    mental effects. Legal Jargon 18: Appendix B-5.

    <u>Federal--Interior</u>. The alteration of the physical, chem-
    ical, or biological properties of the atmosphere or any
    water supply, including change in temperature, taste, color,
    turbidity, or odor, or the discharge of unnatural liquid,
    gas, solid, radioactive, or any other substance which is
    likely to create a nuisance problem or render the air and/
    or water supply injurious to public health or safety, or
    render the air and/or water supply detrimental to wild
    animals, birds, and fish, or other aquatic life. General
    3: p. 63.

    <u>National--Resources</u>. The contamination of soil, water or
    the atmosphere by the discharge of waste or offensive ma-
    terials--gases or chemicals. Energy/Utility 5: p. 10.

Polychlorinated Biphenyls (PCB's). <u>Federal--EPA</u>. A group of tox-
    ic, persistent chemicals used in transformers and capaci-
    tors. Further sale or new use is banned in 1979 law.
    General 1: p. 11.

Polyvinyl Chloride. <u>Federal--EPA</u>. A plastic that releases hydro-
    chloric acid when burned. General 1: p. 11.

Pre-Existing Use. <u>Federal--DOT</u>. A land use that does not con-
    form to a zoning ordinance but which existed before its
    enactment and therefore may not be banned until abandon-
    ment of that land use. Such uses have sometimes been
    banned after the lapse of a reasonable period equivalent
    to a prescribed depreciation period or amortization of the
    use. Transportation 3: p. 54.

Preservation. <u>Federal--Army</u>. The planned protection of natural,
    man-made or cultural resources of a project from destruc-
    tion, overuse, or exploitation due to construction, vis-
    itation, vandalism, etc. Recreation 2: p. A-45.

    <u>Federal--Commerce</u>. To maintain in existing condition;
    protection from permanent alteration by human activity.
    General 6: p. 313.

Pressure Group. <u>Federal--Forest Service</u>. (1) Any special in-
    terest group which actively advocates for special consid-
    eration of its resource use goals, (2) Any politically
    active group with a common set of values about resource
    use allocation. Pressure groups seek to influence de-
    cisions on resource use allocation in excess of their
    proportional representation in the planned-for populace
    by seeking preferential consideration for their resource
    use choices. General 15: p. 161.

Primary Impacts. <u>Federal--DOT</u>. The immediate effects on the
    social, economic and physical environment caused by the
    construction, presence and operation of a highway facil-
    ity. These impacts are usually experienced within the
    right-of-way or in the immediate vicinity of the highway
    facility. Transportation 3: p. 54.

Pristine State. <u>Federal--Army</u>. A state of nature without human
    effect or with negligible human effect. General 9: p. 36.

Probabilistic Model. <u>Federal--Forest Service</u>. A model that makes
    allowances for randomness in one or more of the factors
    that determine the outputs of the model. General 15:
    p. 132.

Process. <u>Federal--DOT</u>. A series of steps or procedures which,
    when taken together, has as a major objective the achieve-
    ment of a particular result. Transportation 3: p. 55.

Process Planning. <u>Federal--Interior</u>. Developing the steps nec-
    essary to achieve a particular goal. Recreation 3: p. 19.

Process Weight. <u>Federal--EPA</u>. The total weight of all materials,
    including fuel, used in a manufacturing process. It is
    used to calculate the allowable rate of emission of pol-
    lutant matter from the process. General 1: p. 12.

Profile View. <u>Federal--Forest Service</u>. An illustration in which
    the features coinciding in space with the plans of section-
    ing are strongly emphasized (usually by thick linework).

Features in the background are usually shown with much
thinner linework and without any reduction in size proportion to their increasing distance from the plane of sec-
tioning (as would occur if they were drawn according to
the rules of linear perspective). Underground features
are seldom shown in "profile" type drawings nor is there
ever any exaggeration of the scale of vertical or horizontal features. General 15: p. 165.

Program. Federal--DOT. A concerted set of activities which attempts to encompass all of an agency's efforts to achieve
a particular objective or set of allied objectives. Transportation 3: p. 55.

Programming. Federal--Forest Service. The process of deciding on
specific courses of action to be followed in carrying out
planning decisions and objectives. It also involves decisions in terms of total inputs required or total costs
to be incurred over a period of years as to personnel, material, and financial resources to be applied in carrying
out programs. General 15: p. 165.

Property Improvement. Federal--DOT. Any improvement which en-
hances the income producing capacity or market value of a
parcel of land. Improvements may include construction of
a building, structural improvements, etc. Transportation
3: p. 55.

Property Tax. Federal--DOT. A levy on the owners of real or
personal property; generally an ad valorem tax. Transportation 3: p. 55.

Prototype. Federal--DOT. The natural (or full-scale) entity simulated by a model. Also, a structure that is yet to be
built and for which model experiments are being conducted.
Transportation 3: p. 55.

Psychological Barrier. Federal--DOT. An obstruction or impediment to access that is perceived in the minds of an individual or group of individuals. This obstruction may
not, however, be an actual physical barrier. Transportation 3: p. 55.

Public Domain. Federal--Interior. Unpatented Federal lands and
waters. Recreation 3: p. 20.

Quadrat. Federal--Interior. A sampling area, most commonly one
square meter, used for analyzing vegetation. General 3:
p. 66.

Rating Curve. <u>National--Soil Conservation</u>. A graphic or some-
     times tabular representation of performance or output under
     a stated series of conditions; for example, a rating curve
     for a flume shows volume of flow per unit time at various
     stages of depths of flow. Earth 4: p. 41g.

Real Property Tax Base. <u>Federal--DOT</u>. The total assessed value
     of land and imporvements within a local political juris-
     diction against which ad valorem real property taxes are
     applied. Transportation 3: p. 57.

Reasonably Available Control Technology (RACT). <u>State--Committee</u>.
     Abbreviation for reasonably available control technology.
     Air 3: no page number.

Receptor. <u>Federal--DOT</u>. Person, structure or location which
     does or would experience water, noise or air pollution;
     point where pollutant concentrations are monitored or pro-
     jected. Transportation 3: p. 57.

Redevelopment. <u>Federal--DOT</u>. The revision or replacement of an
     existing land use and/or population distribution pattern
     through the acquisition of an essentially built-up area,
     it clearance and rebuilding in accordance with a compre-
     hensive plan. Often involves the assembly of small par-
     cels into larger tracts of land for new public and pri-
     vate development. A form of recuperative change in the
     physical city by which outworn or outmoded structures and
     facilities and, in time, whole areas are altered or re-
     placed in response to pressures of economic and social
     change. Transportation 3: p. 57.

Reevaluation. <u>Federal--Army</u>. That aspect of planning which ex-
     amines the conditions, problems, purposes and development
     of an existing project to determine the best overall course
     of action for meeting current and future needs. Employs
     the full test of feasibility. Recreation 2: p. A-47.

Regional Plan. <u>Federal--Forest Service</u>. The plan for a region
     (i.e., an area larger than a single city and whose bound-
     aries have typically been designated according to some
     physiographic, biological, poltical, administrative, eco-
     nomic, demographic, or other criteria). General 15: p.
     175.

Regional Planning Agency (RPA). <u>State--Committee</u>. An organiza-
     tion of local elected officials or their designees set up
     under state law to conduct regional planning activities.
     Air 3: no page number.

Regression Analysis. <u>State--Natural Resources</u>. A statistical
  method of predicting the value of dependent variables from
  given values of independent variables. General 12: p. 338.

Regulatory Agencies. <u>Federal--HEW</u>. Those organizations charged
  with responsibility for developing criteria documents, setting standards, and enforcing those standards in order to
  prevent or control particular problems; as the Environmental Protection Agency is charged with these responsibilities in the areas of air and water pollution. Health 1:
  p. 39.

Rehabilitation. <u>Federal--DOT</u>. The improvement or restoration of
  a predominantly built-up area which, though consistent with
  a comprehensive plan in terms of intensity of development
  and land use patterns, is in a stage of incipient blight.
  Transportation 3: p. 57.

Related Land. <u>Regional--Commission</u>. In connection with studies
  of water and related land resources, "related land" is:
  (1) that land on which projected use and/or management
  practices may cause significant effects on the runoff and/
  or quantity and/or quality of the water resource to which
  it relates; (2) that land the use or management of which
  is significantly affected by or depends on existing and/
  or proposed measures for the management, development, or
  use of the water resource to which it relates. General
  13: p. 591.

Relational Study. <u>Federal--Army</u>. A report which combines results
  of available feasibility studies and site information to
  produce a comparison of project potential with outside com-
  peting or complementary facilities. Recreation 2: p. A-47.

Remote Sensing. <u>Federal--Army</u>. The scientific detection, recognition, inventory and analysis of land and water area by
  the use of distant sensors or recording devices such as
  photography, thermal scanners, radar, etc. Recreation 2:
  p. A-47.

  <u>Federal--Forest Service</u>. Any data or information acquisition technique which utilizes airborne techniques and/or
  equipment to determine the characteristics of an area. Aerial photos are the most common form of remote sensing product, but side scanning radar (SLAR) and other pictoral
  and nonpictoral forms of data collections are also produced
  by some remote sensing techniques. General 15: p. 178.

  <u>Federal--Interior</u>. The acquiring of information or data
  through the use of aerial cameras or other sensing devices

that are situated at a distance from the area being in-
vestigated. Recreation 3: p. 23.

Renewable Resource. Federal--DOT. A resource capable of being
continuously renewed or replaced through such processes as
organic reproduction and cultivation such as those practiced in agriculture, animal husbandry, forestry and fisheries. Transportation 3: p. 57.

Resource. Federal--DOT. A tangible product of the earth or biosphere capable of serving, supplying, or supporting some
human purpose or need. Transportation 3: p. 58.

Federal--GSA. All actions and ideas as well as living and
nonliving materials utilized in the action. Legal Jargon
18: Appendix B-6.

Resource Allocation. Federal--Forest Service. The division of
limited resource capacity or supplies among the compet-
itors for use. General 15: p. 180.

Resource Base Analysis. State--Transportation. A comprehensive,
systematic and simplified approach for the collection and
analysis of social, economic, and environmental information
for involving the public and other agencies at the Systems
Planning Phase. Transportation 1: p. xiii.

Resource Inventory. Federal--Army. An orderly written and/or
graphic collection of elements of natural, man-made and/or
cultural resources which have been acquired through on-site
inspection, remote sensing, or other reports of particular
project or site. Recreation 2: p. A-47.

Revenue-Cost Analysis. Federal--Forest Service. A method of
comparing alternatives by analyzing the monetary income of
each alternative would generate in relation to its cost.
General 15: p. 181.

Right-Of-Way Acquisition. Federal--DOT. The process of obtain-
ing parcels of property for highway right-of-way, usually
through voluntary negotiation or condemnation under the
power of eminent domain. Transportation 3: p. 58.

Scale. Federal--Forest Service. Graphics usage. The proportional relationship (ratio) between the reduced size at which
something is being represented on a map or other type of
drawing and its true distance or size relationship. General 15: p. 185.

Scenario. Federal--Forest Service. A word picture of a fixed
    sequence of future events in a defined environment. Gen-
    eral 15: p. 185.

Scenic Area. Federal--Forest Service. U.S. Forest Service usage.
    A place which has been designated by the Forest Service as
    containing outstanding or matchless beauty which requires
    special management to preserve these qualities. General
    15: p. 185.

Secondary Impacts. Federal--DOT. Those effects that are trig-
    gered by a direct or primary impact. Transportation 3:
    p. 60.

Secondary Pollutants. National--Soil Conservation. Those pollu-
    tants that result from the chemical reactions involving
    primary pollutants or related atmospheric contaminants,
    for example, oxidants from photochemical activity. Earth
    4: p. 45g.

Sensitivity Analysis. Federal--Forest Service. Repetition of an
    analysis with different quantitative values for the vari-
    ables in order to determine their effects on the results
    of the basic analysis. If a small change in an assumption
    results in a proportionately greater change in the results,
    then the results are said to be "sensitive" to that assump-
    tion or parameter. General 15: p. 188.

Separable Cost. National--Soil Conservation. The difference be-
    tween the cost of a multiple-purpose development and the
    cost of the development with the purpose omitted. Earth
    2: p. 16g.

Service Area. Federal--DOT. The geographic area served by a part-
    icular public facility, such as school, library, police
    station, park, etc. Transportation 3: p. 60.

    Regional--River Basin. An area described for planning pur-
    poses whose boundaries would include the future population
    or industrial activities which could logically and function-
    ally obtain water supply and waste disposal services from
    a central or integrated system or where the problems are
    so interrelated that the planning should be done on an in-
    tegrated basis. Energy/Utility 7: p. 257.

Short-Range Planning. Federal--Forest Service. (1) Planning for
    a future less than 5 years distant, (2) U.S. Forest Service
    usage. Information use for planning and control for the
    period covered by specific action plans--i.e., functional
    plans. General 15: p. 189.

Significant. Federal--Commerce. A measurable change in the built
     or natural environment that is cause for concern. General
     6: p. 314.

     Federal--DOT. In relation to environmental analysis, the
     term includes considerations of importance and magnitude,
     primarily the former. Transportation 3: p. 61.

Significant Deterioration. Federal--EPA. Pollution from a new
     source in previously "clean" areas. General 1: p. 14.

Simulation. Federal--Forest Service. A technique for solving
     complex problems that are not amenable to solution using
     formal analytical techniques. Essentially simulation con-
     sists of a representation of a system or organization by
     means of a model. The behavior of the system under various
     possible operational conditions is then analyzed through
     repeated manipulation of the model. General 15: pp. 189-
     190.

     State--Natural Resources. In modeling, reproduction of
     the behavior of a real or hypothetical prototype. Gener-
     al 12: p. 339.

Site. Federal--Commerce. The actual location of the proposed
     development and the appropriate surrounding region that
     may be affected by the development. General 13: p. 21.

Site Analysis. Federal--Army. The examination and evaluation of
     the natural and man-made resources of a particular area
     of land or water to determine its suitability for develop-
     ment or management. Recreation 2: p. A-49.

Site Plan. Federal--Army. A plan which is site specific and
     sufficiently detailed so that each structure, facility and
     connecting route is shown in its correct scale and relation-
     ship for the purpose of determining costs and construction
     techniques and materials. Recreation 2: p. A-49.

Site Specific. Federal--Forest Service. In "land use planning"
     usage this term commonly refers to something only valid
     for, or confined to, a certain given parcel of land and/
     or water. The term may apply to data, studies to obtain
     information, environmental impacts, use restrictions, etc.
     General 15: p. 191.

Slag. Federal--Interior. A molten or solidified ash; a mineral
     substance formed by chemical action and fusion at furnace
     operating temperatures. General 4: p. 9.

Slurry. Federal--Interior. A mixture of a liquid and a solid.
     General 4: p. 10.

Social Analysis. Federal--Forest Service. An analysis of the
     social (as distinct from the economic and environmental)
     effects of a given plan or proposal for action. Social
     analysis includes identification and evaluation of all
     pertinent desirable and undesirable consequences to all
     segments of society, stated in some comparable quantita-
     tive terms--such as persons or percent of population in
     each affected social segment. It also includes a subject-
     ive analysis of social factors not expressable in quanti-
     tative terms. General 15: p. 192.

Social Equity. Federal--Forest Service. The distribution of the
     gains and losses that will accrue to individuals or groups
     (defined according to social criteria), as a consequence
     of land use planning decisions, in a manner which is in
     reasonable conformity to accepted standards of natural
     rights, law and justice, without prejudice, favoritism or
     fraud and without causing undue hardship. General 15:
     p. 195.

Sorption. Federal--EPA. The action of soaking up or attracting
     substances; used in many pollution control processes. Gen-
     eral 1: p. 14.

Special Interest Group. Federal--Forest Service. Any group
     (whether formally organized or not) with a specialized
     set of shared preferences about how resource use should
     be allocated. General 15: p. 203.

Special Management Zone. Federal--Forest Service. U.S. Forest
     Service usage. Areas of unusual public interest or other
     significance. Examples are: wilderness, primitive areas,
     experimental forests, natural areas, scenic areas, and his-
     torical, geological, or archeological areas. Such areas
     are classified or formally designated by the Congress,
     Secretary of Agriculture, Chief of the Forest Service, or
     Regional Foresters. General 15: p. 203.

Specific Cost. National--Soil Conservation. That cost incurred
     solely for a single purpose, for example, a pipeline to
     carry municipal water. Earth 2: p. 16g.

Sprawl. Federal--EPA. Unplanned development of open land. Gen-
     eral 1: p. 14.

Stagnation. Federal--EPA. Lack of motion in a mass of air or
     water, which tends to hold pollutants. General 1: p. 14.

Standard Deviation. State--Natural Resources. A measurement
    showing the differences of values from the arithmetic
    average. General 12: p. 339.

Standard Metropolitan Statistical Area (SMSA). Federal--DOT.
    Generally consists of a county or group of counties con-
    taining at least one city (or twin cities) having a pop-
    ulation of 50,000 or more plus adjacent counties which are
    metropolitan in character and are economically and socially
    integrated with the central city. In New England, towns
    and cities rather than counties are the units used in de-
    fining SMSA's. The name of the central city or cities is
    used as the name of the SMSA. There is no limit to the
    number of adjacent counties included in the SMSA as long
    as they are integrated with the central city nor is an
    SMSA limited to a single state; boundaries may cross state
    lines, as in the case of Washington, D.C./Maryland/Virginia
    SMSA. Transportation 3: p. 63.

Stationary Source. Federal--EPA. A pollution location that is
    fixed rather than moving. One point of pollution rather
    than widespread. General 1: p. 14.

Statistics. Federal--HEW. A branch of mathematics dealing with
    the collection, analysis, interpretation, and presentation
    of masses of numerical data. Health 1: p. 39.

Strategic Planning. Federal--Forest Service. The study of ob-
    jectives and analysis of alternative ways to achieve ob-
    jectives in terms of their accomplishments. General 15:
    p. 210.

Strategy. Federal--Forest Service. A consideration of alterna-
    tive means to reach an objective. General 15: p. 210.

Stratification. Federal--Forest Service. A land use planning
    concept which was probably borrowed from the statistical
    technique of "stratified sampling" which subdivides a study
    area into units which are, more or less, internally homo-
    geneous with respect to the (those) character(s) of inter-
    est. Subsequently, sampling data is taken to characterize
    each unit. This strategy provides maximum information for
    an economy of sampling effort. General 15: p. 210. (Re-
    vised).

Study Area. Federal--DOT. A geographic area isolated from a
    larger context for purposes of examination and analysis
    of the specific phenomena and activities occurring in the
    area. Transportation 3: p. 64.

Subdivision. <u>National--Soil Conservation</u>. The division or re-
    division of a lot, tract, or parcel of land into two or
    more areas either by platting or metes and bounds descrip-
    tion. Earth 2: p. 54g.

Subjective. <u>Federal--GSA</u>. That which cannot be measured accord-
    ing to agreed upon standards or techniqes. Whether or not
    such agreed upon standards or techniques exist is in no way
    related to the importance or significance of an environmen-
    tal impact question. Legal Jargon 18: Appendix B-7.

Substantial. <u>Federal--GSA</u>. In relation to environmental analy-
    sis, the term "substantial" implies an impact which is
    sufficiently great to alter the basic nature or substance
    of an environmental system or element. Legal Jargon 18:
    Appendix B-7.

Substantive Comment. <u>Federal--GSA</u>. A formally submitted response,
    with reasons for the response, to the content of an environ-
    mental impact statement. Legal Jargon 18: Appendix B-7.

Surrounding Region. <u>State--Environmental Protection</u>. At the min-
    imum, the county or counties likely to be affected by the
    proposed facility. Legal Jargon 14: p. 5.

Surveillance System. <u>Federal--EPA</u>. A series of monitoring de-
    vices designed to determine environmental quality. Gen-
    eral 1: p. 15.

Survey. <u>Federal--DOT</u>. A critical examination of facts or con-
    ditions to provide information on a situation. Usually
    conducted by interviews and/or on-site visitations. Trans-
    portation 3: p. 66.

Symbolic Resources Sites. <u>State--Transportation</u>. Refers to those
    cultural or educational areas that are important for their
    symbolic significance to the community. Parameters could
    include: public and private mangement areas, historic
    sites, unique archeological, botanical, cultural and ed-
    ucational areas. Transportation 1: p. xv.

Synergism. <u>Federal--GSA</u>. Combined action of two or more sub-
    stances that is greater than the sum of the individual
    actions. For example, the toxicity of many chemical agents
    is greatly increased by increased temperatures. General
    2: p. IX-20.

    <u>State--Committee</u>. The action of separate substances such
    that the total effect is greater than the sum of the effects

of the substances acting independently. Air 3: no page number.

Synoptic Model. <u>Federal--DOT</u>. A part of the APRAC-1A Model which generates hourly pollutant concentrations as a function of time, for comparison and verification with observed concentrations and for operational applications. Transportation 3: p. 66.

Systematic Sample. <u>National--Soil Conservation</u>. A sample consisting of sampling units selected in conformity with some regular pattern (e.g., the sample formed from every 20th two-chain strip of forest, or from every 10th tree in every 5th row). Earth 4: p. 55g.

Systems Evaluation. <u>Federal--DOT</u>. The continuous determination of the extent to which the planned objectives or capabilities of a system coincide with those actually achieved during systems implementation. Includes the measurement of the change in effectiveness and/or efficiency resulting from the implementation of the system, and the determination of areas where improvement could be made. Transportation 3: p. 66.

Systems Planning Or Systems Analysis. <u>Federal--DOT</u>. A means of organizing elements into an integrated analytic and/or decisionmaking procedure to achieve the best possible results. A comprehensive, rational and precise approach to problem solving which focuses on the structure and interrelationships of elements in complex systems like a city, rather than on individual parts. It is amplified by mathematical, operations research, and decisionmaking techniques, as well as simulation models, where appropriate. In city planning, the systems planning approach is analogous to "comprehensive planning". Transportation 3: p. 66.

Theory. <u>Federal--Interior</u>. The possible explanation of phenomenon for which only a small amount of evidence is available. General 3: p. 76.

Time-Contour Map. <u>Federal--DOT</u>. A map on which the contour lines connect points of equal travel time from a specified central point, given a particular mode of transport. Transportation 3: p. 67.

Trade Area. <u>Federal--DOT</u>. The primary distributing area for the community's retail and wholesale functions. Transportation 3: p. 68.

Trade-Off Analysis. <u>Federal--DOT</u>. The balancing and weighing of
relevant considerations by decisionmakers when choices are
offered between mutually exclusive options. Transportation
3: p. 68.

Transect. <u>National--Soil Conservation</u>. A cross section of an
area used as a sample for recording, mapping, or studying
vegetation and its use. A transect may be a series of
plots, a belt or strip, or merely a line, depending on why
it is being used. Earth 4: p. 57g.

Turbidity. <u>Federal--EPA</u>. Hazy air due to the presence of particulates and pollutants; a similar cloudy condition in water
due to suspended silt or organic matter. General 1: p. 15.

Urban Renewal. <u>Federal--DOT</u>. A form of recuperative change in
the physical city by which dysfunctional or outmoded structures and facilities are replaced in response to pressures
of economic and social change. It occurs by improving urban
environments through public initiative and assistance in
demolishing slums, rehabilitating or conserving existing
structures, providing better housing, commercial, industrial and public buildings, as well as other amenities.
Transportation 3: p. 70.

Use. <u>Federal--DOT</u>. Denotes the specific purpose for which land
or a building is designed or occupied. Transportation 3:
p. 70.

Variance. <u>Federal--EPA</u>. Government permission for a delay or
exception in the application of a given law, ordinance, or
regulation. General 1: p. 16.

  <u>Federal--Interior</u>. Sanction granted by a governing body
  for delay or exception in the application of a given law,
  ordinance, or regulation. General 3: p. 81.

Visual Character. <u>Federal--DOT</u>. The aspects of the appearance
of an area that lend themselves to generalization as typ-
ifying the whole. Transportation 3: p. 71.

Visual Landscape Character. <u>Federal--Forest Service</u>. U.S. Forest Service, Visual Management System usage. The overall
impression created by a landscape's unique combination of
visual features (such as land, vegetation, water, structures) as seen in terms of form, line, color and texture.
General 15: p. 227.

Visual Sensitivity Level. <u>Federal--Forest Service</u>. U.S. Forest
Service, Visual Management System usage. A three-level

rating system used to delineate areas receiving different
amounts of exposure (present and potential) to user groups
with different attitudes toward changes in scenic quality
(such as might occur as a result of management activities).
The system initially classifies all travel routes, special
interest areas and water bodies into areas of primary and
secondary aesthetic management importance on basis of their
national importance, number of users, duration of use and
area size. The system next uses the assumption that aesthetic quality will be of major concern to recreational
users and minor concern to functional users of forest
areas (such as daily commuters and loggers) as the other
basis for classifying the entire planning area into three
sensitivity levels. General 15: p. 228.

Visual Variety Class. Federal--Forest Service. U.S. Forest Service, Visual Mangement System usage. A classification system for establishing three visual landscape categories
according to the relative importance of the visual features,
i.e., landforms, vegetation patterns, stream or lake water
forms and rock formations. The class A ("distinctive")
category contains those landscape features of unusual or
outstanding visual quality, and, that are usually not
common in the visual character type or visual character
subtype. The class B ("common") category contains those
landscape features which tend to be common throughout the
visual character type or subtype and thus are not outstanding in visual quality. The class C ("minimal") category contains those areas with few distinguishing landscape features--including all areas not classed as A or B.
This classification system is based on the premise that
all landscapes have some visual values, but those with the
most variety or diversity of visual features have the
greatest potential for being or attaining high scenic value.
General 15: pp. 228-229.

Zonation. Federal--Army. Distinct, conspicuous layers or belts,
e.g., in soils, vegetation, bodies of water, and on mountains. General 9: p. 50.

Zone. Federal--DOT. A discrete spatial area of analysis such
as a community census tract or transportation zone. Transportation 3: p. 73.

Zoning Ordinance. National--Soil Conservation. An ordinance
based on the police power of government to protect the
public health, safety, and general welfare. It may regulate the type of use and intensity of development of
land and structures to the extent necessary for a public

purpose. Requirements may vary among various geographically defined areas called zones. Regulations generally
cover such items as height and bulk of buildings, density
of dwelling units, off-street parking, control of signs,
and use of land for residential, commercial, industrial,
or agricultural purposes. A zoning ordinance is one of
the major methods for implementation of the comprehensive
plan. Earth 2: p. 61g.

2. LEGAL JARGON

Index

Environmental Impact Analysis
Environmental Impact Appraisal
Environmental Impact Assess-
    ment (EIA)
Environmental Impact Assess-
    ment Report
Environmental Impact Report
    (EIR)
Environmental Impact State-
    ment (EIS)
Environmental Impact State-
    ment Preparation Notice
Environmental Impact State-
    ment Process
Environmental Notification
    Form (ENF)
Environmental Report (ER)
Environmental Resources
Environmental Review
Environmental Statement (ES)
Environmental Statement Prep-
    aration Plan
Environmental Assessment Work-
    sheet (EAW)
Excluded Action
Executive Order
Exempt Classes Of Action
Extramural Project
Feasibility And Planning
    Studies
Feasible
Federal Agency
Federal Finding
Final Environmental Impact
    Statement (FEIS)
Finding Of Inapplicability
Finding Of No Significant
    Impact
Government Action
Governmental Permit
Human Element
Human Environment
Human Life
Impacts
Inadequate EIS
Initial Study
Intramural Project
Joint Environmental Impact
    Statement
Land Pollution

Lead Agency
Lead City Agency
Legislation
Legislative Action
License
Local Agency
Major Action
Major Amendatory
Major Development Project
Major Federal Action
Master EIR
Matter
Ministerial Action
Ministerial Projects
Mitigation
Negative Declaration
Negative Declaration EIS
Negative Declaration Notice
Negative Environmental Dec-
    laration (NED)
NEPA-Associated Documents
NEPA Process
New Source
New Source And Environmental
    Questionnaire (NS/EQ)
Notice Of Completion
Notice Of Determination
Notice Of Exemption
Notice Of Intent
Notice Of Preparation
Opinion
Other Approving Agencies
Owner Of Land Abutting The
    Activity
Participating Agency
Participating City Agency
PER Announcement Sheet
Permit
Permit Condition
Permitee
Permit Pre-Condition
Person
Persons With Special Expertise
Petition
Plans
Potential New Source Applicant
Preliminary Analysis
Preliminary Environmental Report
    (PER)
Preliminary Environmental Review
    (PER)

Prior Finding Affirmation
Private Action
Project
Project Data Statement
Project Sponsor
Program
Program EIR
Programmatic Review
Proponent
Proposal
Proposer
Proprietary Department
Public Agency
Public Entities
Public Hearing
Recirculation
Referring Agency
Relevant A-95 Clearinghouse
Report
Resource Inventory
Resources
Responsible Agency
Responsible Official
Review
Reviewing Agencies
Schedule Of Compliance
Scope
Section 208 Plan
Short-Term Impacts
Significant
Significant Effect
Significant Effect On The En-
    vironment
Significant Environmental
    Effects
Significant Environmental
    Impact
Significant Impact
Significantly
Site
Source
Special-Purpose Unit Of
    Government
Sponsoring Agency
Staged EIR
State EIR Guidelines
Statement Of Nonsignificant
    Impact
Statement Of Overriding
    Considerations

Statement Record
Supplemental Environmental
    Impact Statement
Supporting Data
Terminology
Third Party
Threshold
Threshold Of Determination
Tiering
Typically Associated Environ-
    mental Effect
Undertaking
United States

Terms

Abstract. State--Executive Order. A concise series of state-
    ments, with appropriate graphics, which describes an ac-
    tivity or action; proposed by a state agency, and iden-
    tifies any potentially adverse environmental effects and
    public concerns or controversies that may occur as a re-
    sult of the proposed activity or action for the purpose
    of determining the need to prepare an environmental im-
    pact statement (EIS). Legal Jargon 27: p. 3.

Act. Federal--CEQ. Means the National Environmental Policy Act,
    as amended (42 U.S.C. 4321, et seq.) which is also refer-
    red to as "NEPA". Legal Jargon 30: p. 25244.

Acting Agency. State--Department of Ecology. Means an agency
    with jurisdiction which has received an application for
    a license, or which is proposing an action. Legal Jar-
    gon 37: p. 2.

Action. Regional--Commission. A resolution by the Commission
    approving, disapproving, modifying or otherwise disposing
    of a project, program, legislation or any part thereof.
    Legal Jargon 4: p. 464.

    State--Commission. Any program or project to be initiated
    by any agency or applicant. Legal Jargon 3: p. 2.

    State--Environmental Conservation. Include but are not
    limited to: (1) projects or physical activities such as
    construction or other activities which change the use or
    appearance of any natural resource or structure which (i)
    are directly undertaken by an agency, or (ii) involve fund-
    ing by an agency, or (iii) require one or more permits from
    an agency or agencies; (2) planning activities of an agency
    when such planning substantially commits an agency to a
    definite course of future decisions; (3) policy making act-
    ivities of an agency such as the making, modification or
    establishment of rules, regulations procedures and pol-
    icies; and (4) combinations of the above. Capital projects
    directly undertaken by an agency commonly consist of sev-
    eral distinct sets of activities or steps (e.g., planning,
    design, contracting, construction and operation}. For the
    purposes of this Part, the entire set of activities or
    steps can be considered an "action". If it is determined
    that an environmental impact statement is necessary, only
    one draft and one final environmental impact statement need
    be prepared on the action if the statements address each
    step at a level of detail sufficient for adequate analysis

and disclosure of environmental effects. In the case of
a project or activity involving funding or a permit from
an agency, the entire project shall be considered an "action" regardless of whether such funding or permit relates
to the project as a whole or to a portion or component of
it. Legal Jargon 24: pp. 3-4.

State--Environmental Council. The whole of a project which
will cause physical manipulation of the environment, directly or indirectly. The determination of whether an
action requires environmental documents shall be made by
reference to the physical activity to be undertaken and
not to the governmental process of approving the activity.
"Action" does not include the following: (1) Proposals
and enactments of the Legislature, (2) The rules, orders,
or recommendations of public agencies, (3) Executive Orders of the Govenor, or their implementation by public
agencies, (4) Judicial orders, except orders establishing
judicial ditches pursuant to Minn. Stat. ch. 106(1974),
(5) Submission of proposals to a vote of the people of
the state. Legal Jargon 26: p. 1093.

State--Water Resources. Any activity, pursuit or procedure requiring permission from the department, or any department activity, pursuit or procedure which may affect
the human environment. Legal Jargon 21: p. 1.

City--Office of the Mayor. Any activity of an agency,
other than an exempt action enumerated in section 4 of
this Executive Order, including but not limited to the
following: (1) non-ministerial decisions on physical
activities such as construction or other activities which
change the use or appearance of any natural resource or
structure; (2) non-ministerial decisions on funding activities such as the proposing, approval or disapproval of
contracts, grants, subsidies, loans, tax abatements or
exemptions or other forms of direct or indirect financial
assistance, other than expense budget funding activities;
(3) planning activities such as site selection for other
activities and the proposing, approval or disapproval of
master or long range plans, zoning or other land use maps,
ordinances or regulations, development plans or other plans
designed to provide a program for future activities; (4)
policy making activities such as the making, modification
or establishment or rules, regulations, procedures, policies and guidelines; (5) non-ministerial decisions on li-
censing activities, such as the proposing, approval or disapproval of a lease, permit, license, certificate or other
entitlement for use or permission to act. Legal Jargon
23: pp. 2-3.

Action Significantly Affecting The Quality Of The Human Environ-
      ment. Federal--Transportation. An action in which the
      overall cumulative primary and secondary consequences sig-
      nificantly alter the quality of the human environment, cur-
      tail the choices of beneficial uses of the human environ-
      ment, or interfere with the attainment of long-range human
      environmental goals. Transportation 2: p. 3.

Administrative Action. Federal--Energy. Means a major DOE activ-
      ity, other than a legislative action as defined herein, sig-
      nificantly affecting the quality of the human environment.
      Legal Jargon 32: p. 7235.

Affecting. Federal--CEQ. Means will or may have an effect on.
      Legal Jargon 30: p. 25244.

Agency. State--Commission. Any department, office, board or com-
      mission of the State or County government which is part of
      the executive branch of that government. Legal Jargon 3:
      p. 2.

      State--Environmental Affairs. An agency, department, board,
      commission or authority of the commonwealth, and any author-
      ity of any political subdivision which is specifically cre-
      ated as an authority under special or general law. In cases
      of doubt as to whether a body is an "agency" for purposes
      of these regulations, an opinion of the secretary should
      be sought under MGL, Chapter 30A, Section 8. Legal Jargon
      6: p. 4.

      State--Law. The executive and administrative departments,
      offices, boards, commissions, and other units of the state
      government. Legal Jargon 5: p. 602.

Agency Action. State--Commission. An action proposed by an agency
      which will use State or County lands or funds. Legal Jar-
      gon 3: p. 2.

Agency Official. Federal--Historic Preservation. The head of the
      Federal agency having responsibility for the undertaking or
      a subordinate employee of the Federal agency to whom such
      authority has been delegated. Laws/Regulations 1: p. 3367.

Alter. State--Environmental Quality. Including, but not limited
      to, any one or more of the following actions upon areas
      described in the Act: (a) The removal, excavation or dredg-
      ing of soil, sand, gravel, or aggregate material of any
      kind, (b) The changing of pre-existing drainage character-
      istics, flushing characteristics, salinity distribution,
      sedimentation patterns, flow patterns and flood storage

retention areas, (c) The drainage or distribution of the
water level or water table, (d) The dumping, discharging
or filling with any materials which could degrade the
water quality, (e) The driving of pilings or the erection
of buildings or structures of any kind, (f) The placing
of obstructions whether or not they interfere with the
flow of water, (g) The destruction of plant life, including
the cutting of trees, which could result in harm to an
area described in the Act, (h) The changing of water temperature, biochemical oxygen demand (BOD), and other natural
characteristics of the receiving water, (i) Changes which
may adversely affect shellfish or fisheries. Legal Jargon
28: p. 2.

Alternatives. State--Natural Resources. Other actions or activities which may be reasonably available to achieve the
same or altered purpose(s) of the proposed action. Legal
Jargon 21: p. 1.

An Action Significantly Affecting The Quality Of The Human Environment. Federal--DOT. An action in which the overall cumulative primary and secondary consequences significantly
alter the quality of the human environment, curtail the
choices of beneficial uses of the human environment, or
interfere with the attainment of long-range human environmental goals. Legal Jargon 13: p. 41806.

Applicant. Regional--Commission. Proposed action's sponsor, including the Commission when it sponsors an action. Legal
Jargon 4: p. 464.

State--Commission. Any person that, pursuant to statute,
ordinance, rule, or regulation, officially request approval
from an agency for a proposed action. Legal Jargon 3:
p. 2.

State--Environmental Conservation. Any person making an
application or other request to an agency to provide funding or to grant a permit in connection with a proposed
action. Legal Jargon 24: p. 4.

State--Environmental Protection. Any person requesting a
permit who has submitted an application to the Department.
Legal Jargon 14: p. 1.

State--Resources Agency. A person who proposes to carry
out a project which needs a lease, permit, license, certificate, or other entitlement to use or financial assistance
from one or more public agencies when that person applies

for the governmental approval or assistance. Legal Jargon 15: p. 5.

City--Law. An applicant is a person who proposes to carry
out a project and needs a lease, permit, license, certificate, or other entitlement or use, or who is requesting
financing assistance from one or more public agencies to
carry out a project. Legal Jargon 2: p. II-1.

City--Office of the Mayor. Any person required to file an
application pursuant to this Executive Order. Legal Jargon 23: p. 3.

Applicant's Environmental Report (AER). State--Executive Order.
A series of statements and technical data furnished by an
applicant for state approval or action which describes a
proposed activity and which may be used by the recipient
state agency in determining the need for an EIS or as an
informational base for their preparation of an appropriate
type of EIS. Legal Jargon 27: p. 4.

Approval. State--Commission. For the purposes of these Regulations means a discretionary consent, sanction, or recommendation required for an agency prior to actual implementation of an action, as distinguished from a ministerial
situation. Legal Jargon 3: p. 2.

State--Environmental Council. The issuance of a governmental permit, or any review of a proposed action required
by state or federal law or regulation. Legal Jargon 26:
p. 1093.

State--Resources Agency. The decision by a public agency
which commits the agency to a definite course of action in
regard to a project intended to be carried out by any person. The exact date of approval of any project is a matter
determined by each public agency according to its rules,
regulations, and ordinances. Legislative action in regard
to a project often constitutes approval. In connection
with private activities, approval occurs upon the earliest
commitment to issue or the issuance by the public agency
of a discretionary contract, grant, subsidy, loan, or other
form of financial assistance, lease, permit, license, certificate, or other entitlement for use of the project.
Legal Jargon 15: p. 5.

Approval Of An Action. State--Environmental Conservation. A decision or commitment by an agency to undertake or carry
out a direct action; to provide funding in connection with
a project or activity proposed by another person; to issue

a permit or to otherwise authorize a project or activity
proposed by another person; or to adopt or approve a plan,
policy, rule or regulation. Legal Jargon 24: p. 5.

Assessment. <u>State--Commission</u>. An evaluation by an agency of a
proposed action to determine whether an Environmental Impact Statement is required. Legal Jargon 3: p. 2.

Categorical Exclusion. <u>Federal--CEQ</u>. Means a category of actions
which do not individually or cumulatively have a significant effect on the human environment and which have been
found to have no such effect in procedures adopted by a
Federal agency in implementation of these regulations (Section 1507.3) and for which, therefore, neither an environmental assessment nor an environmental impact is needed.
Any such procedure shall provide for extraordinary circumstances in which a normally excluded action may have a significant environmental effect. Legal Jargon 30: p. 25244.

Categorical Exemption. <u>State--Resources Agency</u>. An exception
from the requirements of CEQA for a class of projects based
on a finding by the Secretary for Resources that the class
of projects does not have a significant effect on the environment. Legal Jargon 15: p. 5.

<u>City--Environmental Section</u>. An exemption from the require-
ments of CEQA (California Environmental Quality Act) based
on a finding by the Secretary for Resources and the Los
Angeles City Council that certain types of projects do not
have a significant effect on the environment. Legal Jargon 2: p. II-2.

Circulation Period. <u>Federal--DOT</u>. Process of sending copies of
both draft and final EIS's to Federal, State, and local
agencies, and the public. The process differs slightly
for draft and final statements. Legal Jargon 35: p. A-1.

Clearinghouse A-95. <u>National--Soil Conservation</u>. An agency at
the state and metropolitan or regional level (usually an
agency responsible for comprehensive planning) designated
to provide a review and comment on projects requesting
federal assistance. The purpose is to assure maximum consistency among projects in different functional areas and
with the comprehensive plans of state, regional, and local
agencies. "A-95" is derived from Circular A-95 issued by
the Office of Management and Budget, which established the
program. Earth 4: p. 12g.

Comments. <u>Federal--GSA</u>. All formal reactions by public and private entities to the proposed action and to the environmen-

tal impact statement. Legal Jargon 16: p. 2.

Federal--HUD. Formal reactions by one Federal agency to
the proposed action and the Draft Environmental Impact
Statement of another Federal agency; also formal reactions
of State and local agencies or of public or private groups
to the proposed action. Each Federal agency, including
HUD, is obliged to take these comments received on its
Draft Environmental Impact Statement into account in the
Final Environmental Impact Statement and in the approval
or disapproval of the proposal. The HUD format for comments on Environmental Impact Statements of other agencies
is set forth in Appendix G. Legal Jargon 11: p. 19183.

Complaint. State--Environmental Control. Any charge, however
informal, to or by the council, that any person or agency,
private or public, is polluting the air, land, or water
or is violating the provisions of sections 81-1501 to
81-1533 or any rule or regulation of the council in respect therof. Legal Jargon 25: p. 1.

Comprehensive. Federal--Army. The conscious consideration of a
study area's environmental resources, problems and needs;
the scope of which emphasizes breadth rather than depth.
Legal Jargon 31: p. 10782.

Conditional Negative Declaration. City--Office of the Mayor.
A written statement prepared by the lead agencies after
conducting an environmental analysis of an action and ac-
cepted by the applicant in writing, which announces that
the lead agencies have determined that the action will not
have a significant effect on the environment if the action
is modified in accordance with conditions or alternatives
designed to avoid adverse environmental impacts. Legal
Jargon 23: p. 3.

Construction. Federal--EPA. As defined in section 306(a)(5) of
the FWPCA means "any placement, assembly, or installation
of facilities or equipment (including contractural obligations to purchase such facilities or equipment) at the premises where such equipment will be used, including preparation work at such premises". Legal Jargon 19: p. 2453.

City--Commission. Any and all physical activity necessary
or incidental to the erection, placement, demolition, as-
sembling, altering, cleaning, repairing, installing, or
equipping of buildings, and other structures, public or
private highways, roads, premises, parks, utility lines, or
property, and shall include land clearing, grading, excavating, filling, and paving. Noise 1: p. 2.

Construction Site. <u>City--Commission</u>. That area within which a
     contractor confines a construction operation. This in-
     cludes defined boundary lines of the project itself plus
     any contractor staging area outside those define boundary
     lines used expressly for the construction. Noise 1: p. 2.

Consulted Agency. <u>State--Department of Ecology</u>. Means any agency
     with jurisdiction or with expertise which is required by
     the lead agency to provide information during a threshold
     determination or predraft consultation or which receives a
     draft environmental impact statement. An agency shall not
     be considered to be a consulted agency merely because it
     receives a proposed declaration of nonsignificance. Legal
     Jargon 37: p. 3.

Cooperating Agency. <u>Federal--CEQ</u>. Means any Federal agency other
     than a lead agency which has jurisdiction by law or special
     expertise with respect to any environmental impact involved
     in a proposal (or a reasonable alternative) for legislation
     or other major Federal action significantly affecting the
     quality of the human environment. The selection and re-
     sponsibility of a cooperating agency are described in Sec-
     tion 1501.6. A State or local agency or similar qualifi-
     cations or, when the effects are on a reservation, an In-
     dian Tribe may by agreement with the lead agency become a
     cooperating agency. Legal Jargon 30: p. 25244.

Council. <u>Federal--CEQ</u>. Means the Council on Environmental Qual-
     ity established by Title II of the Act. Legal Jargon 30:
     p. 25244.

Cumulative Effects. <u>Federal--GSA</u>. Those environmental impacts
     which result when individual actions, taken at the same
     time or over a period of time, are considered collectively.
     These actions may be taken by one or more agencies or other
     parties. The actions may be influenced by other actions.
     Legal Jargon 18: Appendix B-3.

Cumulative Impact. <u>Federal--CEQ</u>. The impact on the environment
     which results from the incremental impact of the action
     when added to other past, present, and reasonably foresee-
     able future actions regardless of what agency (Federal or
     non-Federal) or person undertakes such other actions. Cum-
     ulative impacts can result from individually minor but col-
     lectively significant actions taking place over a period of
     time. Legal Jargon 30: p. 25244.

     <u>Federal--Commerce</u>. Shall mean actions which, in themselves,
     would not involve a significant impact on the quality of

the human environment, but which, nonetheless shall be considered as having a significant impact if they can reasonably be expected to set a precedent for a series of subsequent actions which, when considered cumulatively, would
result in a significant impact. Also, several related
Federal actions in a specific area may result in a significant impact as a result of the aggregated actions. Legal
Jargon 29: p. 3.

State--Resources Agency. Refer to two or more individual
effects which, when considered together, are considerable
or which compound or increase other environmental impacts.
The individual effects may be changes resulting from a
single project or a number of separate projects. Legal
Jargon 15: p. 5.

Damage To The Environment. State--Environmental Affairs. Any
destruction, damage or impairment, actual or probable, to
any of the natural resources of the commonwealth and shall
include but not be limited to air pollution, water pollution, improper sewage disposal, pesticide pollution, excessive noise, improper operation of dumping grounds, impairment and eutrophication of rivers, streams, flood plains,
lakes, ponds, or other surfaces or sub-surface water resources; destruction of seashores, dunes, marine resources,
underwater archaeological resources, wetlands, open spaces,
natural areas, parks or historic districts or sites. Damage to the environment shall not be construed to include
any insignificant damage to or impairment of such resources.
Legal Jargon 6: p. 4.

Days. State--Environmental Affairs. Calendar days; provided,
that in computing time periods such periods shall exclude
the day of the event which starts the period running and
further provided that if the last day of a period falls on
a Sunday, legal holiday or declared state of emergency day,
such period shall be extended to close of business on the
next business day. Legal Jargon 6: p. 4.

State--Environmental Council. In computing any period of
time perscribed or allowed in these Rules, the day of the
act or the event from which the designated period of time
begins to run shall not be included. The last day of the
period so computed shall be included unless it is a Saturday, Sunday, or a legal holiday, in which event the period
runs until the end of the next day which is not a Saturday, Sunday, or a legal holiday. When the period of time
prescribed or allowed is 15 days or less, intermediate Sat-
urdays, Sundays, and legal holidays shall be excluded in
the computation. Legal Jargon 26: p. 1093.

Decision. **Federal--Historic Preservation**. Means the exercise of
     agency auhtority at any stage of an undertaking where al-
     terations might be made in the undertaking to modify its
     impact upon historic and cultural properties. Legal Jar-
     gon 34: p. 3367.

     State--Environmental Protection. A ruling of the Commis-
     sioner on the application for a permit. Legal Jargon 14:
     p. 1.

Decision-Making Body. **State--Resources Agency**. Any person or
     group of people within a public agency permitted by law
     to approve or disapprove the project at issue. Legal Jar-
     gon 15: p. 5.

     City--Environmental Section. The group or individuals hav-
     ing project approval authority. Legal Jargon 2: p. II-3.

Decision Points. **Federal--HUD**. Those points of Federal commit-
     ment in the decision-making process before which prescribed
     environmental clearances must be completed. HUD decision
     points are set forth in Appendix A-1 of this Handbook.
     Legal Jargon 11: p. 19183.

Declaration Of Nonsignificance. **State--Department of Ecology**.
     Means the written decision by the responsible official of
     the lead agency that a proposal will not have a significant
     adverse environmental impact and that therefore no environ-
     mental impact statement is required. A form substantially
     consistent with that in WAC 197-10-355 shall be used for
     this declaration. Legal Jargon 37: p. 3.

Declaration Of Significance. **State--Department of Ecology**. Means
     the written decision by the responsible offical of the lead
     agency that a proposal will or could have a significant ad-
     verse environmental impact and that therefore an environ-
     mental impact statement is required. A form substantially
     consistent with that in WAC 197-10-355 shall be used for
     this declaration. Legal Jargon 37: p. 3.

Detail. **Federal--Army**. Pertains to the depth to which resource
     elements are studied and described. Legal Jargon 31:
     p. 10782.

Determination. **State--Environmental Quality**. Shall be defined
     as follows: (a) A written finding by the conservation com-
     mission whether the proposed work shall or shall not re-
     quire the filing of a Notice of Intent, (b) A written find-
     ing by the conservation commission, after a public hearing,

whether the area on which the proposed work is to be done
is significant to the interests described in the Act, (c)
A written finding by the Commissioner, upon a request made
to him, whether the area on which the proposed work is to
be done is significant to the interests described in the
Act. Legal Jargon 28: pp. 3-4.

Direct Action. State--Environmental Conservation. An action
planned and proposed for implementation by an agency itself for purposes of carrying out its governmental duties,
functions, obligations or responsibilities. A direct action may take various forms, including but not limited to
a capital project, rule making, procedure making and policy making. Legal Jargon 24: p. 5.

Discretionary Project. State--Resource Agency. An activity defined as a project which requires the exercise of judgment,
deliberation, or decision on the part of the public agency
or body in the process of approving or disapproving a particular activity, as distinguished from situations where
the public agency or body merely has to determine whether
there has been conformity with applicable statutes, ordinances, or regulations. A timber harvesting plan submitted to the State Forester for approval under the requie-
ments of the Z'berg-Nejedly Forest Practice Act of 1973
(Pub. Res. Code Section 4511 et seq.) constitutes a discretionary project within the meaning of the California
Environmental Quality Act. Section 21065(c). Legal Jargon 15: p. 5.

Draft Environmental Impact Statement (DEIS). Federal--DOT. This
is the document that represents FAA's evaluation of the
environmental impact of a proposed action when coordination pursuant to Section 102(2)(C) of NEPA is initiated.
The agency makes its own evaluation and assumes responsibility for the draft environmental impact statement. It
is simultaneously distributed by FAA to the Council on Environmental Quality (CEQ), other appropriate Federal agencies, state and local agencies, and to the public. Legal
Jargon 12: p. 32648.

Federal--EPA. The document prepared by EPA, or under EPA
guidance, which attempts to identify and analyze the environmental impacts of a proposed EPA action and feasibility alternatives, and is circulated for public comment prior
to preparation of the final environmental impact statement
(final EIS). Legal Jargon 19: p. 2453.

State--Council on the Environment. A Federal document that
fully addresses the sponsoring agency's consideration of

environmental impacts, alternatives, mitigation measures,
etc. The Draft EIS will include information contained in
the EA when available. Legal Jargon 1: p. 3.

Ecological Impact. Federal--GSA. The total of an environmental
change, either natural or man-made, on the ecology of the
area. Legal Jargon 18: Appendix B-3.

Effects. Federal--CEQ. Effects include: (a) Direct effects,
which are caused by the action and occur at the same time
and place, (b) Indirect effects, which are caused by the
action and are later in time or farther removed in distance,
but are still reasonably foreseeable. Indirect effects may
include growth inducing effects and other effects related
to induced changes in the pattern of land use, population
density or growth rate, and related effects on air and
water and other natural systems, including ecosystems. Effects and impacts used in these regulations are synonymous.
Effects includes ecological (such as the effects on natural
resources and on the components, structures, and function-
ing of affected ecosystems), economic, soical, or health,
whether direct, indirect, or cumulative. Effects may also
include those resulting from actions which may have both
beneficial and detrimental effects, even if on balance the
agency believes that the effect will be beneficial. Legal
Jargon 30: p. 25244.

Element. State--Executive Order. A portion or component of the
natural or human resource complex of the State which can
be identified, either because of its nature, size, location
or value as potentially being altered by a proposed action.
Legal Jargon 27: p. 3.

Emergency. State--Commission. A sudden unexpected occurrence demanding immediate action to prevent or mitigate loss or damage to life, health, property, or essential public services.
Legal Jargon 3: p. 2.

State--Resources Agency. A sudden, unexpected occurrence,
involving a clear and imminent danger, demanding immediate
action to prevent or mitigate loss of, or damage to life,
health, property, or essential public services. Emergency
includes such occurrences as fire, flood, earthquake, or
other soil or geologic movements, as well as such occurrences as riot, accident, or sabotage. Legal Jargon 15:
p. 5.

Emergency Actions. State--Environmental Commission. (a) Projects
undertaken, carried out, or approved by the Board or

Department to repair or restore property or facilities
damaged or destroyed as a result of a disaster when a disaster has been declared by the govenor or other appropriate
government official; (b) Emergency repairs to public services facilities necessary to maintain service; or (c) Projects, whether public or private, undertaken to prevent or
mitigate immediate threats to public health, safety, or
welfare. Legal Jargon 10: p. 2.

Environment. Federal--GSA. The whole complex of physical, social,
cultural, and aesthetic factors which affect individuals
and communities and ultimately determine their form, character, relationship, and survival. Legal Jargon 16: p. 1.

Federal--HUD. Environment is not defined in NEPA or in
the CEQ Guidelines. However, it is clear from section 102
of the Act and elsewhere that the term is meant to be interpreted broadly to include physical, social, cultural,
and aesthetic dimensions. Examples of environmental considerations are: air and water quality, erosion control,
natural hazzards, land use planning, site selection and
design, subdivision development, conservation of flora and
fauna, urban congestion, overcrowding, displacement and re-
location resulting from public and private action or natural
disaster, noise pollution, urban blight, code violations and
building abandonment, urban sprawl, urban growth policy,
preservation of cultural resources, including properties
on the National Register of Historic Places, urban design
and the quality of the built environment, the impact of
the environment on people and their activities. Legal Jargon 11: p. 19183.

Regional--Commission. For the purposes of the regulations
in this part is the major natural, man-made or affected environment as implied by the National Environmental Policy
Act of 1969. Legal Jargon 4: p. 464.

State--Commission. Man's surroundings, inclusive of all
the physical, economic and social conditions which exist
within the area which will be affected by a proposed action
including land, human and animal communities, air, water,
minerals, flora, fauna, ambient noise, and objects of historic or aesthetic significance. Legal Jargon 3: p. 2.

State--Environmental Conservation. The physical conditions
which will be affected by a proposed action, including land,
air, water, minerals, flora, fauna, noise, objects or historic or aesthetic significance, existing patterns of population concentration, distribution or growth, and existing

community or neighborhood character. Legal Jargon 24:
p. 5.

State--Environmental Council. The physical conditions existing in the area which will be affected by the proposed
action, including land, air, water, minerals, flora, fauna,
ambient noise, energy resources available to the area, and
man-made objects or natural features of historic, geologic
or aesthetic significance. Legal Jargon 26: p. 1093.

State--Executive Order. Environment shall be defined as
the natural resources of the state, including air, water,
land, mineral and energy resources and the flora and fauna
(not including human beings). Legal Jargon 27: p. 3.

State--Interagency Council. The natural, social and economic conditions which exist within the area which will be
affected by a proposed project. Legal Jargon 7: p. 6.

State--Law. The physical conditions which will be affected
by a proposed action, including land, air, water, minerals,
flora, fauna, noise, objects of historic or aesthetic significance, existing patterns of population concentration,
distribution, or growth, and existing community or neighborhood character. Legal Jargon 5: p. 602.

City--Environmental Section. Environment, for purposes of
implementing CEQA (California Environmental Quality Act),
is the physical conditions which exist within the area
which will be affected by a proposed project including
land, air, water, minerals, flora, fauna, ambient noise,
and objects of historic or aesthetic significance. Legal
Jargon 2: p. II-3.

Environmental Analysis. City--Office of the Mayor. The lead
agencies' evaluation of the short and long term, primary
and secondary environmental effects of an action, with
particular attention to the same areas of environmental
impacts as would be contained in an EIS. It is the means
by which the lead agencies determine whether an action under consideration may or will not have a significant effect
on the environment. Legal Jargon 23: p. 4.

Environmental Appraisal (EA). Federal--EPA. A document, based on
the environmental review, which supports a negative declaration. Legal Jargon 19: p. 2454.

Environmental Assessment (EA). Federal--CEQ. (a) Means a public
document for which a Federal agency is responsible that
serves to: (1) Briefly provide sufficient evidence and

analysis for determining whether to prepare an environmental impact statement or a finding of no significant impact,
(2) Aid an agency's compliance with the Act when no environmental impact statement is necessary, (3) Facilitate
preparation of such a statement when one is necessary. (b)
Shall include brief discussions of the need for the proposal, of alternatives as required by sec. 102(2)(E), of
the environmental impacts of the proposed action and alternatives, and a listing of agencies and persons consulted.
Most environmental impacts of the proposed action and alternatives, and a listing of agencies and persons consulted. Most environmental assessments do not exceed several
pages in length. Legal Jargon 30: p. 25244.

Federal--Energy. Means a document prepared by DOE which
assesses whether a proposed DOE action would be "major"
and would "significantly affect" the quality of the human
environment, and which serves as the basis for a determination as to whether an environmental impact statement (EIS)
is required. Legal Jargon 32: p. 7235.

Federal--EPA. A written analysis submitted to EPA by the
grantees or contractors describing the environmental impacts of proposed actions undertaken with the financial
support of EPA. For facilities or section 208 plans as
defined in Section 6.102 (j) and (k), the assessment must
be an integral, though identifiable, part of the plan submitted to EPA for review. Legal Jargon 33: p. 16815.

Federal--GSA. An evaluation occurring early in the approval process of the potential environmental impact of a project or activity. Legal Jargon 16: p. 1.

Regional--Commission. An analysis by the Commission prior
to the preparation of an environmental impact statement, of
an applicant's environmental report or of a Commission-spon-
sored action to determine whether the action proposed will
have a significant effect involving the quality of the human environment. Legal Jargon 4: p. 464.

State--Council on the Environment. The term "environmental assessment" refers to the actual assessment of environmental impacts as well as to the documentation of the assessment. The latter is a Federal document that contains
minimal information about environmental impacts and the
significance of the impacts. The EA can serve as the basis
for determining the necessity of the preparation of an EIS.
Legal Jargon 1: p. 3.

State--Environmental Board. A cursory assessment of the
probable environmental effect of a proposed action, determined in accordance with the provisions of Reg. EMB-2.
Legal Jargon 9: p. 2.

State--Natural Resources. A brief assessment of environmental impacts of a proposed action by a state or federal
agency which also evaluates the need for an EIS. Legal
Jargon 20: p. 2.

Environmental Assessment Form. City--Office of the Mayor. A written form completed by the lead agencies, designed to assist
their evaluation of actions to determine whether an action
under consideration may or will not have a significant
effect on the environment. Legal Jargon 23: p. 4.

Environmental Decision Memorandum. Federal--DOT. This is an FAA
staff memorandum transmitting the negative declaration or
environmental impact statement and the proposed Federal
Finding to the responsible official. The memorandum sets
forth the action to be taken, discusses the key issues,
and indicates any factors requiring special consideration.
Legal Jargon 12: p. 32648.

Environmental Documents. Federal--CEQ. Includes the documents
specified in Sections 1508.9, 1508.11, 1508.13, and 1508.21.
Legal Jargon 30: p. 25244.

State--Resources Agency. Draft and Final EIRs, Initial
Studies, Negative Declarations, Notices of Completion, and
Notices of Determination. Legal Jargon 15: p. 6.

Environmental Elements. State--Executive Orders. Environmental
elements are identified in three categories: biological,
physical, chemical and include such things as water sheds,
water bodies (all or in part), forest, recreation, or habitat areas, environmental zones (shore lands, etc.) or any
other definable location, space, mass or resource complex
that would be directly or indirectly impacted by a proposed action. Legal Jargon 27: p. 3.

Environmental Impact. Federal--EPA. Shall refer to both the adverse and the beneficial impacts associated with a new
source. Legal Jargon 19: p. 2453.

Federal--HUD. Any alteration of environmental conditions
or creation of a new set of environmental conditions, adverse or beneficial, caused or induced by the action or
set of actions under consideration. Assessment of the
significance of the environmental impact generally involves

two major elements: a quantitative measure of magnitude
and a qualitative measure of importance. Such a determination is a matter of agency judgment and consensus; at
the project level, this judgment shall be governed by HUD,
environmental policies and standards. Legal Jargon 11:
p. 19183.

State--Commission. An effect of any kind, whether immediate or delayed, on any component or the whole of the environment. Legal Jargon 3: p. 2.

Environmental Impact Analysis. Federal--GSA. The orderly and
logical process by which the potential impact of a proposed
development project on its immediate and more distant environments is analyzed. Types of analyses may range from
impact on animal and plant life to impact on urban economy
or health, depending on the nature and location of the development project. Legal Jargon 18: Appendix B-3.

Environmental Impact Appraisal. Federal--EPA. Based on an environmental review and supports a negative declaration.
It describes a proposed EPA action, its expected environmental impact, and the basis for the conclusion that no
significant impact is anticipated. Legal Jargon 33:
p. 16815.

Environmental Impact Assessment (EIA). Federal--EPA. The report,
prepared by the applicant for an NPDES permit to discharge
as a new source, which identifies and analyzes the environmental impacts of the applicant's proposed source and fea-
sibile alternatives as provided in Section 6.908 of this
Part. Legal Jargon 19: p. 2453.

Federal--GSA. A document, occurring early in the planning
process, for evaluating the potential environmental impacts
of a project or activity. An assessment covers the same
topical areas as an impact statement, but with less detail.
An assessment results in a decision that an environmental
impact statement is necessary, or that the proposed action
will have no significant effect and therefore a negative
declaration can be made. Legal Jargon 18: Appendix B-3.

Environmental Impact Assessment Report. Federal--DOT. This is
the report, prepared by the sponsor of an action, analyzing
the environmental impact of a proposed action for which Federal financial assistance is being requested or for which
a Federal authorization is required. This report may serve
as the basis, in whole or in part, for the FAA's draft environmental impact statement or negative declaration. Legal
Jargon 12: p. 32648.

Environmental Impact Report (EIR). State--Resources Agency. A
    detailed statement setting forth the environmental effects
    and considerations pertaining to a project as specified in
    Section 21100 of the California Environmental Quality Act,
    and may mean either a draft or a final EIR. (a) Draft EIR
    means an EIR containing the information specified in Sec-
    tions 15141, 15142 of these Guidelines. Where a Lead Agency
    consults with Responsible Agencies in the preparation of a
    draft EIR, the draft EIR shall also contain the information
    specified in Section 15144. (b) Final EIR means an EIR con-
    taining the information contained in the draft EIR, comments
    either verbatim or in summary received in the review pro-
    cess, a list of persons commenting, and the response of the
    Lead Agency to the comments received. The final EIR is dis-
    cussed in detail in Section 15146. Legal Jargon 15: p. 6.

    City--Environmental Section. A concise statement setting
    forth the environmental effects and considerations pertain-
    ing to a project as specified in Section 21100 of the Cal-
    ifornia Environmental Quality Act. Legal Jargon 2: p. II-4.

Environmental Impact Statement (EIS). Federal--CEQ. Means a de-
    tailed written statement as required by Sec. 102(2)(C) of
    the Act. Legal Jargon 30: p. 25244.

    Federal--GSA. A detailed statement which, pursuant to sec-
    tion 102(2)(C) of the NEPA, to the fullest extent possible,
    identifies and analyzes, among other things, the antici-
    pated environmental impact of a proposed GSA action and
    discusses how the adverse effects will be mitigated. Legal
    Jargon 16: p. 1.

    Regional--Commission. A document prepared by the Commission
    which identifies and analyzes in detail the environmental
    impacts of a major action by the Commission having signif-
    icant effects involving the quality of the human environ-
    ment. Legal Jargon 4: p. 464.

    State--Commission. An informational document prepared in
    compliance with Chapter 343, Hawaii Revised Statutes, ap-
    plicable rules, and these Regulations, and which discloses:
    the environmental effects of a proposed action, the effects
    of a proposed action on the economic and social welfare of
    the community and State, the effects of the economic activ-
    ities arising out of the proposed action, the measures pro-
    posed to minimize adverse effects, and the alternatives to
    the action and their environmental effects. Legal Jargon
    3: p. 3.

State--Council on the Environment. A documentation by an
agency sponsoring a project, (which includes funding, li-
censing, or permitting in the case of Federal agencies),
of that project's potential impacts on the environment.
An EIS includes discussion of alternatives to the proposal,
environmental impacts and measures proposed to avoid adverse impacts and mitigate unavoidable adverse impacts.
Federal and State environmental impact statements can be
classified according to the phase of the process during
which the documents are prepared. Legal Jargon 1: pp. 2-3.

State--Environmental Board. A detailed report on the environmental impact of a proposed action, listing adverse
environmental effects which cannot be avoided should the
action be implemented, alternatives to the proposed action,
any irreversible and irretrievable commitments of resources
which would be involved, the growth-inducing aspects of the
proposed action, effects of the proposed action of the use
and conservation of energy resources, the rationale for
selecting the final proposed action, and other information
as further herein specified. Legal Jargon 9: p. 2.

State--Interagency Council. A formal document presenting
a comprehensive, detailed discussion of the proposed project which describes the type, magnitude and importance of
the consequential interrelated actions and reactions on the
environment. The EIS will be required for projects which
are significant and/or have adverse environmental effects
(impact). Legal Jargon 7: p. 6.

State--Law. A detailed statement setting forth the matters
specified in Section 34-A-9-7. It includes any comments on
a draft environmental statement which are received pursuant
to Section 34-A-9-8, and the agency's response to such comments, to the extent that they raise issues not adequately
resolved in the draft environmental statement. Legal Jargon 5: p. 602.

State--Natural Resources. A detailed statement describing
a proposed action, its environmental setting, potential environmental effects and alternatives. An EIS is prepared
by a state or federal agency following guidelines issued by
CEQ, for every action significantly affecting the quality of
the human environment. Legal Jargon 20: p. 1. (Revised).

State--Resources Agency. An environmental impact document
prepared pursuant to the National Environmental Policy Act
(NEPA). The Federal Government uses the term EIS in place
of the term EIR which is used in CEQA. Legal Jargon 15:
p. 6.

City--Environmental Section. An environmental impact report prepared pursuant to the National Environmental Policy
Act (NEPA). Legal Jargon 2: p. II-5.

Environmental Impact Statement Preparation Notice. State--Com-
mission. A document informing the Commission of an agency
determination, after an assessment, that the preparation
of an Environmental Impact Statement is required. Legal
Jargon 3: p. 3.

State--Environmental Council. A written statement by the
Responsible Agency or Responsible Person which requires an
EIS to be prepared. Legal Jargon 26: p. 1093.

Environmental Impact Statement Process. State--Council on the
Environment. A formal procedure that has been established
to organize the preparation and review of environmental
impact statements. Legal Jargon 1: p. 4.

Environmental Notification Form (ENF). State--Environmental Affairs. An environmental notification form. Legal Jargon
6: p. 5.

Environmental Report (ER). Federal--Energy. Means a document submitted to DOE by an applicant in support of an undertaking
which identifies the environmental impacts of the proposed
undertaking and its alternatives. Legal Jargon 32: p. 7235.

Regional--Commission. A document to be submitted by ap-
plicants proposing an action which requires an environmental assessment. Legal Jargon 4: p. 464.

City--Office of the Mayor. A report to be submitted to the
lead agencies by a non-agency applicant when the lead agencies prepare or cause to be prepared a draft EIS for an action involving such an applicant. An environmental report
shall contain an analysis of the environmental factors specified in section 9 of this Executive Order as they relate
to the applicant's proposed action and such other information as may be necessary for compliance with this Executive Order, including the preparation of an EIS. Legal
Jargon 23: p. 4.

Environmental Resources. Federal--Army. Those elements, features,
conditions and areas valued by man that can be characterized
as physiographic, biological, cultural, and aesthetic. Legal Jargon 31: p. 10782.

Environmental Review. Federal--EPA. A formal evaluation undertaken by EPA to determine whether a proposed EPA action may

have a significant impact on the environment. The environmental assessment is one of the major sources of information
used in this review. Legal Jargon 33: p. 16815.

Federal--EPA. The evaluation undertaken by EPA to determine whether the issuance of a new source NPDES permit may
have a significant impact on the environment. The environmental impact assessment is one of the major sources of information used in this review. The environmental review
shall be completed with the issuance by the responsible
official of either a notice of intent or a negative declaration unless the review is reopened by the responsible
official because of additional relevant information. Legal Jargon 19: p. 2453.

Environmental Statement (ES). Federal--Interior. A document analyzing the environmental impact of proposed or recommended
actions and its alternatives on the human environment. The
document is referred to, or characterized, by the following levels of preparation. (1) Preliminary Draft; environmental statement (PDES) prepared in the preliminary stages
of the draft environmental statement preparation process
for review by the Bureau and other agencies with expertise
or jurisdiction, (2) Draft; environmental statement (DES)
filed with the Council on Environmental Quality (CEQ) and
distributed for public comment, (3) Preliminary Final; environmental statement (PFES) prepared in the preliminary
stages of the final environmental statement preparation
process for review by the Bureau and other agencies with
expertise or jurisdiction, (4) Final; environmental statement (FES) filed with CEQ incorporating the necessary
changes after analysis of public comment. Legal Jargon 17:
no page number.

Environmental Statement Preparation Plan. Federal--Interior.
A written plan showing the purpose of action, ES level,
scope of ES, organizational level and arrangements of the
preparation work flow and responsibilities, issues and
problems associated with the proposal and ES, ES outline,
public involvement arrangements, and preparation schedule.
Legal Jargon 17: no page number.

Environment Assessment Worksheet (EAW). State--Environmental
Council. A worksheet provided by the Council to determine
whether an EIS is required. Legal Jargon 26: p. 1094.

Excluded Action. State--Environmental Conservation. An action
which has been approved prior to the effective dates set
forth in SEQR and in section 617.12 of this Part. Legal
Jargon 24: p. 6.

Executive Order. <u>Federal--Historic Preservation</u>. Means Executive
     Order 11593, May 13, 1971, "Protection and Enhancement of
     the Cultural Environment", 36 FR 8921, 16 U.S.C. 470.
     Legal Jargon 34: p. 3367.

Exempt Classes Of Action. <u>State--Commission</u>. Exceptions from
     the requirements of Chapter 343, Hawaii Revised Statutes
     for a class of actions, based on a determination that the
     class of actions will probably have a minimal or no sig-
     nificant effect on the environment. Legal Jargon 3: p. 3.

Extramural Project. <u>Federal--EPA</u>. A project undertaken by grant
     or contract. Legal Jargon 33: p. 16815.

Feasibility And Planning Studies. <u>City--Environmental Section</u>.
     Activities involving only studies for possible future ac-
     tions which the agency, board or commission has not ap-
     proved, adopted or funded. Legal Jargon 2: p. II-5.

Feasible. <u>State--Resources Agency</u>. Capable of being accomplished
     in a successful manner within a reasonable period of time,
     taking into account economic, environmental, social, and
     technological factors. Legal Jargon 15: p. 6.

Federal Agency. <u>Federal--CEQ</u>. Means all agencies of the Federal
     Government. It does not mean the Congress, the Judiciary,
     or the President, including the performance of staff func-
     tions for the President in his Executive Office. Legal
     Jargon 30: p. 25244.

Federal Finding. <u>Federal--DOT</u>. This is a determination by the
     responsible official signifying approval or disapproval of
     a negative declaration or a final environmental impact
     statement. Legal Jargon 12: p. 32648.

Final Environmental Impact Statement (FEIS). <u>Federal--DOT</u>. The
     detailed statement on a major action which significantly
     affects the quality of the human environment, as required
     by section 102(2)(C) of the National Environmental Policy
     Act of 1969, 42 U.S.C. 4332 (2)(C). It contains the same
     supporting information required in the draft EIS with ap-
     propriate revisions to reflect the comments received from
     circulation of the draft EIS and the public hearing process.
     Legal Jargon 13: p. 41806.

     <u>Federal--DOT</u>. This is the document that represents FAA's
     final evaluation of the environmental impact of a proposed
     major Federal action. The final environmental impact state-
     ment will usually consist of the draft environmental impact

statement as amended if necessary, comments thereon, responses thereto, the decision memorandum, and a Federal
Finding. Reports cited as a reference in the statement
need not be included in the documentation. The environmental impact statement is the vehicle for considering the
environmental impacts of a proposed Federal action. This
document must accompany each proposed action through the
Federal decision-making process. Legal Jargon 12: p. 32648.

Federal--EPA. Any document prepared by EPA or under EPA
guidance which identifies and analyzes in detail the environmental impacts of a proposed EPA action and incorporates comments made on the draft EIS. Legal Jargon 19:
p. 2453.

State--Council on the Environment. A Federal document that
fully addresses all considerations of environmental impacts,
etc. The Final EIS is based on the findings of the sponsoring agency as documented in the Draft EIS and on the review and comments that the initiating agency receives on
the Draft EIS. Legal Jargon 1: pp. 3-4.

State--Environmental Commission. A document summarizing
or, if necessary, including the major conclusions and supporting information of a draft environmental impact statement and specifically including the Board's or Department's
response to all substantive comments or objections raised
by the public or other agencies since issuance of the draft
environmental impact statement. Legal Jargon 10: p. 2.

Finding Of Inapplicability. Federal--HUD. A determination by an
authorized HUD official, under the authority of the HUD
Guidelines approved by CEQ, which are issued in accordance
with CEQ Guidelines implementing NEPA, that no Environmental Impact Statement is required for the proposed HUD action under consideration. The Finding of Inapplicability
for project level actions is part of the Special Environmental Clearance. The format for the Finding of Inapplicability for policy actions is set forth in Appendix F.
Legal Jargon 11: p. 19183.

Finding Of No Significant Impact. Federal--CEQ. Means a document
by a Federal agency briefly presenting the reasons why an
action, not otherwise excluded (Section 1508.4), will not
have a significant effect on the human environment and for
which an environmental impact statement therefore will not
be prepared. It shall include the environmental assessment
or a summary of it and shall note any other environmental
documents related to it (1501.7 (a)(5)). Legal Jargon 30:
pp. 25244-25245.

Government Action. State--Environmental Council. An action pro-
    posed to be undertaken by a public agency directly or an
    action supported or licensed, in whole or in part, by a
    governmental permit issued by a public agency. Legal Jar-
    gon 26: p. 1094.

Governmental Permit. State--Environmental Council. A lease, per-
    mit, license, certificate, variance, or other entitlement
    of use, or the commitment to issue or the issuance of a
    discretionary contract, grant, subsidy, loan, or other form
    of financial assistance, by a public agency to another pub-
    lic agency or to a private person. Legal Jargon 26:
    p. 1094.

Human Element. State--Executive Order. Human elements are iden-
    tified in two categories: social and economic, including
    public health, demographic factors, communities, neighbor-
    hoods, and economic gains and losses that would be directly
    or indirectly caused by a proposed action. Legal Jargon
    27: p. 3.

Human Environment. Federal--CEQ. Shall be interpreted compre-
    hensively to include the natural and physical environment
    and the interaction of people with that environment. (See
    the definition of "effects" Section 1508.8). This means
    that exclusively economic or social effects are not intend-
    ed by themselves to require preparation of an environmental
    impact statement. When an environmental impact statement
    is prepared and economic or social and natural or physical
    environmental effects are interrelated, then the environ-
    mental impact statement will discuss all of these effects
    on the human environment. Legal Jargon 30: p. 25245.

    Federal--DOT. This is the aggregate of all external con-
    ditions and influences (ecological, biological, economic,
    social, cultural, historical, aesthetic, etc.) that affects
    the life of a human. Legal Jargon 12: p. 32648.

    State--Environmental Commission. Includes but is not lim-
    ited to biological, physical, social, economic, cultural,
    and aesthetic factors that interrelate to form the environ-
    ment in which Montanans live. Legal Jargon 10: p. 2.

    State--Natural Resources. The totality of conditions and
    influences, both natural and man-made, which surround and
    affect all organisms, including man. Legal Jargon 21: p. 1.

Human Life. State--Executive Order. The citizenry, possessing
    unique demographic, socio-economic and cultural character-
    istics, socially organized and possessing a repertoire of

techniques (technology) to obtain sustenance from the environment. Legal Jargon 27: p. 3.

Impacts. <u>Federal--Army</u>. Impacts, (also known as "effects") are
the economic, social, and environmental consequences expected to result from alternative plans. The impacts of
a plan are the measured changes between with the plan and
"without" conditons. Legal Jargon 31: p. 10782.

<u>State--Executive Order</u>. Any alteration or change in some
element of the human or natural environment. Determination of change would be based on an evaluation of the existing state of the environmental element in question.
The element of the human or natural environment could be
any of three categories: biological, including such subcategories as human, animal, plant, and aquatic; physical
and chemical, which would include factors associated with
impacts on water, air, land and energy resources; and social, including impacts on community as well as individuals.
Impact includes State, interstate, or international effects.
Legal Jargon 27: p. 3.

Inadequate EIS. <u>State--Environmental Council</u>. An EIS that fails
sufficiently to examine potential environmental effects,
alternatives, or desirable modifications, or an EIS not
prepared in compliance with the Act and these Rules. Legal Jargon 26: p. 1094.

<u>State--Natural Resources</u>. An EIS that fails to reasonably
examine possible and real environmental effects, alternatives, modifications, procedural requirements, and other
factors required and further described in WEPA and this
chapter. Legal Jargon 21: p. 1.

Initial Study. <u>State--Resources Agency</u>. A preliminary analysis
prepared by the lead agency pursuant to Section 15080 to
determine whether an EIR or a Negative Declaration must be
prepared. Legal Jargon 15: p. 6.

<u>City--Environmental Section</u>. A comprehensive analysis of
those aspects of the environment which could potentially
affect a project or be affected by a project conducted to
determine whether a project may have a significant effect
on the environment. Legal Jargon 2: p. II-5.

Intramural Project. <u>Federal--EPA</u>. An in-house project undertaken
by EPA personnel. Legal Jargon 33: p. 16815.

Joint Environmental Impact Statement. <u>State--Environmental Com-
mission</u>. An environmental impact statement prepared jointly

by more than one agency, state and/or federal, when such
agencies are involved in the same or closely related proposed actions. Legal Jargon 10: p. 2.

Land Pollution. State--Environmental Control. The presence upon
or within the land resources of the state of one or more
contaminants or combinations thereof, including, but not
limited to, refuse, garbage, rubbish, or junk, in such quantities and of such quality as will or are likely to (a)
create a nuisance; (b) be harmful, detrimental or injurious
to public health, safety or welfare; (c) be injurious to
plant and animal life and property; or (d) be detrimental
to the economic and social development, the scenic beauty
or the enjoyment of the natural attractions of the state.
Legal Jargon 25: p. 3.

Lead Agency. Federal--CEQ. Means the agency or agencies which
have prepared or have taken primary responsibility to prepare the environmental impact statement. Legal Jargon 30:
p. 25245.

Federal--GSA. The Federal agency which has primary authority for committing the Federal Government to a course of
action with significant environmental impact. Legal Jargon 16: p. 2.

State--Environmental Commission. The agency of the state
that has primary authority for committing the government to
a course of action having significant environmental impact,
or is the agency chosen to supervise the preparation of a
joint environmental impact statement where more than one
agency is involved in the action. Legal Jargon 10: p. 2.

State--Interagency Council. A single agency designated to
assume supervisory responsibility for preparation of an impact statement for an action which involves multiple agency
jurisdictions, interdependent functions, or geographic proximity. Legal Jargon 7: p. 6.

State--Resources Agency. The public agency which has the
principal responsibility for carrying out or approving a
project. The Lead Agency will prepare the environmental
documents for the project either directly or by contract.
Criteria for determining which agency will be the Lead Agency for a project are contained in Section 15065. Legal
Jargon 15: p. 7.

Lead City Agency. City--Environmental Section. The City department, bureau, division, section, office, officer or agency
which has the principal responsibility for carrying out a

project which is subject to provisions of CEQA, or has the
principal responsibility for processing the application for
a lease, permit, license, or other entitlement for use for
a project which is subject to the provisions of CEQA. If
more than one City Agency meets the Lead Agency criteria,
the Lead City Agency shall be the City Agency that normally
acts first on such projects. Legal Jargon 2: p. II-5/6.

Legislation. Federal--CEQ. Includes a bill or legislative proposal to Congress developed by or with the significant cooperation and support of a Federal agency, but does not include requests for appropriations. The test for significant cooperation is whether the proposal is in fact predominantly that of the agency rather than another source.
Drafting does not by itself constitute significant cooperation. Proposals for legislation include requests for
ratification of treaties. Only the agency which has primary responsibility for the subject matter involved will
prepare a legislative environmental impact statement.
Legal Jargon 30: p. 25245. (Revised).

Legislative Action. Federal--Energy. Means a DOE recommendation
or report on DOE proposals for legislation significantly
affecting the quality of the human environment. Legal Jargon 32: p. 7235.

License. State--Department of Ecology. Means any form of written
permission given to any person, organization or agency to
engage in any activity, as required by law or agency rule.
A license includes all or part of any agency permit, certificate, approval, registration, charter, or plat approvals or rezones to facilitate a particular project. The
term does not include a license required solely for revenue
purposes. Legal Jargon 37: p. 3.

Local Agency. State--Environmental Council. Any general or special purpose unit of government of the state with less than
state-wide jurisdiction, including but not limited to regional development commissions, counties, municipalities,
townships, port authorities, housing authorities, and all
agencies, committees, and boards thereof. Legal Jargon
26: p. 1094.

State--Resources Agency. Any public agency other than a
state agency, board or commission. Local agency includes,
but is not limited to, cities, counties, charter cities
and counties, districts, school districts, special districts, redevelopment agencies, local agency formation commissions and any board, commission, or organizational subdivision of a local agency when so designated by order or

resolution of the governing legislative body of the local
agency. Legal Jargon 15: p. 7.

City--Environmental Section. Any public agency other than
a State agency, board or commission. The City of Los
Angeles constitutes a single local agency. Legal Jargon 2:
p. II-6.

Major Action. Federal--DOT. An action of superior, large and considerable importance, involving substantial planning, time,
resources or expenditures. Legal Jargon 13: p. 41806.

State--Natural Resources. An action of magnitude and complexity which will notably or seriously affect the quality
of the human environment. Legal Jargon 21: p. 1.

Major Amendatory. Federal--HUD. For the purposes of this Handbook, a significant change in the nature, magnitude or extent of the action from that which was originally evaluated
and which may have a significant affect on the quality of
the human environment, such as an environmentally significant change in location or site, area covered, size or
design in which case they require environmental clearance.
An increase or decrease in cost is considered a major amendatory only when the increase or decrease reflects such an
environmentally significant change in the project. Legal
Jargon 11: p. 19183.

Major Development Project. State--Council. Shall include but is
not limited to shopping centers, subdivisions and other
housing developments, and industrial and commercial projects, but shall not include any projects of less than two
contiguous acres in extent. Legal Jargon 22: p. 3.

Major Federal Action. Federal--CEQ. Includes actions with effects that may be major and which are potentially subject
to Federal control and responsibility. Major reinforces
but does not have a meaning independent of significantly
(Section 1508.25). Legal Jargon 30: p. 25245. (Revised).

Master EIR. City--Environmental Section. An EIR covering a geographical area that may involve cumulative environmental
impacts from a number of separate projects within the geographical area. Legal Jargon 2: p. II-4.

Matter. Federal--CEQ. Includes for the purposes of Part 1504:
(a) With respect to Environmental Protection Agency, any
proposed legislation, project, action or regulation as
those terms are used in Section 309(a) of the Clean Air
Act (42 U.S.C. 7609), (b) With respect to all other agen-

cies, any proposed major federal action to which Section
102(2)(C) of NEPA applies. Legal Jargon 30: p. 25245.

Ministerial Action. <u>City--Office of the Mayor</u>. An action performed upon a given state of facts in a prescribed manner
  imposed by law without the exercise of any judgment or discretion as to the propriety of the action, although such
  law may require, in some degree, a construction of its
  language or intent. Legal Jargon 23: p. 5.

Ministerial Projects. <u>State--Resources Agency</u>. As a general rule
  include those activities defined as projects which are undertaken or approved by a governmental decision which a
  public officer or public agency makes upon a given state
  of facts in a prescribed manner in obedience to the mandate
  of legal authority. With these projects, the officer or
  agency must act upon the given facts without regard to his
  own judgment or opinion concerning the propriety or wisdom
  of the act although the statue, ordinance, or regulation
  may require, in some degree, a construction of its language
  by the officer. In summary, a ministerial decision involves only the use of fixed standards or objective measurements without personal judgment. Legal Jargon 15:
  p. 7.

Mitigation. <u>Federal--CEQ</u>. Includes: (a) Avoiding the impact
  altogether by not taking a certain action or parts of an
  action, (b) Minimizing impacts by limiting the degree of
  magnitude of the action and its implementation, (c) Rec-
  tifying the impact by repairing, rehabilitating, or restoring the impacted environment, (d) Reducing or eliminating the impact over time by preservation and maintenance operations during the life of the action, (e) Com-
  pensating for the impact by replacing or providing sub-
  stitute resources or environments. Legal Jargon 30:
  p. 25245.

Negative Declaration. <u>Federal--DOT</u>. A document supporting a
  determination that a proposed major action will not have
  a significant impact upon the quality of the human environment of a magnitude to require the processing of an
  EIS. Legal Jargon 13: p. 41806.

  <u>Federal--DOT</u>. This is the document that constitutes FAA's
  evaluation that a particular action will not significantly
  alter the airport's impact on its surrounding environment
  and the action is not highly controversial on environmental grounds. Coordination and review pursuant to Section
  102 (2)(C) of NEPA are not required. Legal Jargon 12:
  p. 32648.

Federal--EPA. The written announcement, prepared subsequent to the environmental review, which states that EPA
has decided not to prepare a draft environmental impact
statement. Legal Jargon 19: p. 2454.

Federal--EPA. A written announcement, prepared after the
environmental review, which states that EPA has decided
not to prepare an EIS and summarizes the environmental impact appraisal. Legal Jargon 33: p. 16815.

Federal--GSA. An official administrative decision stating
that an analysis of the environmental assessment has been
made and that the proposed action is not considered a major
GSA action having a significant impact on the environment,
and, therefore, will not require the preparation of an environmental impact statement. (The declaration must also
include a summary of any known environmental impacts.)
Legal Jargon 16: p. 1.

Regional--Commission. A determination by the Executive
Director, based upon an environmental assessment, that a
proposed action will not require an environmental impact
statement. Legal Jargon 4: p. 464.

State--Commission. A determination by an agency that a
given action does not have a significant effect on the environment and therefore does not require the preparation
of an EIS. Legal Jargon 3: p. 3.

State--Resources Agency. A written statement by the Lead
Agency briefly describing the reasons that a proposed project, although not otherwise exempt, will not have a significant effect on the environment and therefore does not
require the preparation of an EIR. Legal Jargon 15: p. 7.

City--Office of the Mayor. A written statement prepared
by the lead agencies after conducting an environmental
analysis of an action which announces that the lead agencies have determined that the action will not have a significant effect on the environment. Legal Jargon 23:
p. 5.

Negative Declaration EIS. State--Executive Order. A concise
series of statements, with appropriate graphics, which
describes an activity or action proposed by a State agency,
identifies any potentially adverse environmental effects
and public concerns or controversies that may occur as a
result of the proposed activity or action, defines the significance of the environmental effects and public concerns
or controversies and indicates that the activities or action

does not warrant the preparation of an environmental impact statement (EIS). Legal Jargon 27: p. 4.

Negative Declaration Notice. State--Environmental Council. A written statement by the Responsible Agency or Responsible Person that a proposed action does not require the preparation of an EIS. Legal Jargon 26: p. 1094.

Negative Environmental Declaration (NED). State--Interagency Council. A formal document indicating the lack of project significance and/or adverse environmental effects. Legal Jargon 7: p. 6.

NEPA-Associated Documents. Federal--EPA. Any one or combination of: notices of intent, negative declarations, exemption certifications, environmental impact appraisals, news releases, EIS's, and environmental assessments. Legal Jargon 33: p. 16815.

NEPA Process. Federal--CEQ. Means all measures necessary for compliance with the requirements of Section 2 and Title I of NEPA. Legal Jargon 30: p. 25245.

New Source. Federal--EPA. As defined in section 306(a)(2) of the FWPCA, means "any source, the construction of which is commenced after the publication of proposed regulations prescribing a standard of performance under this section which will be applicable to such source, if such standard is thereafter promulgated in accordance with this section." Legal Jargon 19: p. 2453.

New Source And Environmental Questionnaire (NS/EQ). Federal--EPA. An initial document submitted by an applicant for a new source NPDES permit. This document will furnish information on the status of the proposed source that will allow determination of whether the facility is a new or existing source. In addition, the NS/EQ will also furnish information on the potential environmental impacts of the proposed source. It is the Agency's intention that in the case of sources which will probably have insignificant environmental impacts, the NS/EQ will normally provide sufficient information to fulfill the requirements for an environmental impact assessment. Legal Jargon 19: p. 2453.

Notice Of Completion. State--Resources Agency. A brief notice filed with the Secretary for Resources by a Lead Agency as soon as it has completed a draft EIR and is prepared to send out copies for review. The contents of this notice are explained in Section 15085(c). A copy of this notice appears in Appendix C. Legal Jargon 15: p. 7.

Notice Of Determination. <u>State--Resources Agency</u>. A brief notice
to be filed by a public agency after it approves or determines to carry out a project which is subject to the requirements of CEQA. The contents of this notice are explained in Sections 15083(d) and 15085(h). A copy of this
notice appears in Appendix D. Legal Jargon 15: p. 7.

<u>City--Environmental Section</u>. A notice to be filed by a
Lead City Agency after a project subject to the provisions
of CEQA has been approved. Legal Jargon 2: p. II-7.

<u>City--Office of the Mayor</u>. A written statement prepared
by the lead agencies after conducting an environmental analysis of an action which announces that the lead agencies
have determined that the action may have a significant effect on the environment, thus requiring the preparation of
an EIS. Legal Jargon 23: p. 5.

Notice Of Exemption. <u>State--Resources Agency</u>. A brief notice
which may be filed by a public agency when it has approved
or determined to carry out a project, and it has determined
that it is ministerial, categorically exempt or an emergency
project. Such a notice may also be filed by an applicant
where such a determination has been made by a public agency
which must approve the project. The contents of this notice are explanined in Section 15074(a) and (b). A copy
of this notice appears in Appendix E. Legal Jargon 15:
p. 7.

<u>City--Environmental Section</u>. A notice which may be filed
by a Lead City Agency after the Lead City Agency has approved a project and has determined that it is a ministerial, categorically exempt, or emergency project. Legal
Jargon 2: p. II-7.

Notice Of Intent. <u>Federal--CEQ</u>. Means a notice that an environmental impact statement will be prepared and considered,
the notice shall briefly: (a) Describe the proposed action
and possible alternatives, (b) Describe the agency's proposed scoping process including whether, when, and where
any scoping meeting will be held, (c) State the name and
address of a person within the agency who can answer questions about the proposed action and the environmental impact statement. Legal Jargon 30: p. 25245.

<u>Federal--EPA</u>. The written announcement to Federal, State
and local agencies, and to interested persons, that a draft
environmental impact statement will be prepared. The notice
shall briefly describe the EPA action, its location, the
issues involved (see Exibit 1). The purpose of the notice

is to involve other governmental agencies and interested
persons as early as possible in the planning and evaluation
of actions which may have significant environmental impacts.
This notice should encourage public input in the preparation
of a draft EIS and assure environmental values will be identified and weighed from the outset, rather than accommodated
by adjustments at the end of the decision-making process.
Legal Jargon 19: p. 2453.

Federal--EPA. A memorandum, prepared after the environmental review, announcing to Federal, regional, State, and local agencies, and to interested persons, that a draft EIS
will be prepared. Legal Jargon 33: p. 16815.

Regional--Commission. An announcement to other Federal,
State and local agencies and to the public that the Commission will be preparing an environmental impact statement for a given action. Legal Jargon 4: p. 464.

State--Environmental Quality. A written description of
any activity proposed to be performed, to be submitted to
the conservation commission with copies to the Commissioner
in a form annexed hereto, marked "B" in the appendix, including a completed Environmental Data form. Legal Jargon 28: p. 5.

Notice Of Preparation. State--Resources Agency. A brief notice
sent by a Lead Agency by certified mail to notify the Responsible Agency that the Lead Agency plans to prepare an
EIR for the project. The purpose of the notice is to sol-
icit guidance from the Responsible Agencies as to the scope
and content of the environmental information to be included
in the EIR. A sample form letter is included as Appendix
J to these Guidelines. Public agencies are free to develop
their own formats for this notice. Legal Jargon 15: p. 8.

Opinion. State--Environmental Protection. A written decision by
the Commissioner, embodying an analysis, findings, and conclusions. Legal Jargon 14: p. 3.

Other Approving Agencies. State--Environmental Council. All public agencies other than the Responsible Agency that must
approve a project for which environmental documents are
prepared. Legal Jargon 26: p. 1094.

Owner Of Land Abutting The Activity. State--Environmental Quality.
The owner of record of land directly beside the property
limits of the site of the proposed activity in any direction, including lands located across a street, way or
waterway. Legal Jargon 28: p. 5.

Participating Agency. State--Environmental Affairs. An agency
to which an application for a permit or for financial assistance for a project has or will be made. Legal Jargon
6: p. 5.

Participating City Agency. City--Environmental Section. A City
department, bureau, division, section, office, officer, or
agency which is required by Charter or action of the City
Council to review a particular class of projects and makes
comments or recommendations to the Lead City Agency. Legal Jargon 2: p. II-7.

PER Announcement Sheet. State--Environmental Resources. A brief
document announcing the availability of a PER for public
review. PER Announcement Sheets are sent to all individuals or organizations that are known or thought to have an
interest in a proposal for which an EIS is being prepared
by the Department. Legal Jargon 20: p. 2.

Permit. State--Environmental Affairs. A permit determination,
order or other action, including the issuance of a lease,
deed, license, permit, certificate, variance, approval, or
other entitlement for use, granted to any person, firm or
corporation, including trusts, voluntary association or
other forms of business organizations by an agency for a
project but shall not include a general entitlement to a
person to carry on a trade or profession or to operate
mechanical equipment which does not depend upon the location of such trade or operation. Legal Jargon 6: p. 5.

State--Environmental Conservation. A permit, lease, license, certificate or other entitlement for use or permission to act that may be granted or issued by an agency.
Legal Jargon 24: p. 8.

State--Environmental Protection. Any legal instrument,
constituting permission to construct a facility in the
coastal area, that is issued by the Commissioner pursuant
to N.J.S.A. 13:19-1 et seq. Legal Jargon 14: p. 3.

Permit Condition. State--Environmental Protection. A requirement of the permit after issuance, applicable during, and
after construction. Legal Jargon 14: p. 3.

Permitee. State--Environmental Protection. Any person issued a
permit. Legal Jargon 14: p. 4.

Permit Pre-Condition. State--Environmental Protection. A requirement which must be met before a permit is issued. Legal
Jargon 14: p. 4.

Person. State--Commission. Any individual, partnership, firm,
       association, trust, estate, private corporation, or other
       legal entity other than agencies. Legal Jargon 3: p. 3.

       State--Environmental Affairs. A private person, firm or
       corporation, or any governmental entity which is not an
       agency. Legal Jargon 6: p. 5.

       State--Environmental Protection. Corporations, companies,
       associations, societies, firms, partnerships and joint
       stock companies as well as individuals and governmental
       agencies. Legal Jargon 14: p. 4.

       State--Interagency Council. Any person, partnership, as-
       sociation, organization, business, trust, bank, corporation,
       professional corporation, or company. Legal Jargon 7:
       p. 6.

       State--Natural Resources. Any person, firm, partnership,
       joint venture, joint stock company, association, public or
       private corporation, the state of Wisconsin and all polit-
       ical subdivisions, cooperatives, estate, trust, receiver,
       executor, administrator, fiduciary, and any representative
       appointed by order of any court or otherwise acting on be-
       half of others. Legal Jargon 21: p. 2.

       State--Resources Agency. Person includes any person, firm
       association, organization, partnership, business, trust,
       corporation, company, district, county, city and county,
       city, town, the State, and any of the agencies' political
       subdivisions of such entities. Legal Jargon 15: p. 8.

Persons With Special Expertise. City--Environmental Section.
       Persons with special expertise are those individuals with
       experience, knowledge or formal education in disciplines
       germane to specific contents of environmental documents
       that will allow them to offer authoritative information and
       opinions for the preparation and review of such documents.
       These persons may be members of the general public, or em-
       ployees of private companies or governmental agencies. Le-
       gal Jargon 2: p. II-8.

Petition. State--Environmental Council. A document that contains
       at least 500 signatures and requests the preparation of an
       EIS. Legal Jargon 26: p. 1094.

Plans. State--Environmental Quality. Any engineering drawings
       and data deemed necessary for regulating the proposed ac-
       tivity and for determining whether the Act is applicable.
       Legal Jargon 28: p. 5.

Potential New Source Applicant. <u>Federal--EPA</u>. The prospective
    owner, operator, or designee, of an anticipated point
    source, as defined in Section 502 (14) of the FWPCA, which
    falls within a proposed standard of performance category
    and would require an NPDES permit and further meets the
    definition of a new source set forth in section 6.902 (c)
    above. Legal Jargon 19: p. 2454.

Preliminary Analysis. <u>State--Environmental Protection</u>. An in-
    formal, initial appraisal of a permit application that is
    prepared by the Division for review by the applicant and
    interested persons prior to the Commissioner's decision on
    a permit application. Legal Jargon 14: p. 4.

Preliminary Environmental Report (PER). <u>State--Natural Resources</u>.
    Required by WEPA and Executive Order No. 26 and is equiv-
    alent to a Draft EIS required under NEPA. This is a draft
    of the EIS circulated to the public and concerned agencies
    to elicit comments. Legal Jargon 20: p. 1.

    <u>State--Natural Resources</u>. It is a draft of the environmen-
    tal impact statement. Legal Jargon 21: p. 1.

Preliminary Environmental Review (PER). <u>State--Environmental Com-
    mission</u>. A written analysis of a proposed action to deter-
    mine whether the action might significantly affect the qual-
    ity of the human environment and therefore require a draft
    environmental impact statement. Legal Jargon 10: p. 2.

Prior Finding Affirmation. <u>Federal--DOT</u>. This is a finding by
    the responsible official that a proposed action is within
    the scope of a previously approved environmental impact
    statement or negative declaration. Affirmation of a prior
    finding establishes the continued validity of a previous
    environmental determination with respect to a currently
    proposed Federal action. Legal Jargon 12: p. 32648.

Private Action. <u>State--Environmental Council</u>. An action proposed
    to be undertaken by a private person that does not require
    a governmental permit. Legal Jargon 26: p. 1094.

Project. <u>State--Environmental Affairs</u>. Work, project, or activ-
    ity either directly undertaken by an agency, or if under-
    taken by a person, which seeks the provision of financial
    assistance by an agency, or requires the issuance of a per-
    mit by an agency but shall not include a grant in aid for
    medical services or personal support, such as welfare or
    unemployment funds, to an individual or a third party on
    behalf of an individual. Legal Jargon 6: p. 5.

State--Interagency Council. (1) New and continuing non-
federal projects directly undertaken by an agency; or supported in whole or in part through contracts, grants, subsidies, loans, or other forms of assistance from one or
more agencies, (2) Those non-federal projects of significant environmental impact requiring permits, licenses, cer-
tificates, approval or other entitlement by State agencies
prior to implementation. Legal Jargon 7: pp. 6-7.

State--Resources Agency. (a) Project means the whole of
an action, which has a potential for resulting in a physical change in the environment, directly or ultimately,
that is any of the following: (1) An activity directly
undertaken by any public agency including, but not limited
to, public works construction and related activities, clearing or grading of land, improvements to existing public
structures, enactment and amendment of zoning ordinances,
and the adoption and amendment of local General Plans or
elements thereof pursuant to Government Code Sections
65100-65700. (2) An activity undertaken by a person which
is supported in whole or in part through public agency contracts, grants, subsidies, loans, or other forms of assistance from one or more public agencies. (3) An activity
involving the issuance to a person of a lease, permit, license, certificate, or other entitlement for use by one or
more public agencies. (b) Project does not include: (1)
Anything specifically exempted by state law. (2) Proposals
for legislation to be enacted by the State Legislature other
than requests by state agencies for authorization or funding for projects independently from the Budget Act. (3)
Continuing administrative or maintenance activities, such
as purchases for supplies, personnel-related actions, emergency repairs to public service facilities, general policy
and procedure making (except as they are applied to specific instances covered above), feasibility or planning studies. (4) The submittal of proposals to a vote of the people
of the State or of a particular community. (c) The term
"project" refers to the activity which is being approved and
which may be subject to several discretionary approvals by
governmental agencies. The term "project" does not mean
each separate governmental approval. Legal Jargon 15: p. 8.

Project Data Statement. City--Office of the Mayor. A written submission to the lead agencies by an applicant on a form prescribed by the lead agencies, which provides an identification of and information relating to the environmental impacts of a proposed action. The project data statement is
designed to assist the lead agencies in their evaluation of
an action to determine whether an action under consideration
may or will not have significant effect on the environment.

Legal Jargon 23: pp. 5-6.

Project Sponsor. <u>City--Environmental Section</u>. The private applicant, City agency, or other public entity that proposes to carry out the project. Legal Jargon 2: p. II-8.

Program. <u>Federal--Energy</u>. Means the aggregate of projects which share a common objective or purpose and are so interrelated that planning or decisonmaking with respect to any one component is likely to significantly affect planning or decisionmaking with respect to any other component. Legal Jargon 7235.

Program EIR. <u>City--Environmental Section</u>. An EIR covering environmental factors that are common to a particular type of project. Legal Jargon 2: p. II-4.

Programmatic Review. <u>State--Environmental Commission</u>. A general analysis of related agency-initiated actions, programs or policies, or the continuance of a board policy or program which may involve a series of future actions. Legal Jargon 10: pp. 2-3.

Proponent. <u>State--Environmental Affairs</u>. A person or agency which seeks to undertake or has a major role in the undertaking of a project or its designee, and shall include the successor in interest to such persons or agency, but shall not include a participating agency except with the approval of the secretary. Legal Jargon 6: p. 6.

Proposal. <u>Federal--CEQ</u>. Refers to that stage in the development of an action when an agency subject to the Act has a goal and is actively considering one or more alternatives means of accomplishing that goal and the effects can be meaningfully evaluated. Preparation of an environmental impact statement on a proposal should be timed (Section 1502.5) so that the final statement may be completed in time for the statement to be included in any recommendation or report on the proposal. A proposal may exist in fact as well as by agency declaration that one exists. Legal Jargon 30: p. 25245.

<u>State--Department of Ecology</u>. Means a specific request to undertake any activity submitted to, and seriously considered by, an agency or a decision-maker within an agency, as well as any action or activity which may result from approval of any such request. The scope of a proposal for the purposes of lead agency determination, the threshold determination, and impact statement preparation is further defined in WAC 197-10-060. Legal Jargon 37: p. 4.

Proposer. State--Environmental Council. The private person or
     public agency that will undertake an action or that will
     direct or authorize others to undertake the action. Legal
     Jargon 26: p. 1094.

Proprietary Department. City--Environmental Section. A depart-
     ment having control over its own special funds. Legal
     Jargon 2: p. II-8.

Public Agency. State--Environmental Council. A federal, state,
     regional, or local agency, board, commission, or other
     special purpose unit of government. "Public Agency" in-
     cludes all public educational institutions but does not
     include the courts of this State. Legal Jargon 26: p. 1094.

     State--Resources Agency. Includes any state agency, board
     or commission and any local or regional agency, as defined
     in these Guidelines. It does not include the courts of the
     State. This term does not include agencies of the federal
     government. Legal Jargon 15: p. 9.

     City--Environmental Section. Any State agency, board, or
     commission, or any local agency as defined in these Guide-
     lines. The courts of the State and agencies of the federal
     government are not public agencies for purposes of CEQA com-
     pliance. The City of Los Angeles constitutes a single pub-
     lic agency. Legal Jargon 2: p. II-9.

Public Entities. Federal--GSA. Any Federal, State, or local of-
     fices and legislatures, and any public or semipublic agen-
     cies. Legal Jargon 16: p. 1.

Pubic Hearing. Federal--DOT. UMTA (Urban Mass Transportation Ad-
     ministration) requires that the applicant hold a public
     hearing on each EIS to solicit the views of those who have
     a significant economic, social, or environmental interest
     in the project. The Draft Environmental Impact Statement
     should be available to the public before the hearing and
     good faith efforts should be made to inform all interested
     parties about the hearing. A public notice should be print-
     ed in the newspaper at least 30 days in advance. Legal Jar-
     gon 35: p. A-2.

Recirculation. City--Environmental Section. The act of circu-
     lating for public review a previously reviewed environmen-
     tal document for the purpose of examining significant new
     information or data not contained in the original document.
     Legal Jargon 2: p. II-9.

Referring Agency. Federal--CEQ. Means the federal agency which

has referred any matter to the Council after a determination that the matter is unsatisfactory from the standpoint
of public health or welfare or environmental quality. Legal Jargon 30: pp. 25245-25246.

Relevant A-95 Clearinghouse. Federal--GSA. Clearinghouse(s) listed in OMB Circular No. A-95 (Revised) clearinghouse direct-
ory for the geographical area in which the GSA action is to
take place. (OMB through its Circular A-95 (Revised) established this system of clearinghouses to facilitate intergovernmental and intragovernmental communication.)
Legal Jargon 16: p. 1.

Report. State--Environmental Board. An environmental assessment
or an environmental assessment followed by an environmental
impact statement as further herein defined. Legal Jargon
9: p. 2.

Resource Inventory. Federal--Army. The locating and accounting
of environmental resources of a given geographic area.
Legal Jargon 31: p. 10782.

Resources. Federal--GSA. All actions and ideas, as well as living
and nonliving materials devoted to the action. Legal Jargon 16: p. 1.

State--Natural Resources. Financial, cultural and natural
matter and forms as well as labor and materials used and
affected by a proposed action if permitted. Legal Jargon
21: p. 1.

Responsible Agency. State--Resources Agency. A public agency
which proposes to carry out or approve a project, for which
a Lead Agency has prepared the environmental documents.
For the purposes of CEQA, the term "responsible agency"
includes all public agencies other than the lead agency
which have discretionary approval power over the project.
Legal Jargon 15: p. 9.

City--Environmental Section. A public agency, such as a
city or county, which proposes to carry out or has approval
power over a project, but is not the Lead Agency for the
project. Legal Jargon 2: p. II-9.

Responsible Official. Federal--DOT. This is the official responsible for making the final determination as to whether the
environmental requirements for a proposed Federal action
have been satisfied. Legal Jargon 12: p. 32648.

Review. State--Natural Resources. The study of and comment upon

the EIR, PER, or EIS by agencies which have jurisdiction
by law or special expertise with regard to environmental
effects. Legal Jargon 21: p. 2.

Reviewing Agencies. State--Environmental Council. All public
agencies which have either jurisdiction by law or special
expertise with regard to the environmental effects of an
action for which an EIS is prepared. All agencies that
are members of the Council shall be considered reviewing
agencies. Legal Jargon 26: p. 1094.

Schedule Of Compliance. State--Environmental Control. A schedule
of remedial measures including an enforceable sequence of
actions or operations leading to compliance with an effluent
limitation, other limitation, prohibition, or standard. Legal Jargon 25: p. 3.

Scope. Federal--CEQ. Scope consists of the range of actions, alternatives, and impacts to be considered in an environmental impact statement. Legal Jargon 30: p. 25246.
(Revised).

Section 208 Plan. Federal--EPA. An areawide waste treatment management plan prepared under section 208 of the Federal Water
Pollution Control Act (FWPCA), as amended, under 40 CFR
Part 126 and 40 CFR Part 35, Subpart F. Legal Jargon 33:
p. 16815.

Short-Term Impacts. Federal--GSA. Short-term impacts are those
project effects that will extend beyond the construction
phase but for only a short period of time. For example,
erosion will continue to occur after construction is completed until vegetation has grown enough to hold the soil
in place, often resulting in short-term impacts on water
quality. Legal Jargon 18: Appendix B-6.

Significant. Federal--GSA. In relation to environmental analysis,
the term includes considerations of importance and magnitude, primarily the former. Legal Jargon 18: Appendix B-6.

State--Environmental Quality. That standard to be used by
the conservation commission or the Commissioner in determining what condition, if any, they deem necessary to protect the public interest under the Act. The standard that
shall be considered to establish significance in order to
protect the said public interest shall include one or more
of the following factors: (a) Any actual or potential contamination to public, private or ground water supply including aquifers or recharge areas, land containing shellfish and fisheries including the biological life necessary

to support either a freshwater or coastal wetland ecosystem, (b) Any reduction of the flood storage capacity of a
freshwater wetland, river, stream or creek, (c) Any alteration of a river, stream or creek that results in any increase in the volume or velocity of water which may cause
flooding, (d) Any actions which shall remove, fill, dredge
or alter any area subject to the Act and will result in
any threat to the health, welfare and safety of the individual or the community, (e) An area consisting of "very
poorly drained soil" as described by the National Cooperative Soils Survey of the U.S. Department of Agriculture,
Soil Conservation Service, (f) An area which would be flood-
ed as a result of a 100-year storm as that storm is defined
in the U.S. Department of Commerce Technical Report No. 40
or has been designated in a town by-law as part of a 100-
year flood plain. Legal Jargon 28: p. 7.

State--Executive Order. Having an impact with regard to
any part of the human or natural resources of the state
that may notably and adversely affect humans, use for humans, for wildlife and fish populations, for scientific
study, or may notably and adversely affect biotic communities. Significance is usually, but not exclusively, associated with largeness of scale, uniqueness or scarcity
of resources, with the duration of adverse effects and with
the rate of chemical, biological or physical alteration,
but is not synonymous with only permanent or irreversible
modifications. In considering the significance of any particular proposed activity, consideration must be given to
the number and cumulative importance of other similar activities, present and proposed, so that the total effect
on the environment is the focus of attention, and not the
effect of any individual activity considered in isolation.
Significant characterizes the scale of an action either in
the size or importance of an element of the environment
with regard to maintaining the structural integrity and the
behavioral stability of the element and of the biologic system of which it is a part. The determination of significance should relate to the size of the influence to the
size of the element and system affected. Legal Jargon 27:
p. 2.

Significant Effect. Regional--Commission. That degree of impact
upon the quality of the human environment determined by the
Commission or the Executive Director as sufficient to jus-
tify an environmental impact statement. Legal Jargon 4:
p. 464.

State--Commission. The sum of those effects that affect
the quality of the environment, including irrevocable

commitment of a natural resource, curtailment of the range
of beneficial uses of the environment, conflict's with the
State's environmental policies or long-term environmental
goals and guidelines as established by Chapter 342 and 344,
Hawaii Revised Statutes or any revisions thereof, or amendments thereto, or adverse effects upon the economic or social welfare. Legal Jargon 3: p. 3.

State--Natural Resources. Considerable and important impacts of major state actions which have long-term effects
on the maintenance of the human environment. Legal Jargon
21: p. 1.

City--Environmental Section. A substantial, or potentially
substantial, adverse change in any of the physical conditions within the area affected by the activity including
land, air, water, minerals, flora, fauna, ambient noise,
and objects of historic or aesthetic significance. Legal
Jargon 2: p. II-9.

Significant Effect On The Environment. State--Resources Agency.
A substantial, or potentially substantial, adverse change
in any of the physical conditions within the area affected
by the activity including land, air, water, minerals, flora,
fauna, ambient noise, and objects of historic or aesthetic
significance. Legal Jargon 15: p. 9.

Significant Environmental Effects. Federal--GSA. Socioeconomic
and physical effects which may be beneficial and/or detrimental to the environment, even if the net is believed to
be beneficial. (These effects may be influenced by the
geographical location of the subject project or action.
Significant detrimental effects include those that degrade
the environment, curtail its range of uses, or sacrifice
its long-term productivity to serve only man's short-term
needs.) Legal Jargon 16: p. 2.

Significant Environmental Impact. Federal--HUD. For the purposes
of this Handbook, the term "significant environmental impact" is used to describe the consequences of an action significantly affecting the quality of the human environment.
This term is defined to some extent in paragraph 5 of the
CEQ Guidelines, but the definition in general is a matter
of discretion delegated to agency heads for formulation of
guidelines subject to the approval of CEQ. Legal Jargon 11
p. 19183.

Significant Impact. Federal--Commerce. Shall include impacts of
both beneficial and detrimental effects, even if, on balance, they are believed to have beneficial effect. A

significant environmental impact may exist depending upon
the extent to which there is a potential for: (1) The alteration of an ecosystem; or (2) Measurably affecting existing or future populations of man or other forms of life.
Legal Jargon 29: p. 3.

Significantly. Federal--CEQ. As used in NEPA requires considerations of both context and intensity: (a) Context. This
means that the significance of an action must be analyzed
in several contexts such as society as a whole (global,
national), the affected region, the affected interests, and
the locality. Significant varies with the setting of the
proposed action. For instance, in the case of a site-specific action, significance would usually depend upon the effects in the locale rather than in the world as a whole.
Both short- and long-term effects are relevant. (b) Intensity. This refers to the severity of impact. Responsible officials must bear in mind that more than one agency
may make decisions about partial aspects of a major action.
Legal Jargon 30: p. 25246. (Revised).

Site. State--Environmental Protection. The land or geographic
area owned or leased or which is held under a contract of
sale or any other legal or equitable interest by the applicant or his agent, and upon which the proposed facility
is to be constructed. Legal Jargon 14: p. 4.

Source. Federal--EPA. As defined in section 306(a)(3) of the
FWPCA, means "any building, structure, facility or installation from which there is or may be the discharge of pollutants." Legal Jargon 19: p. 2453.

Special-Purpose Unit Of Government. State--Council. Includes any
special district or public authority. Legal Jargon 22:
p. 3.

Sponsoring Agency. State--Council on the Environment. That Federal or State agency responsible for preparing environmental impact statements for a proposed action. Legal Jargon
1: p. 4.

Staged EIR. City--Environmental Section. An EIR that covers in
general terms an entire project that will be subject to a
number of discretionary approvals over time, evaluating
with specificity only that aspect of the project before
the Decision-Making Body for consideration. Legal Jargon
2: p. II-4.

State EIR Guidelines. City--Environmental Section. The State of
California Guidelines for the implementation of the Calif-

ornia Environmental Quality Act and are contained in Title
14, Division 6 of the California Administrative Code. Legal Jargon 2: p. II-9.

Statement Of Nonsignificant Impact. State--Natural Resources.
A completed environmental assessment screening worksheet
which indicates that the proposed action is not a major
action which will significantly affect the quality of the
human environment and that no EIS is required. Legal Jargon 21: p. 1.

Statement Of Overriding Considerations. City--Environmental Section. A statement identifying other public objectives that,
in the opinion of the Decision-Making Body, warrant approval
of a project notwithstanding its substantial adverse impact
on the environment. Legal Jargon 2: p. II-9.

Statement Record. Federal--Interior. A file containing approved
draft and final environmental statement, meeting or hearing
records, and all written comments. Legal Jargon 17: no
page number.

Supplemental Environmental Impact Statement. Federal--DOT. A
Supplemental EIS is prepared if substantial project changes
are made which significantly affect the environment or where
significant new information regarding its environmental impacts is found after completion of the Final EIS. The Supplemental EIS is prepared in the same manner as the original Draft and Final EIS. Legal Jargon 35: p. A-2.

Supporting Data. City--Environmental Section. The information
and analysis gathered or prepared for the Initial Study
that supports the conclusions relative to the project's
possible environmental impact. Legal Jargon 2: p. II-10.

Terminology. State--Resources Agency. The following words are
used to indicate whether a particular subject in the Guidelines is mandatory, advisory, or permissive: (a) "Must"
or "shall" identifies a mandatory element which all public
agencies are required to follow, (b) "Should" identifies
guidance provided by the Secretary for Resources based on
policy considerations contained in CEQA, in the legislative
history of the statute, or in federal court decisions which
California courts can be expected to follow. Public agencies are advised to follow this guidance in the absence of
compelling, countervailing considerations, (c) "May" identifies a permissive element which is left fully to the discretion of the public agencies involved. Legal Jargon 15:
p. 4.

Third Party. <u>Federal--EPA</u>. A method for preparing EPA's environmental impact statement whereby the applicant retains a consultant, the responsible official exercises a concurrence review, and then the responsible official supervises the approved consultant in the preparation of the EIS. This method is optional and requires approval of both the new source applicant and the responsible official prior to the execution of an agreement to prepare the EIS. Generally, the preparation of the EIS under the third party method would be initiated prior to the preparation of the environmental impact assessment by the applicant and would thereby serve the purpose of any such environmental assessment analyses. Legal Jargon 19: pp. 2453-2454.

Threshold. <u>Federal--HUD</u>. A criterion of size or of environmental impact above which a Special Environmental Clearance is always required. The threshold is designed to screen out the more important HUD actions for special attention. HUD thresholds are set forth in Appendix A-1 of this Handbook. Legal Jargon 11: p. 19183.

Threshold Of Determination. <u>State--Department of Ecology</u>. Means a decision by a lead agency whether or not an environmental impact statement is required for a proposal. Legal Jargon 37: p. 4.

Tiering. <u>Federal--CEQ</u>. Refers to the coverage of general matters in broader environmental impact statements (such as national program or policy statements) with subsequent narrower statements or environmental analyses (such as regional or basinwide program statements or ultimately site-specific statements) incorporating by reference the general discussions and concentrating solely on the issues specific to the statement subsequently prepared. Legal Jargon 30: p. 25246. (Revised).

Typically Associated Environmental Effect. <u>City--Office of the Mayor</u>. Changes in one or more natural resources which usually occur because of impacts on other such resources as a result of natural interrelationships or cycles. Legal Jargon 23: p. 6.

Undertaking. <u>Federal--Energy</u>. Means a proposed initiative of a private person or non-Federal governmental entity which may result in an action. Legal Jargon 32: p. 7235.

<u>Federal--Historic Preservation</u>. Means any Federal action, activity, or program, or the approval, sanction, assistance, or support of any other action, activity or program. Legal Jargon 34: p. 3367. (Revised).

United States. <u>Federal--Executive Order</u>. All of the several
        States, the District of Columbia, the Commonwealth of
        Puerto Rico, American Samoa, the Virgin Islands, Guam,
        and the Trust Territory of the Pacific Islands. Legal
        Jargon 38: p. 363.

3. LAWS AND REGULATIONS

Index

Terms

A-95 Directive. State--Transportation. From the U.S. Office of
    Management and Budget Circular A-95 Directive which pro-
    vides for a review of proposed Federally assisted highway
    projects by interested and affected State and local agen-
    cies to determine the compatibility of the projects with
    the interests of the reviewing agencies. Transportation
    1: p. vi.

Antiquities Act Of 1906 (36 Stat. 225, 16 U.S.C. 431 et seq.)
    Federal--GSA. The earliest legislation enacted to protect
    cultural resources, this act presents the basic principles
    of Federal protection and preservation, affirms the public
    availability of the nation's historic and prehistoric arch-
    aeological resources, and provides for Federal control of
    all archaeological resources on federally owned or control-
    led land. General 2: p. XIII-35.

Archaeological And Historic Preservation Act Of 1974 (88 Stat.
    174, 16 U.S.C. 49 et seq.) Federal--GSA. This act ex-
    tends the requirements of the Reservoir Salvage Act of 1960
    to all Federal or federally assisted or licensed construc-
    tion projects. The act places coordinating responsibility
    with the Secretary of the Interior and, for the first time,
    authorizes all Federal agencies to seek appropriations, ob-
    ligate available funds, or reprogram existing appropriations
    for the recovery, protection, and preservation of signifi-
    cant scientific, pre-historic, or archaeological resources.
    Agencies can either undertake these protective efforts them-
    selves or transfer one percent of the total amount author-
    ized to be appropriated for each project (over $50,000) to
    the Secretary of the Interior for such purposes. (These
    obligations are an important aspect of any measures to min-
    imize harm to cultural resources.) General 2: p. XIII-36.

Clean Air Act (42 U.S.C. 1857 et seq.) Federal--GSA. This act
    formally adopts national ambient air quality standards and
    State implementation plans, which provide for the implemen-
    tation, maintenance, and enforcement of those standards.
    The Environmental Protection Agency has also issued stan-
    dards of performance for new stationary sources, national
    emission standards for hazardous air pollutants, and motor
    vehicle emission standards. General 2: p. IV-19. (Revised).

    State--Board. Common reference to the Clean Air Act Amend-
    ments of 1970 (Public Law 91-604) which are the basis for
    the federal air pollution control effort. The Act estab-
    lishes broad federal authority to combat air pollution and

requires the states to perform a variety of actions to ensure that healthful levels of air quality are achieved and
maintained. Air 2: no page number.

State--Committee. An act of the U.S. Congress passed in
1963 and amended in 1967, 1970, 1973, 1974 and 1977. The
most important amendments required the U.S. Environmental
Protection Agency to set national air quality standards for
air pollutants and insure that these standards are met.
Air 3: no page number.

Clean Air Act, As Amended in 1974 (PL 91-604). Federal--GSA.
Each state is required to develop plans for the implementation, maintenance, and enforcement of federally established air quality standards in every air quality region
within its boundaries. Any transportation controls necessary to reduce carbon monoxide emissions (e.g., reduction
of motor vehicle traffic through expansion of or promotion
of the use of mass transportation or car pools) must be included in the plans. General 2: p. III-21.

Clean Air Amendments Of 1970 (PL 91-604). Federal--HUD. The Clean
Air Act, as amended in 1970, established national primary
and secondary ambient air quality standards and required
each state to adopt implementation plans (SIPs) providing
means for the attainment, maintenance, and enforcement of
the national standards. The Act delegated to the state primary responsibility for carrying out implementation plans
within the designated air quality control regions of each
state. States may, however, authorize a local agency to
administer or develop certain portions of a plan. General
11: p. 41.

Regional--Commission. Established National Ambient Primary
and Secondary Air Quality Standards for six major air pollutants: particulate matter, sulfur oxides, carbon monoxide, photochemical oxidants, hydrocarbons, and nitrogen oxides. These standards have been established by the Federal
Government as the levels of air quality necessary to protect
the public health and welfare. General 10: p. IV-10.

Coastal Zone Management Act of 1972 (PL 92-583). Federal--GSA.
The most comprehensive incorporation of protective measures
relating to geology/soils/hydrology, this act requires that
designated coastal zone states develop management plans and
regulations governing critical environmental areas (e.g.,
wetlands, estuaries, barrier beaches). General 2:
p. VIII-26.

Federal--GSA. Section 307(c)(3) requires a permit for any
activity affecting land or water uses in the State coastal
zone. The applicant must certify compliance with the State
coastal zone management program. General 2: p. IX-14.

Federal--GSA. This act provides matching funds for coastal
states to encourage the development and administration of
State coastal zone management programs, and to promote the
wise use of the marine, estuarine, wetland, and upland areas
bordering the country's shores. General 2: p. X-10.

Federal--HUD. The purpose of the Act, administered by the
Department of Commerce, is to foster more effective and beneficial management, use, and protection of the U.S. coastal
zone. The Act authorizes provision of funds to coastal
states to prepare and administer coastal zone management
programs "to achieve wise use of the land and water resources of the coastal zone, giving full consideration to
ecological, cultural, historic, and aesthetic values as well
as to needs for economic development"(303(b)). General
11: p. 27.

Endangered Species Act Of 1973 (PL 93-204; 87 Stat. 884). Fed-
eral--GSA. The Secretary of the Interior is to maintain
a list of threatened and endangered species of plants and
animals, and all Federal departments and agencies are to
ensure that their actions do not jeopardize the continued
existence of those species. General 2: p. XI-26.

Regional--Commission. Requires federal agencies to ensure
that their programs do not jeopardize endangered species or
destroy or modify habitats critical to their existance.
The Fish and Wildlife Service of the Department of the Interior has compiled a list of rare and endangered species
within the U.S. In addition, the Fish and Wildlife Service
is currently determining the "critical habitat" for 108
endangered species of the United States and Puerto Rico.
General 10: p. IV-22.

Environmental Quality Improvement Act of 1970 (83 Stat. 852; 42
U.S.C. 4371-4374). Federal--Forest Service. Assures that
each Federal department and agency conducting or supporting
public work activities which effect the environment shall
implement the policies established under existing law; and
authorizes an Office of Environmental Quality, which shall
provide the professional and administrative staff for the
Council on Environmental Quality. General 15: p. 73.

Executive Order 11296, Evaluation Of Flood Hazard In Locating Fed-
erally-Owned Or Financed Buildings, Roads, And Other

Facilities, And In Disposing Of Federal Lands And Properties (31 FR 10663), August 11, 1966. Federal--GSA. This
order directs the Federal agencies to avoid the location
of new Federal facilities and federally financed or supported improvements in flood plains. General 2: p. VII-25.

Executive Order 11593, May 13, 1971, For The Protection And Enhancement Of The Cultural Environment (36 F.R. 8921, 16
U.S.C. 470). Federal--GSA. The second most important
legal document aimed at ensuring the protection of cultural
resources, the order requires all Federal agencies to survey properties under their jurisdiction and nominate appropriate candidates to the National Register of Historic
Places. General 2: p. XIII-37. (Revised).

Executive Order 11752, Prevention, Control, And Abatement Of Environmental Pollution At Federal Facilities, December 17,
1973. Federal--GSA. This order stipulates that "heads
of Federal agencies shall ensure that all facilities under
their jurisdiction are designed, constructed, managed, operated, and maintained so as to conform to the following
requirements: Federal, State, interstate, and local air
quality standards and emission limitations adopted in accordance with or effective under the provisions of the
Clean Air Act, as amended." General 2: p. IV-19.

Federal--GSA. Executive Order 11752 and OMB Circular A-106
require that "heads of Federal agencies shall ensure that
all facilities under their jurisdiction are designed, constructed, managed, operated, and maintained so as to conform to...Guidelines for solid waste recovery, collection,
storage, separation, and disposal systems issued by the
EPA Administrator pursuant to the Solid Waste Disposal Act,
as Amended." Federal facilities are not, however, required
to comply with State or local administrative procedures relating to solid waste pollution abatement and control. General 2: p. VI-10.

Federal Water Pollution Control Act Amendments of 1972 (PL 92-500,
33 U.S.C. 1251 et seq.) Federal--GSA. These amendments
restrict the operation of existing facilities and construction of new facilities that adversely affect hydrology
(particularly on-site and adjacent water supply and water
quality). General 2: p. VIII-25.

Federal--GSA. Under section 313 of this act, all properties, facilities, and activities of the Federal Government
are to comply with Federal, State, interstate, and local
water pollution requirements, unless specifically excluded
by the President. The President, however, may not exempt

a Federal source from either national performance standards
(section 306) or toxic and pretreatment standards (section
307). Section 402 authorizes the National Pollutant Discharge Elimination System (NPDES), which requires an EPA
permit for any pollutant discharge that does not meet Federal standards. Effluent limitations and pretreatment
standards for specific industry classifications are published in the Federal Register and are available from EPA.
General 2: p. IX-14.

Federal--HUD. PL 92-500, administered by EPA, greatly ex-
pands the Federal emphasis on water quality management
planning, particularly coordination among water pollution
control activities at the different levels of government.
General 11: p. 35.

Fish And Wildlife Coordination Act (PL 85-624, 16 U.S.C. 661-666).
Federal--GSA. To ensure consideration of wildlife conservation in water resources development plans, this act requires consultation with the U.S. Fish and Wildlife Service,
Department of the Interior. General 2: p. IX-14.

Regional--Commission. Provides for the coordination of
wildlife conservation with water resources development programs. Any water resources development project carried
out by a federal agency or licensed by a federal agency
must consult with the U.S. Fish and Wildlife Service in
regard to the wildlife resources of the proposed project
area. General 10: p. IV-22.

Fishery Conservation And Management Act Of 1976 (FCMA)(PL 94-265).
Federal--Commerce. The FCMA provides for a national program for the conservation and management of fishery resources within a declared Fishery Conservation Zone (FCZ),
contiguous to the territorial sea of the United States.
Plant/Animal 2: p. 106. (Revised).

Flood Disaster Protection Act Of 1973 (PL 93-234). Federal--HUD.
Under this act, HUD is required to identify all communities
that contain areas that are either flood-prone (i.e., one
percent possibility of flooding in any given year--the 100
year floodplain) or subject to erosion and mudslides. HUD
is to issue local hazard boundary maps to the identified
communities, which are then required to adopt and enforce
floodplain management measures. The act precludes construction of new public facilities in flood-prone areas unless
adequate damage control measures are incorporated into the
land use plans by disqualifying such facilities for flood
insurance. General 2: p. VIII-25.

Historic Sites Act Of 1935 (49 Stat. 666, 16 U.S.C. 46 et seq.).
Federal--GSA. This act declares as national policy the
preservation of historic (including prehistoric) sites,
buildings, and objects of national significance. The act
directs the National Park Service to establish a mechanism
for cataloging and identifying historic and archaeological
properties. Three programs were developed as a result of
this directive: The Historic American Building Survey,
Historic American Engineering Record, and the National
Historic Landmarks Program. General 2: p. XIII-36.

Land And Water Conservation Fund Act (78 Stat. 897, as amended;
16 U.S.C. 4601-4 to 4601-11, 23 U.S.C. 120). Federal--
Forest Service. Provides funds for and authorizes Federal
assistance to the states in planning, acquisition, and development of needed land and water areas and facilities
and provides funds for the Federal acquisition and development of outdoor recreation resources. General 15: p. 106.

Marine Protection, Research And Sanctuaries Act Of 1972 (PL 92-
532). Federal--GSA. Title I regulates ocean dumping and
title II authorizes the Department of Commerce to regulate
any activities in designated marine sanctuaries. General
2: p. IX-14.

Multiple-Use-Sustained Yield Act (74 Stat. 215; 16 U.S.C. 528-
531). Federal--Forest Service. Authorizes and directs
that the National Forests be managed under principles of
multiple use for outdoor recreation, range, timber, watershed, and wildlife and fish purposes, and to produce a sustained yield of products and services, and for other purposes. This Act does not affect the use or administration
of the mineral resources of National Forest lands or the
use or administration of Federal lands not within National
Forests. General 15: p. 134.

National Environmental Policy Act (42 U.S.C. 4341, as amended by
PL 94-52, and PL 94-83). Federal--HUD. Public concern
over the deteriorating state of the environment heightened
considerably in the 1960's and culminated at the end of
this decade in passage of the National Environmental Policy
Act, which became law on January 1, 1970. This far-reach-
ing piece of legislation made systematic consideration of
all aspects of environmental quality--natural, built, and
social--a matter of national policy. General 11: p. 18.

State--Natural Resources. Establishes a: (1) National
policy for the environment (not a protection act), (2)
Requires Federal agencies to become aware of environmental
ramifications of their proposed actions, (3) Requires

disclosure to the public of proposed major federal actions
and provides a mechanism for public input into the federal
decision-making process, (4) Requires Federal agencies to
prepare an Environmental Impact Statement (EIS) for every
major action significantly affecting the quality of the
human environment. Legal Jargon 20: p. 1.

City--Environmental Section. The federal law requiring an
environmental assessment for federal actions that involve
impacts on the environment. NEPA is set forth in 42
U.S.C.A. 4321 et seq. Legal Jargon 2: p. II-6.

National Flood Insurance Act Of 1968 (PL 90-448, 82 Stat. 476).
Federal--GSA. This act, administered by the Department of
Housing and Urban Development (HUD), requires that all
flood-prone areas interested in participating in the National Flood Insurance Program complete flood insurance surveys
and studies by August, 1983. General 2: pp. VIII-24 and
VIII-25.

National Historic Preservation Act Of 1966 (80 Stat. 915, 16
U.S.C., 470 et seq.). Federal--GSA. This act is the single
most important law governing the policies of the Federal
agencies toward historic preservation. In addition to ex-
panding the national policy of historic preservation at the
Federal level, it encourages preservation on the State and
private levels, in part through grants for preparing comprehensive statewide surveys and plans for historic preservation. Section 101(a) of the act authorizes the Secretary of the Interior to expand and maintain a National Register of Historic Places for cultural resources. Section
106 outlines specific actions required of Federal agencies
to protect cultural resources, defined as any district,
site, building, structure or object included in the National
Register. To satisfy these requirements, the EIS must consider the effects of a proposed Federal undertaking on the
resources. The act establishes the Advisory Council on
Historic Preservation to comment on Federal actions having
an effect on cultural resources. The Advisory Council has
implemented procedures to facilitate compliance with this
section. General 2: p. XIII-37.

Federal--Historic Preservation. An act to establish a
program for the preservation of additional historic properties throughout the Nation and for other purposes. Laws/
Regulations 1: p. 3367.

National Trails System Act (82 Stat. 919; 16 U.S.C. 1241-1249).
Federal--Forest Service. Institutes a national system of
recreation and scenic trails. Designates the Appalachian

Trail and the Pacific Crest Trail as the initial component
of that system and prescribes the methods by which, and
standards according to which, additional components may be
added to the system. Additional trails should be established primarily near urban areas and secondarily within established, more remotely located scenic areas. General 15:
p. 137.

New Source Performance Standards (40 CFR 60). Regional--Commis-
sion. Regulate the emission rates for specific pollutants
such as: fossil-fuel steam generators, incinerators, port-
land cement plants, petroleum refineries, and coal preparation plants. General 10: p. IV-10.

Noise Control Act Of 1972 (PL 92-574). Federal--GSA. Section 4
of the Noise Control Act of 1972 requires that GSA "comply
with Federal, State, interstate, and local requirements
respecting control and abatement of environmental noise to
the same extent that any person is subject to such requirements." General 2: p. V-18.

Federal--HUD. The Act delegated to the U.S. Environmental
Protection Agency responsibility for coordinating federal
research on noise and for establishing noise standards for
products distributed in common (source control). General
11: p. 45. (Revised).

Regional--Commission. The Environmental Protection Agency
has assumed responsibility for establishing and implementing source standards for the manufacture of major noise
sources such as construction equipment, transportation
equipment, motors and electrical equipment. General 10:
p. IV-61.

Office Of Management And Budget, Circular A-95 Of January 2, 1976
(revised). Federal--GSA. Circular furnishing guidance to
Federal agencies for cooperation with State and local governments in the evaluation, review, and coordination of Federal and federally assisted programs and projects. General
2: p. III-22.

Procedures For The Protection Of Historic And Cultural Properties
Advisory Council On Historic Preservation (41 F.R. 5902,
CFR Title 36, Chapter 8, Part 800). Federal--GSA. Issued
by the Advisory Council on Historic Preservation, these procedures represent the most important document needed to perform an analysis of the environmental impact of a project
on cultural resource. Pursuant to Section 106 of the National Historic Preservation Act and Sections 1(3) and
2(b) of the Executive Order 11593, the Advisory Council

set forth the procedures required for an adequate review
by the Council of Federal undertakings that might affect
either federally owned or non-federally owned historic and
cultural resources eligible for or listed in the National
Register of Historic Places. General 2: p. XIII-38. (Revised).

Reservoir Salvage Act Of 1960 (74 Stat. 220, U.S.C. 469-469c.).
Federal--GSA. To protect any undiscovered archaeological
resources of importance, this act requires any agency of
the United States to notify the Secretary of the Interior
of any plans to construct a dam or issue a license for the
construction of a dam larger than 500 acre-feet or 40 surface areas of capacity. If any archaeological resources
will be affected by the dam, the provisions of the act apply regardless of the size of the reservoir. The Secretary of the Interior will then approve an archaeological
research and/or survey as expeditiously as possible. General 2: p. XIII-36.

Federal--Forest Service. The Reservoir Salvage Act of 1960
provides for the recovery and preservation of "historical
and archaeological data (including relics and specimens)"
that might be lost or destroyed as a result of the construction of dams, reservoirs, and attendant facilities
and activities. General 15: p. 179.

Resource Recovery Act Of 1970. National--Resources. This Act
extended the Solid Waste Disposal Act of 1965 for another
three years and extensively amended it to: (1) redirect
Federal emphasis from disposal to recycling; (2) expand
the grant and demonstration program with the Federal government paying as much as 75 percent of the total costs of programs of interstate solid waste management agencies; and
(3) create a National Commission on Materials Policy to
recommend means for extraction, development, and use of
materials susceptible to recycling, reuse, or self-destruc-
tion. Congress passed one year extensions of the Act in
1973 and in 1974 which kept in effect all provisions of
the 1970 legislation (although the National Commission on
Materials Policy fulfilled its mandate and is no longer in
existence). Energy/Utility 5: p. 12.

Safe Drinking Water Act Of 1974 (PL 93-523). Federal--GSA. Any
federally owned or maintained public water system is required to comply with all national primary drinking water
regulations. General 2: p. IX-15.

Federal--HUD. The Act authorizes the Environmental Protection Agency to issue national drinking water regulations

applicable to all public water supplies. The states are
charged with primary enforcement responsibility and, to
protect underground sources of drinking water from contamination, they are required to establish underground injection control programs which include a permit system governing all such injections. In the interim, section 1424
(e) of the Act prohibits commitment of federal financial
assistance to any project that might endanger an aquifer
if the aquifer is an area's sole or primary drinking water
source. General 11: p. 40.

Solid Waste Disposal Act Of 1965 (PL 89-272) As Amended By Resource Recovery Act Of 1970 (PL 91-512). Federal--GSA.
Sections 209 and 211 stipulate that all Federal agencies
are to comply with the solid waste recovery, collection,
separation, and disposal systems guidelines formulated by
the EPA. General 2: p. VI-10.

National--Resources. With this Act, the Federal government assumed a supporting role in the area of solid waste
management, moving away from the traditional view of solid
waste as the responsibility of local governments. The Act
made the Federal government responsible for research, training, demonstrations of new technology, technical assistance,
and grants-in-aid for state and interstate solid waste planning programs. Energy/Utility 5: p. 13.

Standards For Hazardous Air Pollutants (40 CFR 61). Regional--
Commission. Apply to any stationary source for which a
standard is prescribed. To date, standards have been established for the hazardous substances of asbestos, mercury,
and beryllium. Also, Regulations for the Control of Air
Pollution from New Motor Vehicles and Vehicle Engines (40
CFR 85, 40 CFR 86) establish standards for auto emissions
control in the U.S. General 10: p. IV-10. (Revised).

Physical Glossaries

4. EARTH

Index

Cut-And-Fill
Delta
Deposition
Detritus
Development Of Potential And
    Future Land Use
Differential Settlement
Dip-Slip Fault
Duff
Earthflow
Effective Soil Depth
Elevation
Eluviation
Embankment
Embayment
Epicenter
Erode
Erosion
Erosion Class
Erosion Potential
Erosion Signature
Escarpment
Estuaries
Estuarine Area
Estuarine Sanctuary
Estuary
Existing Land Use
Farm
Fault
Federal Lands
Fen
Fertilizer
Field Capacity
Field Moisture Capacity
Fill
Filling
Finish Elevation
Fjord Estuaries
Floor Area Ratio (FAR)
Fluvial Deposit
Fold
Foreshore
Frail Lands
Geological Areas
Geological Terrace
Geologic Erosion
Geology
Geomorphology
Grade
Grading

Greenbelt
Green Space
Greenway
Gully
Gully Erosion
Hardpan
Homogeneous Estuary
Homogeneous Ground
Horizon
Humus
Hydraulic Fill
Hypersaline Estuary
Igneous Rocks
Impermeable Layer
Impervious Material
Impervious Soil
Inactive Fault
Incompatible Uses
Infiltration
Infiltration Capacity
Infiltration Rate
Infiltration Velocity
Intensity Index
Interbedded
Interstices
Intrusion
Irrigated Land
Irrigated Rangeland
Joint
Key Bed
K Values
Lacustrine Deposits
Land Capability Class
Land Capability Classification
Land Classification
Landform
Land Reclamation
Land Resource Area
Land Resource Region
Land Resources
Landslide
Land Use
Land Use Plan
Land Use Planning
Land Use Type
Late Pleistocene
Lenses
Liquefaction
Liquefication
Liquid Limit (LL)

Lithology
Litter
Littoral
Littoral Drift
Loamy
Loess
Marsh
Mechanical Analysis
Metals
Metamorphic Rocks
Microrelief
Mineral
Mineral Resources
Monocline
Moraine
Morphology
Muck
Muck Soils
Mulch
Multiple Use
National Forest Lands
Natural Area
Natural Area Preserve
Natural Environment
Natural Setting
Neutral Soil
Nonmetals
Non-Renewable Resource
Normal Erosion
Open Space
Organic Deposits
Orographic
Ortstein
Outwash
Overburden
Pan
Parent Material
Peat
Percolation
Percolation Test
Permeability
Physiographic Region
Physiographic Units
Physiography
Planosol
Plasticity
Plasticity Index (PI)
Plastic Limit
Plastic Soil
Porosity

Public Domain Lands
Public Land
Quaternary Age
Range Condition Class
Rangeland
Raw Land
Relief
Residual Geologic Materials
Richter Scale
Rill Erosion
Saline Soil
Salt Marsh
Salt Water Estuary
Sand
Sand Lens
Saturation Point
Scour
Sediment
Sedimentary Rocks
Sedimentation
Seismicity
Shear
Shear Strength
Sheet Erosion
Shore Erosion
Shoreline Erosion
Silt
Site Preparation
Slick Spots
Slip
Slope
Slope Characteristics
Slope Stability
Soil
Soil Aeration
Soil Amendments
Soil Association
Soil Complex
Soil Conditioner
Soil Conservation
Soil Depth
Soil Elasticity
Soil Erosion
Soil Fertility
Soil Horizon
Soil Improvement
Soil K-Factor
Soil Loss Equation
Soil Management
Soil Map

Soil Mapping Unit
Soil Moisture
Soil Morphology
Soil Permeability
Soil Porosity
Soil Probe
Soil Profile
Soil Resource Group
Soil Series
Soil Structure
Soil Structure Types
Soil Survey
Soil Texture
Solum
Spoil
Stability
Stream Bank Erosion
Stream Bed Erosion
Stream Incision
Strike And Dip
Strike-Slip Fault
Structure
Subgrade
Subsoil
Substrata
Substratum
Surface Erosion
Surface Soil
Swale
Syncline
Talus
Tectonic Estuaries
Terrain Analysis
Tidal Flats
Tidal Marsh
Toe
Topographic Map
Topography
Transitory Range
Truncated Soil Profile
Ultra Basic Rock
Unconsolidated Sediment
Undercutting
Uniform Soil
Upland
Urban Areas
Urbanized Area
Urban Land
Virgin Areas
Weathering

Wetlands
Wilderness
Wilderness Area
Wilderness Study Area
Wind Erosion
Xeric
Zone Of Aeration
Zone of Saturation

Terms

AASHTO Classification. National--Soil Conservation. The official
    classification of soil materials and soil aggregate mix-
    tures for highway construction used by the American Assoc-
    iation of State Highway Transportation Officials. Earth 4:
    p. 5g.

ABC Soil. National--Soil Conservation. A soil that has a profile
    including A, B, and C master horizons. Earth 4: p. 5g.

Abrasion. National--Soil Conservation. The wearing away by fric-
    tion, the chief agents being currents of water or wind la-
    den with sand or other rock debris and glaciers. Earth 4:
    p. 5g.

Accelerated Erosion. National--Engineering. Erosion of soil ma-
    terial at a rate more rapid than that of natural, normal or
    geological erosion. Accelerated erosion occurs as a re-
    sult of destruction of vegetal cover or some activity of
    man. Water 1: no page number.

    National--Soil Conservation. Erosion much more rapid than
    normal, natural, or geologic erosion, primarily as a result
    of the influence of the activities of man or, in some cases,
    of other animals or natural catastrophies that expose base
    surfaces, for example, fires. Earth 4: p. 21g.

Accretion. Federal--Army. The process of growth or enlargement
    by external accumulation. General 9: p. 1.

    National--Soil Conservation. The gradual addition of new
    land to old by the deposition of sediment carried by a
    stream. Earth 4: p. 5g.

Acid Soil. Regional--River Basin. A soil giving an acid reaction
    (precisely, below pH 7.0; practically, below pH 6.6)
    throughout most or all of the portion occupied by roots.
    Earth 1: p. 371.

Active Fault. National--Soil Conservation. A fault that has un-
    dergone movement in recent geologic time (the last 10,000
    years) and may be subject to future movement. Earth 4:
    p. 5g.

Adobe. National--Engineering. Term used to describe a structural
    condition of soils, usually of high clay content, which
    crack into roughly cubical blocks when dried. Water 1:
    no page number.

A Horizon. Federal--Interior. The stratum of soil consisting of
     one or more of the following layers: A_o horizon, partly de-
     composed or matted plant remains lying on the top of the
     mineral soil (duff); A_{oo} horizon, the relatively fresh
     leaves and other plant debris, generally of the previous
     year, lying on A_o horizon (litter); A_1 horizon, the sur-
     face mineral layer, relatively high in organic matter, us-
     ually dark in color; A_2 horizon, below the A_1 horizon, in
     places the surface layer, usually lighter in color than the
     underlying horizon, in which leaching of solutes and sus-
     pended materials occurs; A_3 horizon, transitional to the
     B horizon, more like A than B, sometimes absent. General
     3: p. 3.

Air Drainage. National--Engineering. Renewal of soil air by dif-
     fusion and meteorological factors as soil temperature
     changes, barometric variations and action of wind. Water
     1: no page number.

Alkaline Soil. National--Soil Conservation. A soil that has a
     pH value greater than 7.0, particularly above 7.3, through-
     out most or all of the root zone, although the term is com-
     monly applied to only the surface layer or horizon of a
     soil. Earth 4: p. 6g.

Alkali Soil. National--Engineering. Soil that contains suffi-
     cient exchangeable sodium to interfere with water penetra-
     tion and crop growth, either with or without appreciable
     quantities of soluable salts. Water 1: no page number.

Alluvial. Federal--Interior. Pertaining to the sediments, usu-
     ally mineral or inorganic, deposited by running water.
     General 4: Glossary p. 1.

     Federal--Interior. Material deposited by running water,
     such as clay, silt, sand, and gravel. General 3: p. 4.

Alluvial Cone. National--Engineering. Cone or fan-shaped deposit
     of alluvium. Water 1: no page number.

Alluvial Fan. Federal--ERDA. Rock deposit laid down by streams
     flowing from mountains into lowland regions. General 7:
     p. g-1.

     Federal--Interior. A sloping mass of sand, clay, etc.,
     that widens out like a fan where a stream gradually slows
     down as it enters a plain. General 3: p. 4.

Alluvial Terrace. State--Natural Resources. Flat, generally hor-
     izontal, land surface composed of recent, water deposited,

unconsolidated sediments. General 12: p. 335.

Alluvium. Regional--River Basin. Soil material, such as sand, silt, or clay, that has been deposited by water. Earth 1: p. 371.

Angle Of Repose. National--Soil Conservation. Angle between the horizontal and the maximum slope that a soil assumes through natural processes. Earth 4: p. 7g.

Anisotropic Soils. National--Engineering. Soils not having the same physical properties when the direction of measurement is changed. Commonly used in reference to permeability changes with direction of measurement. Water 1: no page number.

Anticlinal. Federal--Interior. Of or relating to a geological anticline, which is an arch of stratified rock in which the layers bend downward in opposite directions from the crest. General 4: Glossary p. 1.

Anticline. Federal--ERDA. An uparched fold in which the rock strata dip away from the fold's axis; the opposite of syncline. General 7: p. g-1.

Federal--Interior. A geological structure or arch formed by strata from opposite sides dipping from a common line. General 3: p. 7.

State--Natural Resources. A configuration of folded, stratified rocks which dip in two directions from a crest. General 12: p. 335.

Arable Land. Federal--Army. Land fit for cultivation. General 9: p. 3.

Attitude. National--Soil Conservation. The relation of some directional features to a rock in a horizontal surface. Earth 4: p. 8g.

Available Soil Moisture. National--Engineering. That part of soil moisture available for use by plants. Water 1: no page number.

Available Water Holding Capacity. Regional--River Basin. The capacity of a soil to hold water in a form available to plants. Amount of moisture held in soil between field capacity, or about one-third atmosphere of tension, and the wilting coefficient, or about 15 atmospheres of tension. Earth 1: p. 371.

Azonal. Federal--Interior. Designating or of zones of soil that
          cannot be sharply distinguished from one another because
          for example, they are of recent formation. General 3:
          p. 10.

Backfill. Federal--EPA. The material used to refill an excava-
          tion, or the process of doing so. General 1: p. 2.

          National--Soil Conservation. The material used to refill
          a ditch or other excavation, or the process of doing so.
          Earth 4: p. 8g.

Backfilling. Federal--Interior. A reclamation technique which
          returns the rock and soil of the overburden to mined cuts
          or pits. This leaves the land in a configuration similar
          to the original form. General 4: p. 1.

Bank. National--Engineering. Margin of raised ground bordering
          a watercourse or lake. Water 1: no page number.

Bank Erosion. Federal--Army. Pertains to the wearing away of
          soil and rock along a stream or river due to water velocity,
          water levels, scouring, or surface runoff. Recreation 2:
          p. A-34.

          Regional--River Basin. Destruction of land areas border-
          ing rivers or water bodies by the cutting or wearing action
          of waves or flowing water. Water 5: p. 1013.

Bar-Built Estuaries. Federal--DOT. Estuaries with shallow basins,
          often partly exposed at low tide, enclosed by a chain of
          offshore or barrier islands, broken at intervals by inlets.
          Transportation 3: p. 15.

Beach. Federal--Army. Depositional area at the shore of an
          ocean or lake covered by mud, sand, gravel or larger rock
          fragments and extending into the water for some distance.
          General 9: p. 5.

Beach Erosion. Regional--River Basin. The retrogression of the
          shore line of large lakes and coastal waters caused by wave
          action, shore currents, or natural causes other than sub-
          sidence. Water 5: p. 1014.

Bedding. City--Planning. The arrangement of rocks or sediments
          in layers, strata, or beds. Usually means the same as
          stratification. General 5: p. 53.

Bedrock. Federal--Army. The solid rock at the surface or under-
          lying other surface materials. General 9: p. 5.

Federal--DOT. The more or less solid, undisturbed rock in place either at the surface or beneath superficial deposits of gravel, sand or soil. Transportation 3: p. 15.

City--Planning. Solid, undisturbed or unweathered rock in place. It should be used only in reference to a geologic formation, and not to hard soils or highly weathered rock materials. General 5: p. 53. (Revised).

Bench.   Federal--Interior. A terrace along the bank of a body of water, often marking a former shoreline. General 4: Glossary p. 1.

Berm.   Federal--Interior. A ledge or shoulder, as along the edge of a road or canal. General 3: p. 12.

National--Engineering. Strip or area of land, usually level, between the spoil bank and edge of a ditch or canal. Water 1: no page number.

Bog.   Federal--Army. A quagmire or wet, spongy ground; often a filled-in lake; composed primarily of dead plant tissues (peat); principally mosses. General 9: p. 7.

Federal--DOT. A standing water body, having only slight flow-through of water, characterized by peat formation, low pH, and water colored with tannin and lignin from vegetation. Transportation 3: p. 17.

Federal--EPA. Wet, spongy land usually poorly drained, highly acid and rich in plant residue, the result of lake eutrophication. General 1: p. 3.

Bottom Lands. Federal--Interior. Those lands immediately adjacent to, and slightly higher than an existing stream channel. General 3: p. 14.

Regional--River Basin. Low land formed by alluvial deposits along a river or stream. Earth 1: p. 371.

State--Soil and Water Commission. A term often used to define lowlands adjacent to streams. Earth 3: p. G-5.

Buffer Strips. Federal--EPA. Strips of grass or other erosion-resisting vegetation between or below cultivated strips or fields. General 1: p. 3.

Buffer Zone. Federal--Army. An allocation of land between functional areas to preserve the integrity of each of those areas and to prevent physical, visual, or aesthetic encroachment on the respective areas. Recreation 2: p. A-31.

Federal--GSA. Area of open space separating two different
land uses. General 2: p. X-12.

Federal--Interior. An area of land designated or managed
for the purpose of separating and insulating two or more
land areas whose uses conflict or are incompatible. Recreation 3: p. 5.

Built Environment. Federal--DOT. This term is used interchangeably with the term, man-made environment. Transportation
3: p. 18.

Built-Up Urban Areas. Federal--Commerce. Land areas already in-
tensely developed for housing, commerce, industry, etc.
General 6: p. 310.

Calcareous Soil. Regional--River Basin. Soil containing sufficient calcium carbonate (often with magnesium carbonate)
to effervesce visibly to the naked eye when treated with
hydrochloric acid. Soil of alkaline in reaction, owing
to the presence of free calcium carbonate; may be more or
less cemented, depending upon concentration and time.
Earth 1: p. 371.

Capillary Zone. National--Engineering. Zone of soil essentially
saturated, in which pores become filled as a result of surface tension. Water 1: no page number.

Cemented. National--Soil Conservation. Indurated; having a hard,
brittle consistency because the particles are held together
by cementing substances such as humus, calcium carbonate,
or the oxides of silicon, iron, and aluminum. The hardness
and brittleness persist even when wet. Earth 2: p. 11g.

Clay. National--Engineering. Soil particles less than 0.002
mm in diameter according to USDA classification. Water 1:
no page number.

Regional--River Basin. As a soil separate, mineral soil
particles less than 0.002 millimeter in diameter. As a
soil textural class, soil material that is 40 percent or
more clay, less than 45 percent sand, and less than 40
percent silt. Earth 1: p. 371.

State--Natural Resources. Rock or mineral particles with
a volume less than an equivalent sphere with a diameter of
0.004mm. General 12: p. 336.

Clay Lense. _Federal--Commerce_. A lense-shaped deposit of clay.
    General 6: p. 310.

Claypan. _National--Engineering_. Dense, subsoil horizon high in
    clay content having a sharply defined upper boundary.
    Formed by downward movement of clay or by synthesis of
    clay in place during soil formation. Soil permeability
    is usually low in these claypans, but bulk density may
    not be appreciably different than in lower horizons.
    Water 1: no page number.

    Regional--River Basin. A compact, slowly permeable soil
    horizon that contains more clay than the horizon above and
    below it. A claypan is commonly hard when dry and plastic
    or stiff when wet. Earth 1: p. 371.

Coarse Fragments. _Regional--River Basin_. Fragments coarser than
    very coarse sand. (1) Gravelly--20 to 35 percent by volume
    of the soil mass is fragments up to 3 inches in diameter,
    (2) Very Gravelly--35 to 80 percent by volume of the soil
    mass is fragments up to 3 inches in diameter, (3) Cobbly--
    20 to 35 percent by volume of the soil fragments from 3 to
    10 inches in diameter, (4) Very Cobbly--35 to 80 percent
    by volume of the soil mass is fragments from 3 to 10 inches
    in diameter, (5) Stony--20 to 35 percent by volume of the
    soil mass is fragments more than 10 inches in diameter,
    (6) Very Stony--35 to 80 percent by volume of the soil mass
    is fragments more than 10 inches in diameter. Earth 1:
    p. 372.

Coastal Zone. _Federal--EPA_. Ocean waters and adjacent lands that
    exert an influence on the uses of the sea and its ecology.
    General 1: p. 4.

    Federal--GSA. Coastal waters and adjacent lands that exert
    a measurable influence on the uses of the sea and its ecol-
    ogy. Legal Jargon 18: Appendix B-2.

Colluvial. _City--Planning_. Pertaining to soil and rock material
    transported chiefly by gravity, but often aided by running
    water. General 5: p. 53.

Colluvium. _Federal--Army_. Rock and soil accumulated at the foot
    of a slope from gravitational forces. General 9: p. 10.

    Federal--Interior. A deposit of rock fragments and soil
    material accumulated at the base of steep slopes as a re-
    sult of gravitational forces. General 3: p. 20.

Regional--River Basin. Soil material, rock fragments, or both, moved by creep, slide, or local wash and deposited on slopes. Earth 1: p. 372.

Compaction. National--Soil Conservation. In geology, the changing of loose sediment into hard, firm rock. Earth 2: p. 14g. (Revised).

Compatible Uses. Federal--Forest Service. Land uses which can exist together, so that no one use improves or detracts from the quality of another. In practice this definition is usually relaxed to include uses which can coexist and conflict only slightly. General 15: p. 47.

Complementary Uses. Federal--Forest Service. Land uses, each of which is improved in quality or quantity by the other uses existing with it in the same area. General 15: p. 48.

Confining Clays. National--Engineering. Impervious or slowly pervious clay layers in the subsoil which restrict or confine the movement of percolating water. Water 1: no page number.

Conflicting Uses. Federal--Forest Service. Land uses by which the quality of each individual use is harmed by the others when they occur together in the same area--due to either competition for limited resources or use byproducts which damage alternative uses. In the extreme when one use prevents another, "conflicting uses" become incompatible uses. General 15: p. 50.

Conglomerate. City--Planning. A sedimentary rock made up of worn or rounded fragments of other rocks, cemented together. The fragments may range from the size of a pea up to large boulders, and may be any kind of rock. General 5: p. 54. (Revised).

Consolidate. National--Soil Conservation. Any or all of the processes whereby loose, soft, or liquid earth materials become firm and hard. Earth 2: p. 15g.

Consolidated Sediment. State--Natural Resources. A well-cemented sediment. General 12: p. 336.

Creep. National--Soil Conservation. Slow mass movement of soil and soil material down relatively steep slopes, primarily under the influence of gravity but facilitated by saturation with water and by alternate freezing and thawing. Earth 4: p. 16g.

Critical Area. <u>National--Soil Conservation</u>. A severely eroded
    sediment producing area that requires special management
    to establish and maintain vegetation in order to stabilize
    soil conditions. Earth 4: p. 16g.

Cropland. <u>Regional--River Basin</u>. Land regularly used for pro-
    duction of crops, except forest land and rangeland. Earth
    1: p. 372.

Cut. <u>Federal--Forest Service</u>. To lower a land surface by removal
    of earth materials. General 15: p. 57.

    <u>National--Engineering</u>. Portion of land surface or area
    from which earth or rock has been removed or will be re-
    moved by excavation; the depth below original ground sur-
    face to excavated surface. Water 1: no page number.

Cut-And-Fill. <u>National--Soil Conservation</u>. Process of earth
    moving by excavating part of an area and using the ex-
    cavated material for adjacent embankments or fill areas.
    Earth 4: p. 17g.

Delta. <u>Federal--Army</u>. The alluvial deposit at the mouth of a
    river. General 9: p. 12.

    <u>National--Soil Conservation</u>. An alluvial deposit formed
    where a stream or river drops its sediment load on enter-
    ing a body of more quiet water; formed largely beneath the
    water surface and in an area often resembling the shape of
    the Greek letter Delta, with the point of entry of the
    stream at one corner. Earth 4: p. 17g.

Deposition. <u>National--Engineering</u>. Transported material depos-
    ited because of decreased velocity of the transporting
    agent--water or wind, examples of which are accumulations
    at the foot of an eroded slope and alluvial fans. Water
    1: no page number.

Detritus. <u>Federal--DOT</u>. (1) Loose fragments, particles or grains
    that have been formed by the disintegration of rocks, (2)
    Any disintegrated matter, debris. Transportation 3: p. 25.

Development Of Potential And Future Land Use. <u>Regional--Commiss-
    ion</u>. In affecting area characteristics, the project may
    also influence development potential and the consequent
    future use of surrounding sites. This is particularly
    true in cases of projects designed to encourage economic
    development. General 10: p. IV-52.

Differential Settlement. <u>National--Soil Conservation</u>. Has reference to deformation of maximum extent in a short horizontal distance. Earth 2: p. 19g.

Dip-Slip Fault. <u>Federal--ERDA</u>. A fault in which one wall has moved up or down the face of the fault relative to the other; contrasts to strike-slip fault. General 7: p. g-3.

Duff. <u>Regional--River Basin</u>. A type of organic surface horizon of forested soils consisting of matted peaty organic matter only slightly decomposed. Earth 1: p. 373.

Earthflow. <u>Federal--DOT</u>. A mass movement landform and process characterized by downslope translation of soil and weathered rock over a discrete basal shear surface (landslide) within well-defined lateral boundaries. Transportation 3: p. 27.

Effective Soil Depth. <u>National--Soil Conservation</u>. The depth of soil material that plant roots can penetrate readily to obtain water and plant nutrients; the depth to a layer that differs sufficiently from the overlying material in physical or chemical properties to prevent or seriously retard the growth of roots. Earth 4: p. 18g.

Elevation. <u>National--Soil Conservation</u>. The variation in the height of the earth's surface; the measure of vertical distance from a known datum plane which on most maps is mean sea level. Earth 2: p. 21g.

Eluviation. <u>State--Water Resources</u>. A process of removal of organic material and clay in solution or in suspension from the soil by percolating waters. It takes place according to the direction of water movement. Energy/Utility 2: Appendix E, p. 11.

Embankment. <u>State--Soil and Water Commission</u>. A man-made deposit of soil, rock, or other material used to form an impoundment. Earth 3: p. G-8.

Embayment. <u>State--Water Resources</u>. (1) A deep depression in a shoreline forming a large open bay, (2) An area within the swing of a bend of a river. Energy/Utility 2: Appendix E, p. 11.

Epicenter. <u>Federal--Interior</u>. That portion of the Earth's crust which is directly above the focus of an earthquake; focus being that point within the Earth which is the center of an earthquake. General 3: p. 30.

Erode. Federal--Interior. To wear away or remove the land sur-
     face by wind, water, or other agents. General 3: p. 31.

Erosion. Federal--Army. The removal of soil or rock by wearing
     away of land surface. General 9: p. 19.

     Federal--Commerce. The wearing away of the land surface
     by running water, wind, or other geological agents. Gen-
     eral 6: p. 311.

     Federal--EPA. The wearing away of land surface by wind
     or water. Erosion occurs naturally from weather or run-
     off but can be intensified by land-clearing practices.
     General 1: p. 6.

     Federal--GSA. Process by which earthly matter or rock is
     loosened and removed from any part of the earth's surface,
     particularly by the movement of water. General 2: p. VII-
     28.

     National--Engineering. Detachment and movement of the
     solid material from the land surface by wind, water and
     ice or by gravity as in landslides. Water 1: no page
     number.

Erosion Class. National--Engineering. Numerical evaluation of
     erosion or erosion potential obtained in a soil conserva-
     tion survey. Water 1: no page number.

Erosion Potential. National--Engineering. Amount of erosion that
     may be expected under given climatic, topographic, soil,
     crop and cultural conditions. Water 1: no page number.

Erosion Signature. Federal--DOT. Any physical evidence which
     identifies slope erosion. Transportation 3: p. 30.

Escarpment. Federal--Interior. A long, inland cliff or steep
     slope, usually high, formed by erosion or possibly by
     faulting. General 3: p. 31.

     National--Soil Conservation. A steep face or a ridge of
     high land; the escarpment of a mountain range is generally
     on that side nearest the sea. Earth 2: p. 22g.

     City--Planning. A cliff, or steep slope of some extent
     formed by a fault or a cliff or steep slope along the mar-
     gin of a plateau, mesa, or terrace. General 5: p. 54.

Estuaries. Federal--EPA. Areas where fresh water meets salt
     water (bays, mouths of rivers, salt marshes, lagoons).

These brackish water ecosystems shelter and feed marine
life, birds, and wildlife. General 1: p. 6.

Estuarine Area. <u>Federal--Forest Service</u>. U.S. Forest Service
usage. The environmental system of an estuary and those
transitional areas which are consistently influenced or
affected by water from an estuary. General 15: p. 75.

Estuarine Sanctuary. <u>Federal--Forest Service</u>. Coastal Zone Management Act usage. A research area which may include any
part or all of an estuary, adjoining transitional areas,
and adjacent uplands, constituting to the extent feasible
a natural unit, set aside to provide scientists and students the opportunity to examine over a period of time the
ecological relationships within the area. General 15:
p. 75.

Estuary. <u>Federal--Commerce</u>. Any confined coastal water body with
a connection to the sea and measurably quantity of marine
salt in the waters; greater than 0.5 parts per thousand
(ppt). General 6: p. 311.

<u>Federal--Forest Service</u>. U.S. Forest Service usage. All
or part of the mouth of a navigable or interstate river or
stream or other body of water having unimpaired natural
connection with the open sea and within which the sea water
is measurably diluted with fresh water derived from land
runoff, as defined in Public Law 89-753, and similar lands
and waters of the Great Lakes, as directed by Public Law
90-454. General 15: p. 76.

<u>Federal--Forest Service</u>. A water body having an open,
natural connection with the sea and within which the sea
water is measurably diluted by freshwater runoff. General
15: p. 76.

<u>Federal--Interior</u>. Shallow coastal water, usually assoc-
iated with the mouth of a river, including adjoining bays,
lagoons, shallow sounds, and marshes where tidal effects
are evident and fresh water and sea water mix. General 4:
p. 5.

<u>Federal--Interior</u>. The zone between the fresh water of a
stream and the salt water of an ocean. General 3: p. 31.

Existing Land Use. <u>Regional--Commission</u>. The immediate affect
of implementing a project which has definite land use impacts is the change that occurs on the project site itself.
This change may or may not be significant when compared to
the present or existing use of the site. However, besides

changing the use of the site (usually from a less intensive to more intensive activity) the project may excit an
influence upon surrounding areas, perhaps to the point of
affecting area character. General 10: p. IV-51.

Farm.   Regional--River Basin. A place operated as a unit of 10
        or more acres from which the sale of agricultural products
        totalled $50 or more annually, or a place operated as a
        unit of less than 10 acres from which the sale of agricul-
        tural products totaled $250 or more annually during the
        previous year. Earth 1: p. 373.

Fault. National--Soil Conservation. A fracture or fracture zone
        of the earth along which there has been displacement of one
        side with respect to the other. Earth 4: p. 22g.

        City--Planning. A break in the continuity of a rock form-
        ation, caused by a shifting or dislodging of the earth's
        crust, such as in an earthquake, in which adjacent surfaces
        are differentially displaced parallel to the plane of the
        break. General 8: p. 2.

        City--Planning. A break in materials of the earth's crust
        along which there has been differential movement. It occurs
        when formations are strained past the breaking point and
        yield along a crack, or series of cracks, usually during an
        earthquake, so that corresponding points on the two sides
        are distinctly offset. A joint is a crack across the bed-
        ding plane of a rock formation. It is not a fault unless
        displacement has occured. General 5: p. 55.

Federal Lands. Regional--River Basin. All classes of land owned
        by the Federal Government, which includes both Public Domain
        land and acquired Federal land. Earth 1: p. 373.

Fen.   Federal--EPA. Low-lying land partly covered with water.
        General 1: p. 7.

Fertilizer. National--Soil Conservation. Any organic or inor-
        ganic material of natural or synthetic origin that is added
        to a soil to supply elements essential to plant growth.
        Earth 2: p. 23g.

Field Capacity. National--Engineering. Amount of water remain-
        ing in a well drained soil when the velocity of downward
        moisture flow has become small. Also called field mois-
        ture capacity. Water 1: no page number.

Field Moisture Capacity. Regional--River Basin. The quantity
        of water which can be permanently retained in the soil

in opposition to the downward pull of gravity. Water 5:
p. 1014.

Fill. National--Soil Conservation. In geology, any sediment deposited by any agent so as to fill or partly fill a channel,
valley, sink, or other depression. Earth 2: p. 23g. (Revised).

Filling. Federal--EPA. Depositing dirt and mud, often raised by
dredging, into marshy areas to create more land for real
estate development. It can destroy the marsh ecology.
General 1: p. 7.

Finish Elevations. State--Environmental Protection. The proposed
elevations of the land surface at the proposed site after
the facility is completed and all sodding, paving, covering,
after earth and fill moving has ceased. Legal Jargon 14:
p. 3.

Fjord Estuaries. Federal--DOT. Deep, U-shaped coastal indentures
gouged out by glaciers and generally with a shallow sill at
their mouths, formed by terminal glacial deposits. Transportation 3: p. 31.

Floor Area Ratio (FAR). Federal--GSA. Maximum allowable Floor
Area Ratio; indicated by numeral in zoning classifications
(e.g., a zoning classification of R-5 means that the land
is restricted to residential use and the square footage of
the building may not exceed half, or .5, of the square
footage of the lot. General 2: p. X-12.

Fluvial Deposit. State--Water Resources. Sediment deposited by
the action of streams. Also called alluvial deposit.
Energy/Utility 2: Appendix E.

Fold. National--Soil Conservation. A bend or flexure in a layer
or layers of rock. Earth 4: p. 24g.

State--Natural Resources. A bend in rock strata. General
12: p. 337.

Foreshore. Federal--Interior. A strip of land on the margin of
a body of water; e.g., a lake or stream. The bottom land
is intermittently exposed when the water level fluctuates.
General 3: p. 36.

Frail Lands. Regional--River Basin. Lands characterized by
either a thin or unstable topsoil, or in some instances
no topsoil whatsoever. Subsoils are normally clays, fine
silts, or sands. Frail lands are ordinarily those on which

the plant cover is sparse or easily injured, leading to
increased runoff or erosion. Many of these areas are geologic parent materials that do not have sufficient soil
development to produce a vegetative cover that would sta-
bilize normal geologic erosion. Some of these areas may
consist of barren rock or shale deposits. Earth 1: p. 374.

Geological Areas. Federal--Forest Service. U.S. Forest Service
usage. A unit of land which has been designated by the
Forest Service as containing outstanding formations or
unique geologic features of the earth's development including caves and fossils. General 15: p. 86.

Federal--Interior. Land areas whose outstanding structural
features clearly reveal phases of the earth's development.
Recreation 3: p. 11.

Geological Terrace. Regional--River Basin. An old alluvial plain,
ordinarily flat or undulating, bordering a river, lake, or
the sea. Stream terraces are frequently called second bottoms, as contrasted to flood plains, and are seldom subject
to overflow. Marine terraces were deposited by the sea and
are generally wide. Earth 1: p. 382.

Geologic Erosion. Federal--Forest Service. The gradual wearing
away of the land surface which occurred prior to the occupancy of a land area by man or the present day wearing away
which is not due to the activities of man. General 15:
p. 74.

Geology. Federal--DOT. The science that treats the origin, composition, structure and history of the earth, especially as
these are revealed by the rocks, and of processes by which
changes in the rocks are brought about. Transportation 3:
p. 33.

Geomorphology. State--Natural Resources. The study of the land
forms of the earth. General 12: p. 337.

Grade. National--Engineering. Degree of slope of a road, channel,
or natural ground. Water 1: no page number.

Grading. State--Soil and Water Commission. Any stripping, cutting, filling, stock piling, or any combination thereof
and shall include the land in its cut and filled condition.
Earth 3: p. G-12.

Greenbelt. Federal--Army. A plot of vegetated land separating or
surrounding areas of intensive residential or industrial

use and devoted to recreation or park uses. General 9:
p. 22.

Federal--Army. A corridor of planted open space which en-
circles a particular area to avoid intrusion by conflicting uses and normally utilized for park and recreation purposes. Recreation 2: p. A-37.

Federal--EPA. Buffer zones created by restricting development from certain land areas. General 1: p. 7.

Federal--Interior. Parklike strips of unoccupied land with
little or no development, usually associated with urban
areas. Recreation 3: p. 11.

Green Space. Federal--Interior. Vegetated land that is 'open'
in character. Recreation 3: p. 11.

Greenway. Federal--Army. A clearly defined and controlled corridor of vegetated land, parklike in character, following
both sides of a water course between designated destination points. Recreation 2: p. A-37.

Gully. State--Soil and Water Commission. An incised channel or
miniature valley cut by concentrated runoff but through
which water commonly flows only during snow. A gully may
be dendritic or branching or it may be linear, rather long,
narrow, and of uniform width. The distinction between
gully and rill is one of depth. A gully is sufficiently
deep that it would not be obliterated by normal tillage
operations, whereas a rill is of lesser depth and would
be smoothed by use of ordinary tillage equipment. Earth
3: p. G-12.

Gully Erosion. National--Engineering. Removal of soil by temporal passage of water in well defined channels. Water 1:
no page number.

Regional--River Basin. The widening, deepening, and head-
cutting of small channels and waterways due to erosion.
Water 5: p. 1014.

Hardpan. National--Engineering. Hardened, compacted or cemented
soil horizon. Water 1: no page number.

National--Soil Conservation. A hardened soil layer in the
lower A or in the B horizon caused by cementation of soil
particles with organic matter or with materials such as
silica, sesquioxides, or calcium carbonate. The hardness
does not change appreciably with changes in the moisture

content, and pieces of the hard layer do not slake in
water. Earth 4: p. 26g.

Regional--River Basin. A hardened or cemented soil horizon, or layer. The soil material may be sandy or clayey,
and it may be cemented by iron oxide, silica, calcium carbonate, or other substances. Earth 1: p. 374.

Homogeneous Estuary. Federal--DOT. An estuary where fresh water
and tidal inflow are more nearly equal, the dominant mixing agent is turbulence, caused by the periodicity of the
tidal action. Transportation 3: p. 35.

Homogeneous Ground. Federal--DOT. Ground of uniform nature in
the area of interest. Transportation 3: p. 35.

Horizon. Federal--Interior. A natural soil layer. General 4:
Glossary p. 6.

Humus. Federal--Interior. Partially decomposed organic material
found in soil and water. General 3: p. 42.

Hydraulic Fill. State--Water Resources. An earth structure or
grading operation in which the fill material is transported and deposited by means of water pumped through a flexible or rigid pipe. Energy/Utility 2: Appendix E, p. 14.

Hypersaline Estuary. Federal--DOT. An estuary in which the inflow of water is small, the tidal amplitude low, the evaporation very high and the salinity of enclosed bays may
rise above that of the sea. Transportation 3: p. 37.

Igneous Rocks. Federal--GSA. Rock formed by solidification from
a molten or partially molten state. General 2: p. VIII-29.

State--Natural Resources. Rocks that have solidified from
the molten state, e.g., granite, basalt, lava. General 12:
p. 337.

Impermeable Layer. National--Engineering. Layer of soil resis-
tant to penetration by water, air, or roots. Water 1: no
page number.

Impervious Material. Federal--Interior. Refers to relatively
waterproof soils, such as clay, through which water will
percolate at about one millionth the rate at which it will
pass through gravel. General 3: p. 43.

Impervious Soil. National--Soil Conservation. A soil through
which water, air, or roots cannot penetrate. No soil is

impervious to water and air all the time. Earth 4: p. 28g.

Inactive Fault. City--Planning Department. Inactive faults are
    geologic faults along which no movement is known to have
    occurred in the last 10,000 years. General 8: p. 3. (Re-
    vised).

Incompatible Uses. Federal--Army. Particular uses which due to
    thier inherent or generated characteristics produce phys-
    ical, visual or aesthetic conflicts with other uses, caus-
    ing need for separation or elimination. Recreation 2:
    p. A-38.

    Federal--Forest Service. Land uses which cannot exist to-
    gether by reason of either competition for limited re-
    sources or use byproducts which prevent alternative use.
    For example, timber harvesting and wilderness preservation
    are incompatible uses for one piece of land. General 15:
    p. 97.

Infiltration. Regional--River Basin. The flow of a fluid into
    a substance through pores or small openings. It cannotes
    flow into a substance in contradistinction to the word per-
    colation, which connotes flow through a porous substance.
    Water 5: p. 1016.

Infiltration Capacity. Regional--River Basin. The maximum rate
    at which the soil, when in a given condition, can absorb
    falling rain or melting snow. Water 5: p. 1016.

Infiltration Rate. National--Soil Conservation. A soil charac-
    teristic determining or describing the maximum rate at
    which water can enter the soil under specified conditons,
    including the presence of an excess of water. Earth 2:
    p. 29g.

Infiltration Velocity. National--Soil Conservation. The actual
    rate at which water is entering the soil at any given time.
    It may be less than the maximum (the infiltration rate) be-
    cause of a limited supply of water (rainfall or irrigation).
    Earth 2: p. 29g.

Intensity Index. Federal--GSA. Ratio of space on all floors or
    land area devoted to a particular class of use to the total
    floor space on all floors or total land area. General 2:
    p. X-12.

Interbedded. City--Planning. Occurring between beds of different
    material. General 5: p. 53.

Interstices. <u>Regional--River Basin</u>. The openings or pore spaces in a rock. In the zone of saturation they are filled with water. Water 5: p. 1017.

Intrusion. <u>State--Natural Resources</u>. A body of igneous rock that invades older rock. General 12: p. 337.

Irrigated Land. <u>Regional--River Basin</u>. Land receiving water by controlled artifical means for agricultural purposes from surface or subsurface sources. Earth 1: p. 374.

Irrigated Rangeland. <u>Regional--River Basin</u>. Land in grass, or other long-term forage growth, of native species to which water is applied by controlled artificial measures. Earth 1: p. 378.

Joint. <u>National--Soil Conservation</u>. A fracture or parting that abruptly interrupts the physical continuity of a rock mass. Earth 2: p. 30g.

Key Bed. <u>National--Utilities</u>. A rock stratum that can be identified over large areas and from which measurements can be taken to determine geologic structure. Energy/Utility 1: p. 33.

K Values. <u>City--Planning</u>. Soil erodability factor--the erosion rater per unit of erosion index for a specific soil in cultivated continuous fallow. Used in the Universal Soil Loss Equation. General 5: p. 56.

Lacustrine Deposits. <u>Regional--River Basin</u>. Stratified materials deposited by lake waters. Earth 1: p. 375.

Land Capability Class. <u>National--Soil Conservation</u>. One of the eight classes of land in the land capability classification of the Soil Conservation Service; distinguished according to the risk of land damage or the difficulty of land use. Earth 4: p. 30g. (Revised).

<u>Regional--River Basin</u>. A group of capability subclasses and units that have the same relative degree of hazards or limitations. The risks of soil damage or limitation in use become progressively greater from Class I to Class VIII. Earth 1: p. 375.

Land Capability Classification. <u>National--Soil Conservation</u>. A grouping of kinds of soil into special units, subclasses, and classes according to their capability for intensive use and the treatments required for sustained use, prepared by the Soil Conservation Service, USDA. Earth 4: p. 31g.

Land Classification. National--Soil Conservation. The arrange-
    ment of land units into various categories based on the
    properties of the land or its suitability for some partic-
    ular purpose. Earth 2: p. 31g.

Landform. Federal--Forest Service. Term used to describe the
    many types of land surfaces which exist as a result of geo-
    logical activity, such as a plateau, plain, basin, mount-
    ain, etc. General 15: p. 109.

    Federal--GSA. Features that together constitute the sur-
    face of the earth. General 2: p. VIII-29.

Land Reclamation. National--Soil Conservation. Making land cap-
    able of more intensive use by changing its general charac-
    ter, as by drainage of excessively wet land; irrigation of
    arid or semiarid land; or recovery of submerged land from
    seas, lakes, and rivers. Large-scale reclamation projects
    usually are carried out through collective effort. Simple
    improvements, such as cleaning sumps or stones from land,
    should not be referred to as land reclamation. Earth 2:
    p. 31g.

Land Resource Area. Federal--Forest Service. U.S. Forest Ser-
    vice usage. Broad geographical areas having similar soil,
    climatic, geologic, vegetative and topographic features.
    General 15: p. 109.

Land Resource Region. Regional--River Basin. Geographically
    associated major land resource areas which divide the
    United States into 20 physiographic regions uniform enough
    to be significant for national planning. Earth 1: p. 375.

Land Resources. Regional--River Basin. An area of land contain-
    ing or supporting all or some of certain resources in some
    combination. The resources include soil, water, timber,
    forage, wildlife, and minerals. Earth 1: p. 375.

Landslide. National--Soil Conservation. A mass of material that
    has slipped downhill under the influence of gravity, fre-
    quently occurring when the material is saturated with water.
    Rapid movement down slope of a mass of soil, rock, or de-
    bris. Earth 2: p. 31g.

Land Use. Regional--River Basin. Primary occupier of a tract of
    land grouped into classes with similar characteristics,
    i.e., cropland, rangeland, forest land, or other. Earth 1:
    p. 375.

Land Use Plan. <u>National--Soil Conservation</u>. The key element of
     a comprehensive plan; describes the recommended location
     and intensity of development for public and private land
     uses such as residential, commercial, industrial, recre-
     ational, and agricultural. Earth 4: p. 31g.

Land Use Planning. <u>National--Engineering</u>. Development of plans
     for the use of land that will, over a long period, best
     serve the general public. Water 1: no page number.

     <u>Regional--Commission</u>. Land use planning and regulations
     are intended to promote orderly growth by channeling de-
     velopment into areas in which the community has made or is
     planning to make capital improvement investments such as
     sewer and water service, schools and roads. A project by
     nature of its type, size, and location, may seriously con-
     flict with these plans. These conflicts arise out of in-
     compatibility of uses, altered development patterns, in-
     duced growth areas not economically serviceable by munici-
     pal facilities, and other project-related impacts. All
     such impacts have significant consequences in terms of
     community planning and development. General 10: p. IV-
     52.

Land Use Type. <u>Federal--Forest Service</u>. A classification system
     based on the purposes for which land is being used--e.g.,
     timber production, range, second home subdivisions, farm-
     ing, etc. "Land use types" are bundles of more or less
     shared land use practices--though the precise types and
     amounts of practices that are utilized in conjunction with
     a specific use type may vary from one specific instance to
     another. General 15: p. 116.

Late Pleistocene. <u>Federal--Interior</u>. Time period preceding the
     Holocene era and extending to 2 or 3 million years ago.
     General 3: p. 47.

Lenses. <u>Federal--Interior</u>. Geologic deposits bounded by conver-
     ging surfaces (at least one of which is curved), thick in
     the middle and thinning out toward the edges, resembling a
     convex lens. General 3: p. 47.

Liquefaction. <u>Federal--EPA</u>. Changing a solid into a liquid form.
     General 1: p. 9.

     <u>City--Planning</u>. Earthquake-induced transformation of a
     stable granular material, such as sand, into a fluidlike
     state, similar to quicksand. General 8: p. 3.

Liquefication. National--Soil Conservation. The sudden decrease
        of the shearing resistance of a cohesionless soil, caused
        by a collapse of the structure from shock or other type of
        strain and associated with a sudden but temporary increase
        in the pore-fluid pressure; a temporary transformation of
        the material into a fluid mass. Earth 4: p. 32g.

Liquid Limit (LL). State--Soil and Water Commission. The mois-
        ture content at which the soil passes from a plastic to a
        liquid state. In engineering, a high liquid limit indi-
        cates that the soil has a high content of clay and low ca-
        pacity for supporting loads. Earth 3: p. G-16.

Lithology. Regional--Commission. The description of the total
        physical characteristics of specified sedimentary rock
        samples or formations, including grain or crystal size,
        mineral constituents, and bedding planes. General 10:
        p. IV-39.

Litter. Federal--Interior. The uppermost organic materials,
        partly or not at all decomposed, on the surface of the
        soil. General 3: p. 48.

Littoral. State--Environmental Protection. Shoreline; related
        to edge of the sea or ocean. General 6: p. 312.

Littoral Drift. State--Environmental Protection. The movement
        of sedimentary material, e.g., sand, silt, gravel, paral-
        lel to shoreline under the influence of wind, waves, and
        currents; commonly used synonymous with longshore trans-
        port. General 6: p. 312.

Loamy. State--Soil and Water Commission. Intermediate in text-
        ure and properties between fine-textured and coarse-text-
        ured soils. Earth 3: p. G-16.

Loess. National--Engineering. Aeolian deposit of predominantly
        silt-sized particles. Water 1: no page number.

        Regional--River Basin. Soil material consisting primarily
        of uniform silt particles that were transported and depos-
        ited by wind. Earth 1: p. 376.

Marsh. Federal--GSA. Area of low-lying wetland, usually charac-
        terized by non-woody vegetation such as grasses and sedges;
        may be either fresh or salt water. General 2: p. XI-28.

Mechanical Analysis. State--Soil and Water Commission. The ana-
        lytical procedure by which soil particles are separated to
        determine the particle size distribution. Earth 3: p. G-17.

Metals. <u>Regional--River Basin</u>. Any of various opaque, fusible,
    ductile, and typically lustrous substances having a chem-
    ical element as distinguished from an alloy, which include
    iron, gold, silver, copper, lead, and zinc. Earth 1: p.
    376.

Metamorphic Rocks. <u>Federal--GSA</u>. Rocks formed in the solid state
    in response to pronounced changes of temperature, pressure,
    and chemical environment. General 2: p. VIII-29.

Microrelief. <u>National--Soil Conservation</u>. Small-scaled local
    differences in topography, including mounds, swales, or
    pits that are only a few feet in diameter and with ele-
    vation differences of up to 6 feet. Earth 2: p. 34g.

Mineral. <u>Federal--Forest Service</u>. Mining laws usage. A sub-
    stance is "mineral" under the general U.S. mining laws:
    (1) if it is scientifically recognized as such, (2) if it
    is classified commercially as such or, (3) if it derives
    from the earth and possesses economic value and utility
    aside from the agricultural purposes of the surface itself.
    General 15: p. 126.

Mineral Resource. <u>Regional--River Basin</u>. Known mineral deposit
    that is regarded as having present or future utility.
    Earth 1: p. 376.

Monocline. <u>Federal--ERDA</u>. A downward flexure in otherwise hor-
    izontal strata without any corresponding upfold to form a
    syncline or anticline. General 7: p. g-7.

Moraine. <u>Federal--Interior</u>. A mass of rocks, gravel, sand, clay,
    etc., carried and deposited directly by a glacier. General
    3: p. 53.

Morphology. <u>Federal--Interior</u>. The scientific study of struc-
    ture. General 3: p. 53.

Muck. <u>Regional--River Basin</u>. Fairly well decomposed organic soil
    material, relatively high in mineral content, dark in color,
    and accumulated under conditions of somewhat poor drainage.
    Earth 1: p. 376.

Muck Soils. <u>Federal--EPA</u>. Earth made from decaying plant mater-
    ials. General 1: p. 9.

Mulch. <u>Federal--EPA</u>. A layer of material (wood chips, straw,
    leaves) placed around plants to hold moisture, prevent
    weed growth, and enrich soil. General 1: p. 9.

National--Engineering. Natural or artificially applied
protective covering of plant residue or other material
such as stones, sand, paper, or brush on the surface of
the soil. Water 1: no page number.

State--Soil and Water Conservation. A natural or artificial layer of plant residue or other materials, covering
the land surface which conserves moisture, holds soil in
place, aids in establishing plant cover, and minimizes
temperature fluctuations. Earth 3: p. G-17.

Multiple Use. Federal--EPA. Harmonious use of land for more than
one purpose; i.e., grazing of livestock, wildlife production, recreation, watershed and timber production. Not
necessarily the combination of uses that will yield the
highest economic return or greatest unit output. General
1: p. 10.

National Forest Lands. Regional--River Basin. Federal lands
which have been designated by Executive Order or Statute
as national forests or purchase units, and other lands
under administration of the Forest Service, including experimental areas and Bankhead-Jones Title III lands. Earth
1: pp. 376-377.

Natural Area. Federal--Army. A designation of project lands
which preserves natural resources for their scientific,
scenic, cultural and/or educational value by limiting development and management practices. Land managed to protect rare and endangered species of flora and fauna will
be designated as natural areas. Recreation 2: p. A-41.

Federal--Army. An area in which natural processes predom-
inate, fluctuations in numbers of organisms are allowed
free play and human intervention is minimal. General 9:
p. 30.

Federal--Interior. An ecosystem protected from human influence. Recreation 3: p. 16.

Natural Area Preserve. Federal--Forest Service. An area which
retains a natural or relatively natural condition. Within
it, wild parks, sanctuaries, refuges or wilderness can be
designated. To qualify as a Natural Area Preserve, a site
should be under protection, usually by a nonprofit agency
or a Government agency. General 15: p. 137.

Natural Environment. Federal--Army. The complex of atmospheric,
geological and biological characteristics found in an area

in the absence of artifacts or influences of a well developed technological, human culture; an environment in
which human impact is not controlling, or significantly
greater than that of other animals. General 9: p. 30.

Natural Setting. Federal--Army. The complex of atmospheric, geological and biological characteristics of an area as they
determine its appearance. General 9: p. 30.

Neutral Soil. Regional--River Basin. A soil that is not significantly acid or alkaline; strictly one have a pH of 7.0;
practically, one having a pH between 6.6 and 7.3. Earth
1: p. 377.

Nonmetals. Regional--River Basin. Naturally occurring mineral
or assemblage of minerals that lack typical metalic properties such as cement minerals, sand and gravel, stone,
lime, clay, phosphate, and potash. Earth 1: p. 377.

Non-Renewable Resource. Federal--Army. Those resources which do
not regenerate themselves or maintain a sustained yield
after being utilized or destroyed, i.e., minerals. Recreation 2: p. A-42.

Normal Erosion. National--Engineering. Erosion which takes place
on the land surface in its natural environment; undistrib-
uted by human activity or influence. Water 1: no page
number.

Open Space. Federal--EPA. A relatively undeveloped green or
wooded area provided usually within an urban development
to minimize feelings of congested living. General 1:
p. 10.

Federal--Forest Service. Land and water areas which are
retained in essentially undeveloped state on a permanent
or semipermanent basis. General 15: p. 145.

Federal--Interior. Undeveloped or predominantly undeveloped land having potential use for recreation and conservation purposes. Recreation 3: p. 17.

Organic Deposits. Federal--GSA. Rocks and other deposits formed
by organisms or their remains. General 2: p. VIII-29.

Orographic. Federal--Interior. Refers to mountains, or to relief characteristics of the land. General 3: p. 57.

Ortstein. Regional--River Basin. Hard, irregularly cemented,
dark-yellow to nearly black sandy material formed by soil

forming processes in the lower part of the solum. Similar
material not firmly cemented is known as orterde. Earth
1: p. 377.

Outwash. Federal--Interior. Mineral materials which have been
carried and sorted by water from high to low elevations.
This would include glacial outwash (materials originally
suspended in glaciers which have been moved and deposited
by melt water). General 3: p. 58.

Overburden. Federal--DOT. Loose soil, gravel, etc. that lie
above bedrock. Transportation 3: p. 50.

Pan. National--Soil Conservation. Horizon or layer in soil that
is strongly compacted, indurated, or very high in clay content. Earth 2: p. 37g.

Parent Material. Federal--GSA. Horizon of weathered rock or
partly weathered soil material from which soil is formed.
General 2: p. VIII-29.

Federal--Interior. The unconsolidated mass from which
the developed soil horizons have originated. As used here,
this term also applies to relatively unconsolidated parent
rock. General 3: p. 59.

Peat. Regional--River Basin. Unconsolidated soil material consisting largely of undecomposed or slightly decomposed organic matter accumulated under conditions of excessive
moisture. Earth 1: p. 377.

Percolation. Regional--River Basin. The movement, under hydrostatic pressure, of water through the interstices of a rock
or soil. Water 5: p. 1018.

Percolation Test. National--Soil Conservation. A measurement of
the percolation of water in soil to determine the suitability of different soils for development including private
sewage systems such as septic tanks and drainfields. Earth
4: p. 38g.

Permeability. Federal--GSA. Characteristic of soil that permits
the transmission of water. Impermeable clay layers in the
soil, for example, may interfere with the downward movement
of water. General 2: p. VIII-29.

National--Engineering. The quality or state of a porous
medium relating to the readiness with which it conducts or
transmits fluids. Water 1: no page number. (Revised).

Regional--River Basin. The quality of a soil that enables
water or air to move through it. The permeability classes
are: (1) Very slow--less than 0.05 inch per hour, (2) Slow
--0.05 to 0.20 inch per hour, (3) Moderately slow--0.20 to
0.80 inch per hour, (4) Moderate--0.80 to 2.50 inches per
hour, (5) Moderately rapid--2.50 to 5.00 inches per hour,
(6) Rapid--5.00 to 10.00 inches per hour, (7) Very rapid--
more than 10.00 inches per hour. Earth 1: p. 377.

City--Planning. Soil permeability is that quality of soil
that enables it to transmit water and air. General 5:
p. 57. (Revised).

Physiographic Region. Federal--Army. A homogeneous land area
which is characterized by a single natural phenomena such
as a mountain range or grassland. Recreation 2: p. A-44.

Physiographic Units. Regional--River Basin. Contrasting areas
in the landscape that have general similarity in the range
of environmental, topographic and physical soil characteristics. Earth 1: p. 377.

Physiography. Federal--Interior. That branch of science that
deals with the physical features of the earth. General
3: p. 62.

Planosol. Regional--River Basin. A group of soils with cemented
or compacted subsoil layers. Earth 1: p. 377.

Plasticity. National--Engineering. Property of a soil which
allows it to be deformed without appreciable volume change
or cracking. Water 1: no page number.

Plasticity Index (PI). State--Soil and Water Commission. The
numerical difference between the liquid limit and the plastic limit; the range of moisture content within which the
soil remains plastic. Earth 3: p. G-18.

City--Planning. The numerical difference between liquid
limit and plastic limit. The plastic limit is the moisture content at which the soil material changes from the
semisolid to plastic state. General 5: p. 57.

Plastic Limit. State--Soil and Water Commission. The moisture
content at which a soil changes from a semisolid to a plastic state. Earth 3: p. G-18.

Plastic Soil. National--Soil Conservation. A soil capable of
being molded or deformed continuously and permanently by
relatively moderate pressure. Earth 2: p. 39g.

Porosity. <u>National--Engineering</u>. Ratio of the space in any porous material, such as a soil, that is not filled with solid
matter, to the total space occupied generally expressed as
a percentage. Water 1: no page number. (Revised).

<u>Regional--River Basin</u>. Porosity is the property of con-
taining openings or interstices. In rock or soil, it is
the ratio (usually expressed as a percentage) of the volume
of openings in that material to the bulk volume of the material. Water 5: p. 1018.

Public Domain Lands. <u>Regional--River Basin</u>. Original Public
Domain lands which have never left Federal ownership; also
includes lands in Federal ownership which were obtained by
the Federal Government in exchange for public lands, or for
timber on public lands. Earth 1: p. 378.

Public Land. <u>Federal--Forest Service</u>. Land for which the title
and control rests with a government--Federal, state, regional, county, or municipal. General 15: p. 167.

Quaternary Age. <u>Federal--Interior</u>. Time period covering the last
2 or 3 million years (consists of the Pleistocene and Hol-
ocene). General 3: p. 66.

Range Condition Class. <u>Regional--River Basin</u>. Range condition
estimates are based on a numerical index, rating the forage
stand and the site and soil mantle. The numerical ratings
have been combined for a comprehensive classification of
range conditions common to range managing agencies in
the following relative terms: Excellent, Good, Fair, Poor,
and Bad. Earth 1: p. 378. (Revised).

Rangeland. <u>Federal--Forest Service</u>. Land on which the natural
plant cover is composed principally of native grasses,
forbs or shrubs valuable for forage. General 15: p. 170.

<u>Regional--River Basin</u>. Land in grass or other long-term
forage growth of native species used primarily for grazing.
It may contain shade trees or scattered timber trees with
less than 10 percent canopy. It includes grassland, land
in perennial forbs, sagebrush land, and brushland other
than sage. The term nonforest range is used to differ-
entiate the nonforest range from the forest range when
both are being discussed. Earth 1: p. 378.

Raw Land. <u>Federal--Forest Service</u>. A term of comparison sometimes used to contrast land which has been built upon
(i.e., "developed") or subdivided in preparation for development with undeveloped and unsubdivided (i.e., "raw")

land. General 15: p. 172.

Relief. <u>Federal--Army</u>. Variations in elevation of the earth's
      surface. General 9: p. 38.

      <u>Federal--GSA</u>. Differences in elevation between an area's
      highest and lowest terrain, generally expressed topograph-
      ically on a contour map. Micro-relief measures small-scale
      differences in elevation within the confines of a specific
      site. General 2: p. VIII-29.

      <u>Federal--Interior</u>. The difference in elevations of the
      land surface. General 3: p. 68.

      <u>Regional--Commission</u>. The elevation or inequalities of a
      land surface. General 10: p. IV-89.

Residual Geologic Materials. <u>Federal--Interior</u>. Bedrock mater-
      ials or weathered or decomposed bedrock materials which
      are still in their place of origin and have not been ap-
      preciably transported by water, wind, or gravity. General
      3: p. 68.

Richter Scale. <u>City--Planning</u>. A logarithmic scale developed by
      Charles Richter to measure earthquake magnitude by the
      energy released, as opposed to earthquake intensity as de-
      termined by effects on people, structures, and earth ma-
      terials. General 8: p. 4.

Rill Erosion. <u>National--Engineering</u>. Erosion causing formation
      of shallow channels that can be smoothed out by normal
      cultivation. Water 1: no page number.

      <u>Regional--River Basin</u>. Removal of soil by running water
      with formation of shallow channels that can be smoothed out
      completely by normal cultivation. Earth 1: p. 379.

Saline Soil. <u>Regional--River Basin</u>. A soil that contains soluable
      salts in amounts that impair growth of plants but that does
      not contain an excess of exchangeable sodium. Earth 1:
      p. 379.

Salt Marsh. <u>Federal--Army</u>. Similar to fresh (grass-dominated)
      marsh, but adjacent to marine areas covered periodically
      (tidally or seasonally) with saline water. General 9:
      p. 39.

Salt Water Estuary. <u>Federal--DOT</u>. An estuary in which a distinct
      layer of dense marine water underlies a surface layer of
      fresh water, thus forming a "wedge". Such a wedge may

extend far upstream, gradually diminishing under the influence of the bottom gradient and the volume of fresh
water in the channel. Transportation 3: p. 59.

Sand.   National--Engineering. Soil particles ranging from 0.05
        to 2.0 mm in diameter. Soil material containing 85 per-
        cent or more particles of this size. Water 1: no page
        number.

        Regional--River Basin. Individual rock or mineral frag-
        ments in soils having diameters ranging from 0.05 to 2.0
        millimeters. Most sand grains consist of quartz, but they
        may be any mineral composition. The textural class name
        of any soil that contains 85 percent or more sand and not
        more than 10 percent clay. Earth 1: p. 379.

Sand Lens. National--Soil Conservation. Lenticular band of sand
        in distinctly sedimentary banded material. Earth 2: p. 44g.

Saturation Point. National--Soil Conservation. In soils, that
        point at which a soil or an aquifer will no longer absorb
        any amount of water without losing an equal amount. Earth
        2: p. 44g.

Scour. National--Soil Conservation. To abrade and wear away;
        used to describe the wearing away of terrace or diversion
        channels or stream beds. Earth 2: p. 44g.

Sediment. Regional--Commission. Any material transported by
        water which will ultimately settle to the bottom after the
        water loses its transporting power. Fine waterborne matter
        deposited or accumulated in beds. General 13: p. 591.

Sedimentary Rocks. Federal--GSA. Rocks formed by the accumula-
        tion of sediment in water or from air. General 2:
        p. VIII-30.

        State--Natural Resources. Rocks, usually layered, laid
        down by water or air processes, e.g., limestone, sandstone.
        General 12: p. 338.

Sedimentation. Federal--GSA. Portion of the metamorphic cycles
        from the separation of the particles from the parent rock,
        to and including their consolidation into another rock.
        General 2: p. VIII-30.

        National--Engineering. Deposition of waterborne sediments
        due to a decrease in velocity and a corresponding reduction
        in the size and amount of sediment which can be carried.
        Water 1: no page number.

Seismicity. Federal--ERDA. The tendency for the occurrence of
        earthquakes. General 7: p. g-9.

        Federal--Interior. The phenomenon of earth movements or
        seismic activity. General 3: p. 70.

        Regional--Commission. The vulnerability of an area to
        earthquakes, which is dependent upon earthquake frequency,
        transmission or dampening properties of consolidated and
        unconsolidated geologic material. General 10: p. IV-39.

        City--Planning. Relative susceptibility to earthquakes.
        General 8: p. 4.

Shear. National--Soil Conservation. A distortion, strain, or
        failure producing a change in form, usually without change
        in volume, in which parallel layers of a body are displaced
        in the direction of thier line of contact. Earth 2: p. 46g.

Shear Strength. National--Soil Conservation. The maximum resis-
        tance of a soil to shearing stresses. Earth 2: p. 46g.

Sheet Erosion. National--Engineering. Removal of a fairly un-
        iform layer of soil or material from the land surface by
        runoff water. Water 1: no page number.

        Regional--River Basin. The removal of a fairly uniform
        layer of soil or materials from the land surface by the
        action of rainfall and runoff water. Earth 1: p. 379.

Shore Erosion. National--Engineering. Removal of sand, soil, or
        rock from the land area adjacent to oceans, seas, lakes, or
        ponds due to the wave action. Water 1: no page number.

Shoreline Erosion. Federal--Army. The active wearing away of
        soil and rock from the edge of a lake or ocean by the ac-
        tion of waves, high water, heavy rains or surface runoff.
        Recreation 2: p. A-35.

Silt. Federal--Commerce. Fine particulate matter suspended in
        water and later deposited on water body bottom. General
        6: p. 314.

        Federal--EPA. Fine particles of soil or rock that can be
        picked up by air or water and deposited as sediment. Gen-
        eral 1: p. 14.

        National--Engineering. Small, 0.05 to 0.002 mm in diameter,
        mineral soil grains intermediate between clay and sand;
        waterborne sediment with diameters of individual grains

approaching that of silt; soil material containing 80 percent or more silt and less than 12 percent clay. Water 1:
no page number.

State--Natural Resources. Rock or mineral particles,
0.002 to 0.02 mm in diameter. General 12: p. 338.

Site Preparation. State--Environmental Protection. That substantial physical activity which is an integral part of a continuous process of land development or redevelopment for
a particular facility and must occur before actual construction of that facility may commence. It does not include preparing any drawing, plan, blueprint, or other
specification; surveying; taking soil borings; performing
percolation tests, or driving test pilings. Further, it
does not include mere denuding of a site, placing temporary access roads and bridges, or construction of demonstration or model dwelling units. Legal Jargon 14: p. 5.

Slick Spots. Regional--River Basin. Small areas in a field that
are slick when wet because they contain excess exchangeable
sodium, or alkali. Earth 1: p. 379.

Slip. National--Engineering. Downhill movement of a mass of
soil. The movement usually is only a short distance and
the soil mass stays relatively intact. A form of landslide. Water 1: no page number.

Slope. Federal--GSA. Inclination of the land surface, generally
described in percentages or degrees. Slope represents the
combination of relief and topography and is thus a function
of an area's rock and soil characteristics. Slope, influences the direction and rate of drainage, runoff, erosion,
and sedimentation, and is important is estimating mass
wasting potential and landslide or mudslide susceptibility.
General 2: p. VIII-30.

National--Engineering. Degree of deviation of a surface
from the horizon, usually expressed in percent or degrees.
Water 1: no page number.

Regional--River Basin. The incline of the land surface,
usually expressed in percentage of slope, which equals the
number of feet of fall per 100 feet of horizontal distance.
(1) 0 to 3 percent--nearly level, (2) 3 to 7 percent--
gently sloping, (3) 7 to 12 percent--moderately sloping,
(4) 12 to 25 percent--strongly sloping, (5) 25 to 40 per-
cent--steeply sloping, (6) 40 to 70 percent--very steeply
sloping, (7) 70 to 100 percent--extremely steeply sloping.
Earth 1: p. 379.

Slope Characteristics. <u>National--Soil Conservation</u>. Slopes may
be characterized as concave (decrease in steepness in lower
portion), uniform, or convex (increase in steepness at
base). Erosion is strongly affected by shape, ranked in
order of increasing erodibility from concave to uniform
to convex. Earth 2: p. 47g.

Slope Stability. <u>Federal--Forest Service</u>. An evaluation (almost
always qualitative and expressed as a probability) of the
tendency for the materials on or constituting a slope
(e.g., rocks, soil, snow) to either remain in place or to
move downhill. General 15: p. 192.

Soil. <u>National--Engineering</u>. Natural body, developed from weathered minerals and decaying organic matter, covering the
earth; the upper layer of the earth in which plants grow.
Water 1: no page number.

<u>Regional--River Basin</u>. A natural three-dimensional body
on the earth's surface that supports plants and that has
properties resulting from the integrated effect of climate
and living matter acting upon parent material, as conditioned by relief over periods of time. Earth 1: p. 379.

<u>State--Soil and Water Commission</u>. The unconsolidated min-
eral and organic material on the immediate surface of the
earth that serves as a natural medium for the growth of
land plants. Earth 3: p. G-20.

Soil Aeration. <u>National--Engineering</u>. Movement of air into the
soil as water is drained out; process by which air and
other gases in the soil are renewed. Water 1: no page
number.

Soil Amendments. <u>Federal--Interior</u>. Refers to materials exclusive of nitrogen, phosphorous, and potassium which are
added to the soil to improve workability, chemical balance,
and yield levels. This would include trace elements and
such other materials as gypsum and sulfur. General 3:
p. 71.

Soil Association. <u>Regional--River Basin</u>. A group of soils, with
or without common characteristics, geographically associated in an individual pattern. Earth 1: p. 379.

Soil Complex. <u>National--Soil Conservation</u>. A mapping unit used
in detailed soil surveys where two or more defined taxonomic units are so intimately intermixed geographically
that it is undesirable or impractical, because of the scale
being used, to separate them. A more intimate mixing of

smaller areas of individual taxonomic units than that described under soil association. Earth 4: p. 49g.

Regional--River Basin. A mapping unit consisting of different kinds of soils that occur in such small individual
areas or in such an intricate pattern that they cannot be
shown separately on a publishable soil map. Earth 1:
p. 380.

Soil Conditioner. Federal--EPA. An organic material like humus
or compost that helps soil absorb water, build a bacterial
community, and distribute nutrients and minerals. General
1: p. 14.

Soil Conservation. National--Engineering. Preservation of soil
against deterioration and loss by using it within its capabilities; application of conservation practices needed
for its protection and improvement. Water 1: no page
number. (Revised).

Soil Depth. Regional--River Basin. The depth of soil material
that plant roots can penetrate readily to obtain water and
nutrients. It is the depth to a layer that, in physical
or chemical properties, differs from the overlying material
to such an extent as to prevent or seriously retard the
growth of roots or penetration of water. The depth classes
are: (1) Very deep--more than 60 inches, (2) Deep--40 to
60 inches, (3) Moderately deep--20 to 40 inches, (4) Shal-
low--10 to 20 inches, (5) Very shallow--0 to 10 inches.
Earth 1: p. 380.

Soil Elasticity. Federal--DOT. A property of soils which causes
the soil to return to its original condition after excita-
tion by vibration. Transportation 3: p. 62.

Soil Erosion. National--Engineering. Detachment and movement of
soil from the land surface by wind or water, including
normal soil erosion and accelerated erosion. Water 1:
no page number.

Soil Fertility. National--Soil Conservation. The quality of a
soil that enables it to provide nutrients in adequate
amounts and in proper balance for the growth of specified
plants, when other growth factors, such as light, moisture,
temperature, and physical condition of soil, are favorable.
Earth 2: p. 49g.

Soil Horizon. National--Engineering. Layer of soil which differs
from the material above and below it. Water 1: no page
number.

National--Soil Conservation. A layer of soil or soil material approximately parallel to the land surface and differing from adjacent genetically related layers in physical, chemical, and biological properties or characteristics, such as color, structure, texture, consistence, kinds and numbers of organisms present, degree of acidity or alkalinity, etc. Earth 4: p. 49g. (Revised).

State--Soil and Water Commission. A layer of soil, approximately paralled to the surface, that has distinct characteristics produced by soil forming factors. Earth 3: p. G-20.

Soil Improvement. National--Engineering. Process cf protecting the soil from excessive erosion and making scil more fertile and productive. Water 1: no page number.

Soil K-Factor. Federal--DOT. An experimentally-determined soil erosion rating based upon physical properties of a soil. Transportation 3: p. 62.

Soil Loss Equation. Federal--GSA. Equation used tc determine the amount of soil which will erode from a unit area over a year's time under varying conditions of rainfall, slope, etc. Legal Jargon 18: Appendix B-6.

Soil Management. National--Soil Conservation. The sum total of all tillage operations, cropping practices, fertilizer, lime, and other treatments conducted on, or applied to, a soil for the production of plants. Earth 2: p. 49g.

Soil Map. National--Soil Conservation. A map showing the distribution of soil types or other soil mapping units in relation to the prominent physical and cultural features of the earth's surface. The following kinds of soil maps are recognized in the U.S.: detailed, detailed reconnaissance, reconnaissance, generalized, and schematic. Earth 4: p. 49g.

Soil Mapping Unit. Regional--River Basin. A portion of the landscape that has similar characteristics and qualities, and whose limits are fixed by precise definitions. Earth 1: p. 380.

Soil Moisture. National--Engineering. Water stored in soils. Water 1: no page number.

Regional--River Basin. Water in the soil zone. Available soil moisture is water easily abstracted by roots of plants.

Unavailable soil moisture is water held so firmly by adhesion and other forces that it cannot usually be absorbed
by plants rapidly enough to produce growth; when soil
moisture falls below the "available" level, a conditon of
"soil-moisture deficiency" is said to occur with respect
to vegetation. Water 2: p. I-23.

Regional--River Basin. Water diffused in the soil, the
upper part of the zone of aeration from which water is
discharged by the transpiration of plants or by soil evaporation. Water 5: p. 1019.

Soil Morphology. National--Soil Conservation. The physical constitution, particularly the structural properties, of a
soil profile as exhibited by the kinds, thickness, and
arrangement of the horizons in the profile, and by the
texture, structure, consistency, and porosity of each horizon. Earth 4: p. 49g. (Revised).

Regional--River Basin. The physical constitution of the
soil, including the texture, structure, porosity, consistence, and color of the various soil horizons, their thickness, and their arrangement in the soil profile. Earth 1:
p. 380.

Soil Permeability. National--Engineering. Capacity of the soil
to transmit fluids. Water 1: no page number.

Soil Porosity. National--Engineering. Percentage of soil volume
not occupied by solids, including all pore space filled
with air and water. Water 1: no page number.

Soil Probe. National--Soil Conservation. A tool having a hollow
cylinder with a cutting edge at the lower end, used for
probing into the soil and withdrawing a small sample for
field or laboratory observation. Earth 4: p. 50g.

Soil Profile. National--Engineering. Vertical section of the
soil from the surface through all its horizons into the
parent material. Water 1: no page number.

National--Soil Conservation. A vertical section of the
soil from the surface through all its horizons, including
C horizons. Earth 4: p. 50g.

Regional--River Basin. A vertical section of the soil
through all its horizons extending to 60 inches or to a
restricting layer. Earth 1: p. 380.

Soil Resource Group. **Regional--River Basin**. A grouping of land
    capability units, or soils that have similar cropping pat-
    terns, yield characteristics, responses to fertilizers,
    management, and land treatment measures. Earth 1: p. 380.

Soil Series. **Regional--River Basin**. A group of soils developed
    from generally similar types of parent material and having
    genetic horizons that, except for texture and the surface
    layer, are similar. Earth 1: p. 380.

    City--Planning. Soils that have profiles almost alike
    make up a soil series. Except for different texture in
    the surface layer, all the soils of one series have major
    horizons that are similar in thickness, arrangement, and
    other important characteristics. Each soil series is named
    for a town or other geographic feature near the place where
    a soil of that series was first observed and mapped. Gen-
    eral 5: p. 57.

Soil Structure. **Federal--Interior**. The arrangement of primary
    soil particles into compound particles or clusters that
    are separated from adjoining aggregates and have proper-
    ties unlike those of an equal mass of unaggregated primary
    soil particles. General 4: Glossary, p. 10.

    National--Engineering. Arrangement of soil particles,
    either single grain or in aggregates, that make up the
    soil mass. Structure may refer to the natural arrange-
    ment of the soil when in place and undisturbed, or the
    soil at any degree of disturbance. The principal types
    of soil structure are platy, prismatic, columnar, blocky,
    nuciform, granular, and crumb. The aggregates in these
    structure types vary in size and degree of stability.
    Water 1: no page number.

Soil Structure Types. **National--Soil Conservation**. A classifi-
    cation of soil structure based on the shape of the aggre-
    gates or peds and their arrangement in the profile. Gen-
    erally the shape of soil structure types is referred to as
    either platy, prismatic, columnar, blocky, granular, or
    crumb. Earth 2: p. 51g.

Soil Survey. **National--Soil Conservation**. A general term for
    the systematic examination of soils in the field and in
    the laboratories; their description and classification;
    the mapping of kinds of soil; the interpretation of soils
    according to their adaptability for various crops, grasses,
    and trees; their behavior under use or treatment for plant
    production or for other purposes; and their productivity
    under different management systems. Earth 2: p. 51g.

Soil Texture. <u>Federal--Interior</u>. Refers to the relative proportions of the various size groups (sand, silt, and clay) of
the individual soil grains in a mass of soil. General 3:
p. 72.

>   <u>National--Engineering</u>. Classification of soil by the pro-
>   portion and graduations of the three size groups of soil
>   grains, i.e., sand, silt and clay, present in the soil.
>   Water 1: no page number.

Solum. <u>National--Soil Conservation</u>. The upper part of a soil
profile, above the parent material, in which the processes
of soil formation are active. The solum in mature soils
includes A and B horizons. Usually the characteristics of
the material in these horizons are quite unlike those of
the underlying parent material. The living roots and other
plant and animal life characteristics of the soil are
largely confined to the solum. Earth 2: p. 53g.

Spoil. <u>Federal--EPA</u>. Dirt or rock that has been removed from
its original location, destroying the composition of the
soil in the process, as with strip-mining or dredging.
General 1: p. 14.

>   <u>National--Engineering</u>. Soil excavated from a canal, ditch,
>   basin, or other site. Also called waste. Water 1: no
>   page number.

Stability. <u>Regional--Commission</u>. The ability of an area to withstand such actions as landslides and subsidence due to
changes in geodynamic processes (e.g., increased pressure
from overlying structure on soil bed causing compression
and sliding). General 10: p. IV-40.

Stream Bank Erosion. <u>National--Engineering</u>. Scouring of material
and the cutting of channel banks by running water. Water
1: no page number.

Stream Bed Erosion. <u>National--Engineering</u>. Scouring of material
and cutting of channel beds by running water. Water 1:
no page number.

Stream Incision. <u>Federal--Interior</u>. The cutting or eroding away
of a streambank. General 3: p. 74.

Strike And Dip. <u>City--Planning</u>. The dip of a stratum or fault
is the angle, in degrees, between the bedding plane or
fault plane and the horizontal, measured at right angles
to the strike. The strike is the direction (azimuth) of
the line formed by intersection of the bedding or fault

plane and a horizontal plane. General 5: p. 58.

Strike-Slip Fault. Federal--ERDA. A fault in which the movement
(offset) has been parallel to the fault's strike; contrasts
to dip-slip fault. General 7: p. g-10.

Structure. Regional--Commission. The composition of rock, rock
mass or whole region of earth's crust including stratification, lamination, unconformity, folds, and faults. General 10: p. IV-39.

Subgrade. Federal--DOT. The surface of the undisturbed soil or
specially prepared, artificially compacted layers of soil
or crushed rock used as a foundation for a highway or air-
field pavement. Transportation 3: p. 65.

Subsoil. National--Engineering. That part of the soil beneath
the topsoil, usually that not having an appreciable organic
matter content. Water 1: no page number.

Regional--River Basin. That portion of the soil profile
below plow depth. It generally is 10 to 32 inches below
the surface. Water 4: p. 343.

Regional--River Basin. Roughly, that part of the soil profile below plow depth. Generally 10 inches to 40 inches
below surface. Earth 1: p. 381.

Substrata. Federal--GSA. Material constituting the bottom of a
stream or lake (e.g., bedrock, boulders and pebbles, gravel,
sand, silt, or clay). General 2: p. XI-29.

Substratum. Regional--River Basin. Any layer beneath the soil
profile. It applies to the parent material and to layers
unlike the parent material that lie below the subsoil.
Earth 1: p. 382.

Surface Erosion. Federal--Forest Service. Erosion which removes
materials from the surface of the land as distinguished from
gully or channel erosion. The two main types of surface
erosion are sheet erosion and rill erosion. General 15:
p. 75.

Surface Soil. National--Engineering. Upper part of the soil ordinarily moved in tillage, or its equivalent in uncultivated soils, about 10 to 20 cm in thickness. Water 1: no
page number.

National--Soil Conservation. The uppermost part of the
soil ordinarily moved in tillage or its equivalent in

uncultivated soils, ranging in depth from about 5 to 8 inches. Frequently designated as the plow layer, the Ap layer, or the Ap horizon. Earth 4: p. 55g.

Regional--River Basin. The soil ordinarily moved in til-lage, or its equivalent in uncultivated soil, about 10 inches in thickness. The plowed layer. Earth 1: p. 382.

Swale. Federal--Commerce. A low-lying or depressed land area commonly wet or moist; an intermittent drainageway. General 6: p. 314.

Syncline. State--Natural Resources. Downfolding of rock layers toward a trough. General 12: p. 339.

Talus. Regional--River Basin. Fragments of rock and soil mater-ial collected at the foot of cliffs or steep slopes, chief-ly as a result of gravitational forces. Earth 1: p. 382.

Tectonic Estuaries. Federal--DOT. Coastal indentures formed by geological faulting or by local subsidence, often with a large inflow of fresh water. Transportation 3: p. 67.

Terrain Analysis. Federal--Army. The identification and inter-pretation of landforms caused by natural processes either by remote sensing or on-site inspection for the purpose of determining feasibility of development or management practices. Recreation 2: p. A-50.

Tidal Flats. Federal--Interior. Areas of nearly flat, barren mud periodically covered by tidal waters. General 3: p. 77.

Tidal Marsh. Federal--Army. Marsh land periodically inundated by tidal oceanic or estuarine water, (i.e., salt marsh). General 9: p. 46.

Federal--EPA. Low, flat marshlands traversed by interlaced channels and tidal sloughs and subject to tidal inundation; normally, the only vegetation present is salt-tolerant bushes and grasses. General 1: p. 15.

Toe. Federal--Interior. Point of intersection of bottom of a slope or embankment with the natural ground or level ground such as upstream or downstream toe of the dam. General 3: p. 77.

Topographic Map. National--Soil Conservation. A representation of the physical features of a portion of the earth's sur-face as a plane surface, on which terrain relief is shown

by a system of lines, each representing a constant elevation above a datum or reference plane. Earth 4: p. 33g.

Topography. Federal--Army. Description or representation of
    natural or artificial features of the landscape; the de-
    scription of any surface, but usually the earth. General
    9: p. 47.

    Federal--EPA. The physical features of a surface area in-
    cluding relative elevations and the position of natural and
    manmade features. General 1: p. 15.

    Federal--GSA. Science of surveying the physical features
    of a district or region and the art of delineating them on
    maps. General 2: p. VIII-30.

    Federal--Interior. The physical shape of the ground sur-
    face. General 3: p. 77.

    Regional--Commission. The topography comprises the geom-
    etry of the ground surface. Its degree of irregularity in-
    fluences the volume of grading or cut-and-fill which a pro-
    ject may entail. The slope and frequency with which these
    irregularities are spaced determines, in part, the effi-
    ciency of runoff from storms and the vulnerability to flash
    floods. General 10: p. IV-89. (Revised).

    State--Board. The configuration of a surface, including
    its relief and the position of its natural and man-made
    features. Air 2: no page number.

Transitory Range. Federal--Forest Service. Land that is suitable
    for grazing use of a nonenduring or temporary nature over a
    period of time. For example, on particular disturbed lands
    grass may cover the area for a period of time before being
    replaced by trees or shrubs not suitable for forage. Gen-
    eral 15: p. 171.

Truncated Soil Profile. National--Soil Conservation. Soil pro-
    file that has been cut down by accelerated erosion or by
    mechanical means. The profile may have lost part or all
    of the A horizon and sometimes the B horizon, leaving only
    the C horizon. Comparison of an eroded soil profile with
    a virgin profile of the same area, soil type, and slope
    conditions, indicates the degree of truncation. Earth 4:
    p. 57g.

Ultra Basic Rock. Regional--River Basin. Very rich in calcium
    and magnesium. Earth 1: p. 382.

Unconsolidated Sediment. <u>State--Natural Resources</u>. Uncemented
sediment, e.g., loose sand. General 12: p. 339.

Undercutting. <u>National--Engineering</u>. Removal of material at the
base of a steep slope, overfall, or cliff by falling water,
a stream, wind erosion, or wave action. This removal
steepens the slope or produces an overhanging cliff.
Water 1: no page number.

Uniform Soil. <u>National--Engineering</u>. Soil having uniform properties throughout its profile. Water 1: no page number.

Upland. <u>Regional--River Basin</u>. The land consisting of material
unworked by water in recent geologic time and generally at
a higher elevation than the alluvial plain or stream terrace; land above the lowlands along rivers or between
hills. Earth 1: p. 382.

Urban Areas. <u>Federal--Forest Service</u>. Areas within the legal
boundaries of cities and towns; suburban areas developed
for residential, industrial, or recreational purposes.
General 15: p. 223.

Urbanized Area. <u>Federal--Interior</u>. A central city, or twin
cities, and surrounding closely settled area. A "central"
city" is a city of 50,000 inhabitants or more in 1970 or
a special census conducted by the Bureau of the Census;
"twin cities" are cities with contiguous boundaries with
a combined population of at least 50,000 inhabitants, with
the smaller of the two cities having a population of at
least 15,000. Recreation 3: p. 26.

Urban Land. <u>National--Soil Conservation</u>. Areas so altered or
obstructed by urban works or structures that identification of soils is not feasible. A miscellaneous land type.
Earth 2: p. 58g.

Virgin Areas. <u>Federal--Interior</u>. Areas of virtually undisturbed
native vegetation. Recreation 3: p. 26.

Weathering. <u>Federal--DOT</u>. The destructive process constituting
that part of erosion whereby earthy and rocky materials
on exposure to atmospheric agents at or near the earth's
surface are changed in character (color, texture, composition, firmness or form) with little or no transport of the
loosened or altered materials. Transportation 3: p. 72.

Wetlands. <u>Federal--Executive Order</u>. Those areas that are inundated by surface or ground water with frequency sufficient
to support and under normal circumstances does or would

support a prevalence of vegetative or aquatic life that
requires saturated or seasonally saturated soil conditions
for growth or reproduction. Wetlands generally include
swamps, marshes, bogs, and similar areas such as sloughs,
potholes, wet meadows, river overflows, mud flats, and
natural ponds. Legal Jargon 39: p. 369.

Wilderness. Federal--Army. A tract or region of land uncultivated and uninhabited by human beings, or unoccupied by
human settlements. General 9: p. 50.

Wilderness Area. Federal--Forest Service. Undeveloped Federal
land retaining its primeval character and influence, without permanent improvements or human habitation, which is
protected and managed so as to preserve its natural conditions and which (1) generally appears to have been affected primarily by the forces of nature, with the imprint
of man's work substantially unnoticeable; (2) has outstanding opportunities for solitude or a primitive and unconfined type of recreation; (3) has at least 5000 acres or
is sufficient size as to make practical its preservation
and use in an unimpaired condition; and (4) may also contain ecological, geological, or features of scientific,
educational, scenic, or historical value. General 15:
p. 231.

Regional--River Basin. A collective term used to describe
all major areas specially classified and set aside for
their primitive and relatively undisturbed esthetic values.
Earth 1: p. 383.

Wilderness Study Area. Federal--Forest Service. An area which
has been placed under formal consideration for possible
inclusion in the wilderness area system. A new study area
in a National Forest is removed from standard management
and planning considerations until its status as a wilderness area is determined. After study such areas will be
classified as either nonselected roadless areas or roadless
and undeveloped areas. General 15: p. 232.

Wind Erosion. National--Engineering. Detachment, transportation,
and deposition of soil by the action of the wind. Water 1:
no page number. (Revised).

Xeric. Federal--Interior. An arid system almost totally lacking
water. General 3: p. 83.

Zone Of Aeration. National--Engineering. Subsurface zone above
the water table in which the soil or permeable rock is not
saturated. Water 1: no page number.

Zone Of Saturation. <u>National--Engineering</u>. Layer of earth
    (may be both soil and underlying strata) containing grav-
    itational or free water. Water 1: no page number.

5.   AIR

Index

Aerosol
Afterburner
Air
Air Basin
Air Contaminant
Air Contaminant Source
Air Mass
Air Monitoring
Air Pollution
Air Pollution Control District
       (APCD)
Air Pollution Control Officer
       (APCO)
Air Pollution Episode
Air Quality Control Region
Air Quality Criteria
Air Quality Maintenance Area
    (AQMA)
Air Quality Maintenance Plan
    (AQMP)
Air Quality Model
Air Quality Standards
Airshed
Ambient
Ambient Air
Ambient Air Quality
Ambient Air Quality Standards
Area Source
Area Source Simulation
Ash
Atmosphere
Atmospheric Area
Atmospheric Pollution

Atmospheric Stability
Atmospheric Stability Classes
Attainment
Automotive Air Pollution
Background Concentrations
Background Level
Background Pollution Levels
Baghouse
Burden Model
Carbon Dioxide (CO_2)
Carbon Monoxide (CO)
Catalytic Converter
Climate
Climatic Year
Coefficient Of Haze (COH)
Combustion
Community Air
Control Strategy
Control Techniques
Critical Year
Cyclone Collector
Deposition Velocity
Deterioration Increment
Diffusion Models
Dispersion Model
Dry Limestone Process
Dust
Dustfall Jar
Electrostatic Precipitator
Emergency Episode
Emission
Emission Accounting Procedure
Emission Ceiling

185

Emission Factor

Emission Inventory

Emission Projections

Emission Standard

Episode

Equivalent Method

Exhaust Analyzer

Exhaust Emission

Fabric Filter

Federal Motor Vehicle Emission
   Control Program (FMVECP)

Flue Gas

Flue Gas Scrubber

Fluorocarbons

Fly Ash

Fugitive Dust

Fume

Gas

Gaussian Model

Grain Loading

Hazardous Air Pollutant

Heat Island Effect

Hi-Volume Sampler

Hydrocarbon (HC)

Hydrogen Sulfide (H_2S)

Indirect Source

Inversion

Isotherm

Lead

Line Source

Major Air Pollutants

Mechanical Turbulence

Meteorological Stability
   Classes

Microclimate

Mist

Mixing Height

Mixing Layer

Mobile Source

Monitoring

National Ambient Air Quality
   Standards (NAAQS)

National Primary Standards

National Secondary Standards

New Source Performance
   Standards (NSPS)

Nitric Oxide (NO)

Nitrogen Dioxide (NO_2)

Nitrogen Oxides (NOx)

Non-Attainment

No Significant Deterioration

Odor

Opacity

Open Burning

Oxidant

Oxide

Ozone (O_3)

Ozone Shield

Particulate Loading

Particulate Matter

Particulates

Peroxyacetyl Nitrate (PAN)

Photochemical Oxidants

Photochemical Process

Plume

Point Source

Pollutant

Pollutant Burden

Precipitators

Prevention Of Significant
   Deterioration (PSD)

Primary Air Pollutants

Primary Effects

Primary Pollutants

Primary Standard

Receptor

Reasonable Receptor Points

Ringelmann Chart

Rubber Cracking Method

Scrubber

Secondary Effects

Secondary Pollutants

Secondary Standard

Smog

Smoke

Soiling Index

Source

Source Review

Stable Air

Stack

Stack Effect

Standard Of Performance

Standard Smoke Scale

State Implementation Plan (SIP)

Sulfates

Sulfur Dioxide (SO_2)

Sulfur Oxides (SOx)

Thermal Turbulence

Total Suspended Particulates
   (TSP)

Transport   Weather
Transportation Control Plan   Wet Scrubber
Troposphere   Wind Direction
Vapor   Wind Rose
Vapor Plumes   Wind Speed

Terms

Aerosol. <u>Federal--Interior</u>. Particulate matter suspended in the
      air. The particulate matter may be in the form of dusts,
      fumes, or mist. General 3: p. 2.

Afterburner. <u>Federal--EPA</u>. An air pollution device that removes
      undesirable organic gases by incineration. General 1:
      p. 1.

Air. <u>State--Committee</u>. Pure air is a mixture of gases contain-
      ing nitrogen, oxygen, carbon dioxide, argon, other inert
      gases and varying amounts of water vapor. Air 3: no page
      number.

Air Basin. <u>State--Board</u>. A region in which the air quality is
      determined by the meteorology and emissions within it,
      with minimal influence on and impact by contiguous re-
      gions. In California, the 14 air basins have boundaries
      established by the Air Resources Board. While the above
      criterion is the major consideration, the boundaries may
      be the borders of counties if such borders are close to
      the boundary of the natural air basin. Air 2: no page
      number.

      <u>City--Planning</u>. A geographical area in which air quality
      tends to behave as a unit because of natural barriers sep-
      arating the air mass from other air masses. General 8:
      p. 1.

Air Contaminant. <u>State--Environmental Control</u>. The presence in
      the outdoor atmosphere of any dust, fume, mist, smoke, va-
      por, gas, or other gaseous fluid, or particulate substance
      differing in composition from or exceeding in concentra-
      tion the natural components of the atmosphere. Legal Jar-
      gon 25: p. 1.

      <u>City--Commission</u>. Any substance in the ambient air space
      and includes, but is not limited to dust, fly-ash, fume,
      mist, odor, smoke, vapor, pollen, micro-organism, radio-
      active material, radiation, heat, gas, any combination
      thereof, or any decay or reaction product thereof. Air 1:
      no page number.

Air Contamination Source. <u>City--Commission</u>. Any source at,
      from, or by reason of which any air contaminant is emitted
      directly or indirectly into the ambient air space. Air 1:
      no page number.

Air Mass. <u>Federal--EPA</u>. A widespread body of air that gains cer-
     tain characteristics while set in one location. The char-
     acteristics change as it moves away. General 1: p. 1.

Air Monitoring. <u>State--Board</u>. Sampling for and measuring of pol-
     lutants present in the atmosphere. Air 2: no page number.

     <u>State--Committee</u>. The continuous sampling for and measur-
     ing of pollutants present in the atmosphere. Air 3: no
     page number.

Air Pollution. <u>Federal--EPA</u>. The presence of contaminant sub-
     stances in the air that do not disperse properly and in-
     terfere with human health. General 1: p. 1.

     <u>Federal--Forest Service</u>. Any substance or energy form
     (heat, light, noise, etc.) which alters the state of the
     air from what would naturally occur. Especially associ-
     ated with those altered states whose physical, chemical,
     biological, psychological or aesthetic impacts have been
     decreed to be undesirable by human value judgments. Gen-
     eral 15: p. 26.

     <u>Federal--GSA</u>. Presence in the atmosphere of substances in
     quantities that adversely affect humans, animals, vegeta-
     tion, or materials. General 2: p. IV-21.

     <u>Federal--GSA</u>. The concentration of contaminants in the
     air when not prevented by the normal dispersive ability of
     the air and when interfering directly or indirectly with
     man's health, safety or comfort or with the full use and
     enjoyment of his property. Legal Jargon 18: Appendix B-1.

     <u>State--Board</u>. Atmospheric substances that are foreign to
     the "natural" atmosphere or are in quantities exceeding
     their natural concentrations. Air 2: no page number.

     <u>State--Committee</u>. Man-made contamination in the atmosphere
     which is harmful to humans, animals or plant life, or which
     damages property. Air 3: no page number.

     <u>State--Environmental Control</u>. The presence in the outdoor
     atmosphere of one or more air contaminants or combinations
     thereof in such quantities and of such duration as are or
     may tend to be injurious to human, plant, or animal life,
     or property, or the conduct of business. Legal Jargon 25:
     p. 1.

Air Pollution Control District (APCD). <u>State--Board</u>. Local gov-
     ernmental agencies on the county level with the legislative

authority to adopt and enforce all rules and regulations
necessary to control non-vehicular sources of air contaminants. Air 2: no page number.

Air Pollution Control Officer (APCO). State--Board. The appoin-
tee of the local APCD with the duties of implementing and
enforcing all rules and regulations as prescribed by the
APCD. Air 2: no page number.

Air Pollution Episode. Federal--EPA. A period of abnormally high
concentration of air pollutants, often due to low winds and
temperature inversion that can cause illness and death.
General 1: p. 1.

Federal--GSA. The occurrence of abnormally high concentrations of air pollutants usually due to low winds and
temperature inversion and accompanied by an increase in
illness and death. Legal Jargon 18: Appendix B-1.

Air Quality Control Region. Federal--EPA. An area designated by
the Federal Government in which communities share a common
air pollution problem, sometimes involving several states.
General 1: p. 1.

Air Quality Criteria. Federal--EPA. The levels of pollution and
lengths of exposure above which adverse effects may occur
on health and welfare. General 1: p. 1.

State--Committee. Varying amounts of pollution and lengths
of exposure at which specific adverse effects to health and
welfare take place. Air 3: no page number.

Air Quality Maintenance Area (AQMA). State--Board. Areas which,
due to current air quality and/or projected growth rates
may have the potential for exceeding any national ambient
air quality standard within the next twenty years. These
areas were identified pursuant to federal regulations (FR
June 18, 1973) and are contained in ARB's Revision 5 of
State Implementation Plan. Air 2: no page number.

City--Planning. An area with the potential for exceeding
any national ambient air quality standard in the period
from 1975 to 1985. General 8: p. 1.

Air Quality Maintenance Plan (AQMP). State--Board. A comprehensive plan to assure the achievement and/or maintenance of
the national air quality standards through the long term
(20-25 years). The plans are to be formulated at the local
level and will integrate direct source controls, land use

plans, and transportation strategies which will be implemented at the State and local level to achieve and maintain the air quality standards. AQMPs are required by the
Clean Air Act and subsequent EPA regulations. Air 2: no
page number.

Air Quality Model. State--Board. A mathematical formulation of
the relationship betwen emissions of air pollutants from
various sources and the concentrations of these pollutants
in the air at various locations. An adequate air quality
model simulates the transport, dispersion, and chemical
reactivity of pollutants within an air basin or sub-basin
area and shows the effects of changes in certain emissions
levels on air quality throughout the air basin or in an
air sub-basin. An air quality model for a particular air
basin requires the introduction of emissions data, meteorological data on seasonal wind patterns and inversion
layers and topographic data. Air 2: no page number.

Air Quality Standards. Federal--EPA. The level of pollutants
prescribed by law that cannot be exceeded during a specified time in a defined area. General 1: p. 1.

Federal--GSA. The prescribed level of pollutants in the
outside air that cannot be exceeded legally during a specified time in a specified geographical area. Legal Jargon 18: Appendix B-1.

State--Board. A quantification of allowable levels of air
pollution. There are two types of National Ambient Air
Quality Standards (NAAQS), the primary standards and secondary standards. Under provision of the Mulford Carrell
Act, the State Air Resources Board is required to set state
air quality standards. However, since there are no specified dates for attainment of the standards, the state air
quality standard are essentially air quality goals. Air 2:
no page number.

Airshed. Federal--Interior. Atmospheric zone potentially influenced by air pollutants from various sources. General 3:
p. 3.

National--Soil Conservation. A common air supply demar-
cated by arbitrary or convenient borders, such as an urban
area. Earth 4: p. 6g.

Ambient. State--Board. As commonly used in air quality discussion, a term descriptive of the condition/state/quality
of the air outside structures, at a particular time and
place. Air 2: no page number.

Ambient Air. Federal--EPA. Any unconfined portion of the atmos-
     phere; open air. General 1: p. 1.

     Federal--Interior. The unconfined space occupied by the
     atmosphere. General 3: p. 5.

     State--Committee. Refers to outdoor air. Air 3: no page
     number.

Ambient Air Quality. Federal--Interior. The prevailing quality
     of the surrounding air in a given area in terms of the type
     and amounts of various air pollutants present. General 4:
     Glossary, p. 1.

Ambient Air Quality Standards. Federal--GSA. Maximum allowable
     concentrations of specified pollutants in the outdoor at-
     mosphere, as established by Federal, State, or local agen-
     cies. General 2: p. IV-21.

Area Source. Federal--EPA. In air pollution, any small individ-
     ual fuel combustion source, including vehicles. A more
     precise legal definition is available in Federal regula-
     tions. General 1: p. 2.

     National--Soil Conservation. Geographic source from which
     air pollution originates. Earth 4: p. 7g.

     Regional--Air. Motor vehicle emissions not sufficiently
     concentrated or of insufficient strength to be considered
     as line sources. Example: parking lots. Air 4: p. 3.

Area Source Simulation. Regional--Air. A system of handling
     parking or other areas as line sources in the acceptable
     (line source) models by means of concentrating the area's
     entire pollutant generation on an appropriate number of
     line segments. The EPA Highway model does this automat-
     ically when a pseudo cut-section technique is employed.
     Air 4: p. 3.

Ash. Federal--Interior. Noncombustible mineral matter as con-
     tained in coal. These minerals are generally similar to
     ordinary sand, silt, and clay in chemical and physical
     properties. General 3: p. 8.

Atmosphere. Federal--EPA. The body of air surrounding the Earth.
     General 1: p. 2.

     State--Board. The envelope of air surrounding the earth.
     Air 2: no page number.

Atmospheric Area. <u>National--Soil Conservation</u>. An airshed in
       which all environmental conditions influence the capacity
       of the atmosphere to dilute and disperse air pollutants.
       Earth 4: p. 8g.

Atmospheric Pollution. <u>City--Commission</u>. The presence in the
       ambient air space, or portion thereof, of one or more air
       contaminants or combinations thereof in such a concentra-
       tion and of such duration as: (a) to cause a nuisance; or
       (b) to be injurious or tend to be, on the basis of current
       information, injurious to human or animal life, vegetation,
       or to property; or (c) to unreasonably interfere with the
       comfortable enjoyment of life and property or the conduct
       of business. Air 1: no page number.

Atmospheric Stability. <u>Regional--Commission</u>. The tendency of
       the atmosphere to either enhance or damp out vertical mo-
       tions affecting the concentration of air pollutants. Un-
       stable atmosphere disperses pollutants more readily than
       a stable atmosphere. General 10: p. IV-8.

Atmospheric Stability Classes. <u>Federal--GSA</u>. Set of six cate-
       gories of stability defined by surface wind speed, cloud
       cover, and insolation. These classes are used to select
       parameters of vertical and crosswind diffusion in the
       Gaussian diffusion model. General 2: p. IV-21.

Attainment. <u>State--Committee</u>. Refers to areas of the country
       where levels of air pollution are below the national air
       quality standards or will be below the standards by a
       certain deadline date. Air 3: no page number.

Automotive Air Pollution. <u>State--Committee</u>. Air pollution pro-
       duced by motor vehicles rather than industrial sources--
       includes hydrocarbons, carbon monoxide, nitrogen oxides,
       photochemical oxidants. Air 3: no page number.

Background Concentrations. <u>Regional--Air</u>. The pollutant concen-
       tration which would remain if all the emissions from all
       the sources that are quantified for use in the modeling
       or calculation process were fully stopped. It should be
       determined by adequate monitoring, but might be calculated
       from statistical data. Where not known, a background con-
       centration providing adequate safety should be used. Air
       4: p. 3.

Background Level. <u>Federal--EPA</u>. In air pollution, the level of
       pollutants present in ambient air from natural sources.
       General 1: p. 2.

State--Board. Amounts of pollutants present in the ambient
air due to natural sources. Examples, marsh gases, pollen,
wind blown dust. Air 2: no page number.

Background Pollutant Levels. Federal--GSA. Pollutant levels
that are representative of the average conditions in a
given area. Pollutant levels predicted from diffusion
models are added to background values to determine the
impact of specific sources on air quality. General 2:
p. IV-21.

Baghouse. Federal--EPA. An air pollution abatement device used
to trap particulates by filtering gas streams through large
fabric bags usually made of glass fibers. General 1: p. 2.

Burden Model. Federal--DOT. A model used to determine the amount
of air pollution which will be generated by a given roadway or other human activity. Transportation 3: p. 18.

Carbon Dioxide (CO_2). Federal--EPA. A colorless, ordorless,
non-poisonous gas normally part of the ambient air, a result of fossil fuel combustion. General 1: p. 3.

Carbon Monoxide (CO). Federal--EPA. A colorless, odorless,
poisonous gas produced by incomplete fossil fuel combustion. General 1: p. 3.

State--Board. A colorless, odorless, toxic gas produced
by the incomplete combustion of carbon-containing substances. One of the major air pollutants, it is emitted
in large quantities in the exhaust of gasoline-powered
vehicles. Air 2: no page number.

State--Committee. A colorless, odorless, very toxic gas
produced by incomplete burning of carbon-containing fuel
such as oil, coal, gasoline and natural gas. One of the
major air pollutants, it is primarily emitted through the
exhaust of gasoline-powered vehicles, and found especially
in congested traffic situations. Air 3: no page number.

City--Planning. A clear, odorless gas which in high concentrations can cause dizziness, unconsciousness, and even
death. The major source of carbon monoxide is the automobile. High concentrations of carbon monoxide are mainly
a local problem, occurring near areas of heavy auto traffic
when ventilation is poor. General 8: p. 2.

Catalytic Converter. Federal--EPA. An air pollution abatement
device that removes organic contamination by oxidizing them
into carbon dioxide and water. General 1: p. 3.

State--Committee. A pollution control device used by most
car manufacturers since the 1975 model year. The converter
changes hydrocarbons in the exhaust into harmless carbon
dioxide and water and changes carbon monoxide into carbon
dioxide. Air 3: no page number.

Climate. Federal--Army. The average conditions of the weather
over a number of years; macroclimate is the climate representative of relatively large areas; microclimate is the
climate of small areas, particularly that of the living
space of certain species, group or community. General 9:
p. 9.

National--Soil Conservation. The sum total of all atmospheric or meteorological influences, pricipally temperature, moisture, wind, pressure, and evaporation, which
combine to characterize a region and give it individuality
by influencing the nature of its land forms, soils, vegetation, and land use. Earth 2: p. 12g.

Climatic Year. National--Soil Conservation. A continuous 12
month period arbitrarily selected for the analysis and
presentation of climatological or streamflow data, generally beginning March 1 or April 1. Earth 2: p. 13g.

Coefficient Of Haze (COH). Federal--EPA. A measurement of visability interference in the atmosphere. General 1: p. 4.

National--Soil Conservation. Unit of measurement for determining visability based on the amount of particles in
1,000 linear feet. Earth 2: p. 13g.

Combustion. Federal--EPA. Burning, or a rapid oxidation accompanied by release of energy in the form of heat and light,
a basic cause of air pollution. General 1: p. 4.

Community Air. National--Soil Conservation. The ambient environment in the immediate vicinity of a community. Earth 4:
p. 14g.

Control Strategy. National--Soil Conservation. The combination
of measures, such as emission limitations, land use plans,
emission taxes, designed to reduce levels of a specific
pollutant in the ambient air. Earth 4: p. 15g.

Control Techniques. National--Soil Conservation. Methods, equipment, and devices applicable to the prevention and control
of air pollutants at their sources, such as process
changes, fuel use limitations, and plant location, rules,
etc. Earth 2: p. 15g.

Critical Year. <u>Federal--DOT</u>. That year, present or future, when total mobile source emissions from roadways affected by a given project will be the greatest, as used in regard to air pollution analysis. Transportation 3: p. 23.

Cyclone Collector. <u>Federal--EPA</u>. A device that uses centrifugal force to pull large particles from polluted air. General 1: p. 4.

> <u>National--Soil Conservation</u>. A device used to collect large-sized particulates from polluted air by centrifugal force. Because of its simplicity and effectiveness, the cyclone collector is widely used in feed mills, sawmills, and other manufacturing operations that generate dust and particulates. Earth 2: p. 17g.

Deposition Velocity. <u>National--Soil Conservation</u>. The deposition rate divided by the ambient concentration of a specific air pollutant. Earth 2: p. 18g.

Deterioration Increment. <u>State--Board</u>. Within the context of the Air Conservation Program, the specific amount that air quality will be allowed to deteriorate from existing ambient levels in specific areas. The amounts of deterioration from specific projects/developments are computed on a cumulative basis. Therefore, once a deterioration increment has been fully utilized, there can be no additional source of pollution unless modifications are made to the size of the increment for that area. Air 2: no page number.

Diffusion Models. <u>Federal--GSA</u>. Mathematical statements that use emission information and meteorological parameters to predict atmospheric concentrations of pollutants. General 2: p. IV-21.

Dispersion Model. <u>State--Board</u>. A mathematical representation and solution of the physical processes that result in the spread of pollutants in the atmosphere. The phenomenon of direct interest in predicting the dispersion of pollutants is turbulent diffusion. Consequently dispersion models are sometimes called diffusion models. However, turbulent diffusion and diffusion models are misnomers. The phrases refer to the spreading of a cloud of particles in a turbulent fluid at a rate many orders of magnitude greater than that from molecular diffusion alone. The spreading is really not due to a "diffusion" phenomenon such as results from molecular collisions but rather is a result of the rapid, irregular motion of lumps of fluid (called eddies) in turbulence. Air 2: no page number.

Dry Limestone Process. Federal--EPA. An air pollution control
     method that uses limestone to absorb the sulfur oxides in
     furnaces and stack gases. General 1: p. 5.

Dust. State--Board. Solid particulate matter. Air 2: no page
     number.

Dustfall Jar. Federal--EPA. An open container used to collect
     large particles from the air for measurement and analysis.
     General 1: p. 5.

Electrostatic Precipitator. Federal--EPA. An air pollution con-
     trol device that imparts an electrical charge to particles
     in a gas stream causing them to collect on an electrode.
     General 1: p. 6.

Emergency Episode. National--Soil Conservation. An air pollu-
     tion incident in a given area caused by a concentration
     of atmospheric pollution reacting with meteorological con-
     ditions that results in a significant increase in illness
     or deaths. Earth 4: p. 21g.

Emission. Federal--EPA. Like effluent but used in regard to air
     pollution. General 1: p. 6.

     Federal--Interior. A discharge of pollutants into the
     atmosphere, usually as a result of burning or the oper-
     ation of internal combustion engines. General 4: Glos-
     sary, p. 4.

     State--Board. Commonly used in air quality discussions to
     describe any gaseous pollutant. Air 2: no page number.

     City--Commission. A discharge or release to the ambient
     air space of any air contaminant. Air 1: no page number.

Emission Accounting Procedure. State--Board. A process that con-
     sists of maintaining a log of all polluting sources and
     amounts of emissions per source for a specified area. Such
     a process is designed to show the additional amount of pol-
     lutants that can be emitted before pre-established limits
     are reached. Air 2: no page number.

Emission Ceiling. State--Board. A term used to describe the
     maximum amount of pollutants that will be permitted to
     be emitted within a specified area or from a particular
     source. Air 2: no page number.

Emission Factor. Federal--EPA. The relationship between the
     amount of pollution produced and the amount of raw material

processed. For example an emission factor for a blast
furnace making iron would be the number of pounds of particulates per ton of raw materials. General 1: p. 6.

National--Soil Conservation. Statistical average of the
amount of a specific pollutant emitted from each type of
polluting source in relation to a unit quality of material
handled, processed, or burned. Earth 4: p. 21g.

State--Board. An expression of the estimated emissions of
air pollutants/unit time/activity. Examples of activities
are low density residential land use, vehicle miles traveled for a certain model and year automobile or truck, and
landing-take-off cycles of aircraft. Emissions factors
are being developed by the Environmental Protection Agency
and the Air Resources Board for use in the preparation of
emissions inventories and emissions projections for planning purposes. Air 2: no page number.

Emission Inventory. Federal--EPA. A listing, by source, of the
amounts of air pollutants discharged into the atmosphere
of a community daily. It is used to establish emission
standards. General 1: p. 6.

National--Soil Conservation. The amount of each primary
air pollutant released daily into a community's atmosphere.
Earth 4: p. 21g.

State--Board. An inventory of the spatial distribution of
sources of air polluting emissions and pollutant types and
quantities from each source within a given geographic area
for a given time period. The location of emissions from
stationary sources and mobile sources are presented separately in an emissions inventory. Air 2: no page number.

State--Committee. A list of air pollutants emitted into
a community's air in amounts, commonly tons per day, by
type of source. Air 3: no page number.

Emission Projections. State--Board. Estimates of emissions likely
to be emitted in a given geographic area at a designated
future date. Air 2: no page number.

Emission Standard. Federal--EPA. The maximum amount of discharge
legally allowed from a single source, mobile or stationary.
General 1: p. 6.

State--Board. The maximum amount of a pollutant that is
permitted to be discharged from a polluting source; e.g.,
the number of pounds of dust that may be emitted from an

industrial process. Air 2: no page number.

State--Committee. The maximum amount of a pollutant that is permitted to be discharged from a single source. Air 3: no page number.

Episode. Federal--EPA. An air pollution incident in a given area caused by a concentration of atmospheric pollution reacting with meteorological conditions that may result in a significant increase in illness or deaths. General 1: p. 6.

Equivalent Method. State--Environmental Protection. Any methods of sampling and analysis for an air pollutant which have been demonstrated to the department's satisfaction to have a consistent and quantitatively known relationship to the reference method under specific conditions. Legal Jargon 8: p. 6.

Exhaust Analyzer. State--Committee. An instrument which measures the amount of hydrocarbons and carbon monoxide in a car's exhaust. Air 3: no page number.

Exhaust Emission. National--Soil Conservation. The air pollutants emitted from the exhaust of the internal combustion engine, namely carbon monoxide, nitrogen oxides, and hydrocarbons. Earth 2: p. 22g.

Fabric Filter. Federal--EPA. A cloth device that catches dust and particles from industrial emissions. General 1: p. 6.

Federal Motor Vehicle Emission Control Program (FMVECP). State-- Committee. Program administered by the EPA which requires the reduction in air pollutant emissions from new motor vehicles. Specific limitations are set for hydrocarbons, nitrogen oxides and carbon monoxide and motor vehicle manufacturers must meet them. Air 3: no page number.

Flue Gas. Federal--EPA. The air coming out of a chimney after combustion. It can include nitrogen oxides, carbon oxides, water vapor, sulfur oxide, particles, and many chemical pollutants. General 1: p. 7.

Flue Gas Scrubber. National--Soil Conservation. A type of equipment that removes fly ash and other objectionable materials from flue gas by the use of sprays, wet baffles, or other means that require water as the primary separation mechanism, also called flue gas washer. Earth 2: p. 24g.

Fluorocarbons. Federal--EPA. A gas used as a propellant in aerosols, thought to be modifying the ozone layer in the stratosphere thereby allowing more harmful solar radiation to reach the Earth's surface. General 1: p. 7.

Fly Ash. Federal--EPA. Noncombustible particles carried by flue gas. General 1: p. 7.

City--Commission. The aerosolized solid component of burned or partially burned fuel. Air 1: no page number.

Fugitive Dust. State--Environmental Protection. Solid airborne particulate matter emitted from any source other than through a stack. Legal Jargon 8: p. 6.

Fume. State--Board. Finely divided solid or liquid particles, usually under one (1) micron in diameter, which have condensed from a metallurgical or similar operation. Examples are fumes from a lead melting pot or an aluminum foundry. This term is generally loosely used, and should be avoided in discussions on air pollution and be replaced by the term "particulate matter". Air 2: no page number.

City--Commission. Any aerosol resulting from chemical reaction, distillation, or sublimation. Air 1: no page number.

Gas. City--Commission. The state of matter having neither independent shape nor independent volume but having a tendency to expand and diffuse infinitely. Air 1: no page number.

Gaussian Model. Federal--GSA. Most atmospheric diffusion models are derived from the mathematical expression for a "normal" (i.e., Gaussian) distribution. General 2: p. IV-21.

Grain Loading. Federal--EPA. The rate at which particles are emitted from a pollution source--measurement is made by the number of grains per cubic foot of gas emitted. General 1: p. 7.

Hazardous Air Pollutant. Federal--EPA. Substances covered by Air Quality Criteria, which may cause or contribute to illness or death; asbestos, beryllium, mercury, and vinyl chloride. General 1: p. 8.

Heat Island Effect. Federal--EPA. A haze dome created in cities by pollutants combining with the heat trapped in the space between tall buildings. This haze prevents natural cooling of air, and in the absence of strong winds can hold high concentrations of pollutants in one place. General 1: p. 8.

National--Soil Conservation. The collection of warm air
in the center of a city; and as the air disperses, it cools
and sinks at the extremities. In time the cooler air from
the edges of the city flows into the center to repeat a
self-contained circulatory pattern. Earth 2: p. 27g.

Hi-Volume Sampler. Federal--EPA. A device used to measure and
analyze suspended particulate pollution. General 1: p. 8.

Hydrocarbon (HC). Federal--EPA. Compounds found in fossil fuels,
that contain carbon and hydrogen and may be carcinogenic.
General 1: p. 8.

Federal--HEW. A vast family of compounds containing carbon
and hydrogen in various combinations found especially in
fossil fuels such as coal, petroleum, natural gas, and
bitumens; some are major air pollutants, some may be carcinogenic, and others contribute to smog. Health 1: p. 37.

Federal--Interior. Any of a vast family of compounds containing carbon and hydrogen in various combinations, found
especially in fossil fuels. Hydrocarbons in the atmosphere
resulting from incomplete combustion are a major source of
air pollution. General 4: Glossary, p. 6.

State--Board. Any of a vast family of compounds containing carbon and hydrogen in various combinations; found especially in fossil fuels. Some of the hydrocarbon compounds are major air pollutants; they may be active participants in the photochemical process or carcinogenic.
Air 2: no page number.

State--Committee. Any of a family of compounds containing
carbon and hydrogen in various combinations found especially in fossil fuels. Although most hydrocarbons are
not harmful by themselves, they react with nitrogen oxides
and form photochemical oxidants which are harmful to human
health. The automobile is a major source of hydrocarbons
by evaporation of gasoline and emission of unburnt hydrocarbons through the tailpipe. Air 3: no page number.

Hydrogen Sulfide (H2S). Federal--EPA. The gas emitted during organic decomposition that smells like rotten eggs. It is
also a byproduct of oil refining and burning can cause
illness in heavy concentrations. General 1: p. 8.

Indirect Source. State--Board. Facilities which are traffic generators and can therefore significantly affect air quality
because of emissions generated by the associated mobile
source activities. The pollutant of primary concern with

indirect sources is carbon monoxide and the formation of localized "hot spots". Examples of indirect source are major highways and airports, large regional shopping centers, major sports complexes and stadiums, large amusement and recreational facilities and major parking facilities. Also known as complex sources. Air 2: no page number.

State--Committee. A facility, building or structure which attracts motor vehicle activity and therefore results in the emission of air pollutants. Indirect sources include roads, parking lots, shopping centers, office buildings, etc. Air 3: no page number.

Inversion. Federal--Army. In meteorology, a condition in which cooler surface air is trapped under an upper layer of warmer air, preventing vertical circulation. General 9: p. 26.

Federal--EPA. An atmospheric condition caused by a layer of warm air preventing the rise of cool air trapped beneath it. This holds down pollutants that might otherwise be dispersed, and can cause an air pollution episode. General 1: p. 9.

Federal--GSA. Vertical temperature profile that shows an increase of temperature with height. This extremely stable condition severely restricts vertical motion. General 2: p. IV-21.

Federal--Interior. The phenomenon in which a layer of cool air is trapped by a layer of warm air above it, precluding the release of the bottom air. General 3: p. 45.

State--Board. The phenomenon of a layer of warm air lying over and trapping cooler air below. A special problem in areas having large sources of emissions because the contaminating substances cannot be dispersed through the layer of warm air and high levels of air pollution result. Air 2: no page number.

Isotherm. Federal--Army. A line on a map connecting points having the same temperature at the same time. General 9: p. 26.

Lead. State--Board. A chemical element existing in numerous compounds many of which are highly toxic to humans. Lead compounds of interest to air pollution control are those produced primarily by combustion of gasoline containing anti-knock lead compounds and by lead smelting facilities. Air 2: no page number.

Line Source. Federal--DOT. As distinguished from a point source
     of pollutant emissions such as a stack, a linear config-
     uration (e.g., a highway) from which pollutants are assumed
     to emanate at the same rate from any point along the line.
     Transportation 3: p. 42.

     Federal--GSA. Pollutants generated along a straight line.
     Highways are modeled as successive line segments, and the
     emissions on any given segment are assumed to be constant.
     General 2: p. IV-22.

     Regional--Air. Motor vehicle emissions having a distinct
     linear pattern, definite location, and significant volume.
     Example: streets, roads, freeways. Air 2: p. 4.

Major Air Pollutants. Federal--GSA. The six primary air pollu-
     tants for which ambient air quality standards have been
     promulgated by the Environmental Protection Agency include:
     carbon monoxide (CO), hydrocarbons (HC), nitrogen oxides
     (NO_2), photochemical oxidants (oz), sulfur oxides (SO_2),
     and total suspended particulates (TSP). General 2:
     p. IV-22.

Mechanical Turbulence. Federal--EPA. The erratic movement of
     air caused by local obstructions such as buildings. Gen-
     eral 1: p. 9.

Meteorological Stability Classes. Federal--GSA. Atmospheric
     stability classes. General 2: p. IV-22.

Microclimate. Federal--Army. Variables of humidity, temperature,
     wind, precipitation, etc., which are site specific due to
     topography, vegetation, exposure, drainage, etc., existing
     in the immediate vicinity of that site. Recreation 2:
     p. A-41.

     Federal--GSA. Climatic conditions within a few feet of
     the ground surface. General 2: p. XI-28.

     Federal--Interior. The fine climatic structure of the air
     space which extends from the surface of the earth to a
     height where the effects of the immediate character of the
     underlying surface no longer can be distinguished from the
     general local climate. General 4: Glossary, p. 7.

     Regional--River Basin. Local climatic conditions, brought
     about by the modification of general climatic conditions
     by local differences in elevation and exposure. Earth 1:
     p. 376.

Mist. State--Board. Liquid particles up to 100 microns in diam-
       eter. Air 2: no page number.

       City--Commission. Any liquid aerosol formed by the con-
       densation of vapor or by the atomization of liquids. Air
       1: no page number.

Mixing Height. Regional--Commission. The mixing height is de-
       fined as the height above the surface through which rel-
       atively vigorous vertical mixing occurs. General 10:
       p. IV-8.

Mixing Layer. Federal--DOT. The atmospheric layer wherein pol-
       lutants experience transport. Transportation 3: p. 45.

Mobile Source. Federal--EPA. A moving producer of air pollution,
       mainly forms of transportation--cars, motorcycles, planes.
       General 1: p. 9.

       State--Board. A moving source of air pollutants such as
       a motor vehicle, airplane, train, or ship. Air 2: no page
       number.

       State--Committee. An air pollution source designed to
       move from one location to another or to be portable and
       includes buses, trucks, aircraft, automobiles. Air 3: no
       page number.

Monitoring. State--Board. The sampling and measuring, on a con-
       tinual or continuous basis, of air pollutants. Air 2: no
       page number.

National Ambient Air Quality Standards (NAAQS). State--Board.
       Standards for air quality as set by the EPA Administrator,
       prescribed in terms of maximum concentrations permissible
       for various pollutants and averaging times. Air 2: no
       page number.

       State--Committee. The maximum amount of a specific pol-
       lutant allowed by law in the outdoor air. There are two
       kinds of NAAQS: primary standards are designed to protect
       human health and secondary standards are designed to pre-
       vent damage to materials. Air 3: no page number.

National Primary Standards. State--Board. The levels of air qual-
       ity necessary, with an adequate margin of safety, to pro-
       tect the public health. Each state must attain the pri-
       mary standards no later than three years after the state's
       implementation plan is approved by the Environmental Pro-
       tection Agency (EPA). Air 2: no page number.

National Secondary Standards. State--Board. The levels of air
     quality necessary to protect the public welfare from any
     known or anticipated adverse effects of a pollutant. Each
     state must attain the secondary standards within a "rea-
     sonable time" after implementing plan is approved by the
     EPA. Air 2: no page number.

New Source Performance Standards (NSPS). State--Board. Regula-
     tions promulgated by EPA that establish the maximum per-
     missible emission levels from certain types of new sources.
     Air 2: no page number.

Nitric Oxide (NO). Federal--EPA. A gas formed by combustion un-
     der high temperature and high pressure in an internal com-
     bustion engine. It changes into nitrogen dioxide in the
     ambient air and contributes to photochemical smog. Gen-
     eral 1: p. 10.

Nitrogen Dioxide (NO_2). Federal--EPA. The result of nitric oxide
     combining with oxygen in the atmosphere; a major component
     of photochemical smog. General 1: p. 10.

Nitrogen Oxides (NOx). State--Board. Gases formed from atmo-
     spheric nitrogen and oxygen when combustion takes place
     under conditions of high temperature and high pressure;
     considered major pollutants. Also formed in industrial
     processes, such as the production of nitric acid. Air
     2: no page number.

     State--Board. Compounds which are formed of nitrogen and
     oxygen when combustion takes place under conditions of high
     temperature and high pressure. They are emitted by sources
     such as automobiles, electric generating plants and incin-
     erators. Nitrogen oxides are harmful to human lungs in
     themselves but also combine with hydrocarbons to form pho-
     tochemical oxidants. Air 3: no page number.

Non-Attainment. State--Committee. Refers to regions where air
     quality levels are above the national ambient standards
     or where existing control strategies are not expected to
     lower air pollution to acceptable levels by the deadline
     date. Air 3: no page number.

No Significant Deterioration. State--Board. Pursuant to a court
     decision, EPA adopted regulations for preventing signifi-
     cant deterioration of air quality in those areas of the
     states where air quality is at levels better than NAAQS.
     The regulations apply only to particulate matter and sul-
     fur dioxide. Air 2: no page number.

Odor. City--Commission. That property of gaseous, liquid, or
       solid materials that elicits a physiologic response by
       the human sense of smell. Air 1: no page number.

Opacity. State--Environmental Protection. The degree to which
          emissions reduce the transmission of a light source. Le-
          gal Jargon 8: p. 8.

Open Burning. City--Commission. The practice of burning under
               such conditions that the products of combustion are emit-
               ted directly to the ambient air space and are not conducted
               through a stack, chimney, duct, or pipe, and shall include
               but not be limited to outdoor burning of brush, leaves,
               debris, rubbish, and above or underground smouldering
               fires. Air 1: no page number.

Oxidant. Federal--EPA. A substance containing oxygen that reacts
          chemically in air to produce a new substance; primary
          source of photochemical smog. General 1: p. 11.

          State--Board. Substances in the air such as ozone which
          make available oxygen or oxygenated compounds for chemical
          reaction. Oxidants may be formed from the reaction of
          certain reactive hydrocarbons and nitrogen dioxide, under
          the influences of sunlight. Air 2: no page number.

Oxide. State--Board. A compound of two elements, one of which is
        oxygen. Air 2: no page number.

Ozone (O_3). Federal--EPA. A pungent, colorless, toxic gas that
                contributes to photochemical smog. General 1: p. 11.

                State--Board. A pungent, colorless, toxic gas. As a pro-
                duct of the photochemical process, it is a major air pol-
                lutant. Air 2: no page number.

                State--Committee. A colorless toxic gas which is one com-
                ponent of photochemical oxidants. Air 3: no page number.

Ozone Shield. Federal--HEW. A naturally variable layer of ozone
               in the earth's upper atmosphere which absorbs the most
               lethal wavelengths of ultraviolet radiation from the sun;
               it thus serves as a shield between this radiation and the
               inhabitants of the earth. Health 1: p. 38.

Particulate Loading. Federal--EPA. The introduction of partic-
                      ulates into ambient air. General 1: p. 11.

Particulate Matter. Federal--Interior. The most prevalent atmos-
                     pheric pollutant is suspended particulate matter. Most

particulate measuring devices are designed to measure the
range between 1 to 10 microns. Larger particles settle
out of the air. However, the smaller particles smaller
than 1 micron, are readily respirable, contributing significantly to respiratory diseases and reducing visibil-
ity. General 3: p. 60.

State--Board. A particle, or particles, of solid or liquid matter such as soot, dust, aerosols, fumes and mists.
Air 2: no page number.

State--Environmental Protection. Any material, except
water in uncombined form, that is or has been airborne
and exists as a liquid or a solid at standard conditions.
Legal Jargon 8: p. 8.

City--Commission. Any material that exists in a finely
divided form as a liquid or solid at ambient air temperatures, humidity, and pressures. Air 1: no page number.

Particulates. Federal--EPA. Fine liquid or solid particles such
as dust, smoke, mist, fumes, or smog, found in air or emissions. General 1: p. 11.

Federal--GSA. Solid or liquid particles of variable size
and composition. Sizes range from 0.01 microns for lead
fumes to 100 microns for sawdust. General 2: p. IV-22.

Federal--Interior. Finely divided solid or liquid particles in the air or in an emission. Particulates include
dust, smoke, fumes, mist, spray, and fog. General 4:
Glossary, p. 8.

National--Resources. Suspended small colloidal size particles of ash, charred paper, dust, soot, or other partially
incinerated matter carried in the products of combustion.
Energy/Utility 5: p. 10.

Peroxyacetyl Nitrate (PAN). Federal--EPA. A pollutant created
by the action of sunlight on hydrocarbons and nitrogen
oxides in the air. An ingredient of smog. General 1:
p. 11.

Photochemical Oxidants. Federal--GSA. Oxidizing substances (ranging from ozone to complex organic substances) produces in
the atmosphere through complex reactions involving sunlight, oxides of nitrogen, and hydrocarbons. General 2:
p. IV-22.

State--Committee. Compounds formed by the reaction of hydrocarbons and nitrogen oxides. The reaction is speeded
up by intense sunlight. Motor vehicles are chiefly responsible for high levels of oxidants. Photochemical oxidants
cause irritation of eyes and respiratory tract and can lead
to lung damage. Materials such as rubber and plants such
as tobacco are damaged by oxidants. Air 3: no page number.

Photochemical Process. State--Board. The chemical changes brought
about by the radiant energy of the sun acting upon various
pollutant substances. The products are known as photochemical smog. Air 2: no page number.

Plume. Federal--EPA. Visible emission from a flue or chimney.
General 1: p. 11.

Point Source. Federal--GSA. Pollutants emitted from a single or
closely associated group of stacks. General 2: p. IV-22.

State--Board. A stack or other highly localized pollutant
source. Contrasted to an area source. Air 2: no page
number.

Pollutant. State--Board. In air quality discussion, any one of
the atmospheric substances for which there is a nation and/
or State ambient air quality standard that specifies the
levels of concentration at which undesirable health effects
begin to occur. Air 2: no page number.

Pollutant Burden. Federal--DOT. In reference to air pollution,
the total amounts of pollutants which are (or will be)
generated by a given roadway or other human activity.
Transportation 3: pp. 53-54.

Precipitators. Federal--EPA. Air pollution control devices that
collect particles from an emission by mechanical or electrical means. General 1: p. 12.

Prevention Of Significant Deterioration (PSD). State--Committee.
Provision in the Clean Air Act which provides that air
quality in any region, which is already clean (below the
air quality standards), shall not be allowed to become
dirty. Air 3: no page number.

Primary Air Pollutants. National--Soil Conservation. Emissions
directly evolved from an identifiable source of pollution,
such as fluorides from smoke stacks. Earth 4: p. 40g.

Primary Effects. State--Board. Effects on air quality directly
attributable to a development, project or activity. Primar

effects of a freeway, for example, would be the degradation of air quality directly caused by air polluting emissions from vehicles traveling on the freeway. Air 2: no
page number.

Primary Pollutants. State--Board. Pollutants emitted directly
from sources. Air 2: no page number.

Primary Standard. City--Planning. The levels of air quality necessary, with an adequate margin of safety, to protect the
public health. General 8: p. 3.

Receptors. Federal--GSA. Locations associated with the presence
of segments of the general population considered most susceptible to air pollutants (e.g., hospitals, nursing homes,
schools). General 2: p. IV-22.

State--Board. Commonly used in air quality discussions to
describe any human, animal, plant or material that can be
adversely affected by air pollutants. Air 2: no page
number.

Reasonable Receptor Points. Regional--Air. Such locations where
people might reasonably be expected for time periods consistent with the averaging time specified for the pollutant. For carbon monoxide, the critical time period is
eight hours. Air 4: p. 4.

Ringelmann Chart. Federal--EPA. A series of shaded illustrations
used to measure the opacity of air pollution emissions.
The chart ranges from light grey (number 1) through black
(number 5) and is used to set and enforce emission standards. General 1: p. 13.

Rubber Cracking Method. National--Soil Conservation. A method
for determining levels of air pollution; stretched strips
of unvulcanized rubber are exposed to the air, and the average depth of cracks induced by oxidants determines the
intensity of these air pollutants in the atmosphere. Earth
2: p. 43g.

Scrubber. Federal--Interior. Equipment used to remove pollutants,
such as sulfur dioxides or particulate matter, from stack
gas emissions. General 4: p. 9.

Secondary Effects. State--Board. Effects on air quality indirectly caused by a development project, or activity. For example, the secondary effects of a freeway would be the degradation of air quality caused by the industries attracted

to a site now accessible via the freeway. Air 2: no page
number.

Secondary Pollutants. State--Board. Pollutants formed by chemical and photochemical reactions in the atmosphere. Air 2:
no page number.

Secondary Standard. City--Planning. The levels of air quality
necessary to protect the public welfare from any known or
anticipated adverse effects of a pollutant. General 8:
p. 4.

Smog. Federal--EPA. Air pollution associated with oxidants.
General 1: p. 14.

State--Committee. The irritating haze resulting from the
sun's effect on pollutants in the air. Photochemical oxidants are often referred to as smog or photochemical smog.
Air 3: no page number.

Smoke. Federal--EPA. Particles suspended in air after incomplete
combustion of materials containing carbon. General 1: p.
14.

City--Commission. Means the visible aerosol, which may
contain fly-ash, resulting from combustion of materials
but does not mean condensed water vapor. Air 1: no page
number.

Soiling Index. State--Environmental Protection. A measure of the
soiling properties of suspended particles in air determined
by drawing a measured volume of air through a known area of
Whatman number four filter paper for a measured period of
time, expressed as Coefficient of Haze (COH)/1000 linear
feet, or equivalent. Legal Jargon 8: p. 10.

Source. State--Board. Commonly used in air quality discussions
to describe any process/activity/facility that produces
pollutants. Air 2: no page number.

Source Review. State--Board. A general term used to describe any
one of a number of specific reviews that need to be conducted on proposed facilities that will produce pollutants.
The review is designed to determine the acceptability of
the source with regard to types and amounts of pollutants
that will be produced as the existing pollution controls
and/or strategies within the area of impact. Air 2: no
page number.

Stable Air. Federal--EPA. A mass of air that is not moving norm-
     ally, so that it holds rather than disperses pollutants.
     General 1: p. 14.

Stack. Federal--EPA. A chimney or smokestack; a vertical pipe
     that discharges used air. General 1: p. 14.

     Federal--Interior. A vertical passage through which pro-
     ducts of combustion are conducted to the atmosphere. Gen-
     eral 4: Glossary, p. 10.

Stack Effect. Federal--EPA. Used air, as in a chimney, that moves
     upward because it is warmer than the surrounding atmos-
     phere.

Standard Of Performance. State--Board. An emission limitation
     imposed on a particular category of pollution sources,
     either by EPA, the state, or local APCD. Limitations may
     take the form of emission standards or of requirements for
     specific operating procedures. Air 2: no page number.

Standard Smoke Scale. City--Commission. The scale specified in
     the Standard Method of Test for Smoke Density in the Flue
     Gases from Distillate Fuels, ASTM D2156-65, as published
     by the American Society for Testing and Materials. Air
     1: no page number.

State Implementation Plan (SIP). State--Committee. A program of
     steps which will be taken to ensure that a state will meet
     the national ambient air quality standards by a required
     date. Under federal law each state must have such a plan
     approved by the Environmental Protection Agency and any
     changes in the plan must also be approved by them. Air
     3: no page number.

Sulfates. State--Board. Principally the sulfur oxides SO_3, HSO_4,
     and H_2SO_4 not (SO_2) that are generally found as aerosols
     and particulates. Major sources are power plants, refin-
     eries, chemical industry processes and recently demonstra-
     ted to be a by-product of the catalytic converter installed
     on new automobiles. Air 2: no page number.

Sulfur Dioxide (SO_2). Federal--EPA. A heavy, pungent, colorless
     gas formed primarily by the combustion of fossil fuels.
     This major air pollutant is unhealthy for plants, animals,
     and people. General 1: p. 15.

     Federal--Interior. One of several forms of sulfur in the
     air; an air pollutant generated principally from combustion
     of fuels that contain sulfur. General 4: Glossary, p. 10.

Sulfur Oxides (SOx). Federal--Interior. Compounds of sulfur
        combined with oxygen that have a significant influence
        on air pollution. General 4: Glossary, p. 10.

        State--Board. Pungent; colorless gases formed primarily
        by the combustion of fossil fuels containing sulfur; con-
        sidered major air pollutants; sulfur oxides may damage
        the respiratory tract as well as vegetation. Air 2: no
        page number.

        State--Committee. Pungent, colorless gases formed pri-
        marily by the combustion of fossil fuels which contain
        sulfur. Air 3: no page number.

Thermal Turbulence. National--Soil Conservation. Air mixing
        caused by convection. Earth 2: p. 56g.

Total Suspended Particulates (TSP). State--Committee. Partic-
        ulates consist of solids and liquids such as dust, smoke,
        mist and sprays. They are caused by many types of indus-
        trial sources and the burning of fuel. Air 3: no page
        number.

Transport. State--Board. Generally used in air quality discus-
        sions to describe the physical phenomena whereby pollution
        is carried from the site of emission to another site. Air
        2: no page number.

Transportation Control Plan. State--Committee. A series of strat-
        egies required by the Environmental Protection Agency in
        any region where the national ambient air quality stand-
        ards are exceeded for automotive air pollutants. The
        strategies include reducing vehicle use, decreasing emis-
        sions from individual vehicles and increasing the use of
        public transportation. Air 3: no page number.

Troposphere. Federal--EPA. The portion of the atmosphere between
        seven and ten miles from the Earth's surface, where clouds
        form. General 1: p. 15.

Vapor. City--Commission. The gaseous state of certain substances
        that can exist in equilibrium with their solid or liquid
        states under standard conditions. Air 1: no page number.

Vapor Plumes. Federal--EPA. Fine gases that are visable because
        they contain water droplets. General 1: p. 16.

Weather. National--Soil Conservation. The state of the atmos-
        phere at any given time with regard to precipitation,
        temperature, humidity, cloudiness, wind movement, and

barometric pressure. Earth 2: p. 60g.

Wet Scrubber. National--Soil Conservation. An air cleaning de-
    vice that literally washes out the dust. Exhaust air is
    forced into a spray chamber, where fine water particles
    cause the dust to drop from the air stream. The dust-
    ladden water is then treated to remove the solid material
    and is often recirculated. Earth 2: p. 60g.

Wind Direction. Regional--Commission. The prevailing direction
    from which the wind is blowing expressed in 4, 8, 16, 32
    compass point direction. General 10: p. IV-8.

Wind Rose. Federal--ERDA. A diagram designed to show the distri-
    bution of prevailing wind directions at a given location;
    some variations include wind speed groupings by direction.
    General 7: p. g-11.

    Federal--GSA. Graph illustrating the frequency distribu-
    tion of wind direction and speed at a given location over
    a specified period of time. General 2: p. IV-22.

Wind Speed. Regional--Commission. The speed of the wind which
    is important in dispersal of pollutants expressed in
    knots or miles per hour. General 10: p. IV-8.

6. WATER

Index

Deep Percolation
Density Stratification
Depletion
Depletion Curve
Desalinization
Detention Dam
Detergent
Digotrophic
Dike
Dike Embankment
Dilution Ratio
Discharge
Dissolved Oxygen (DO)
Dissolved Solids
Distillation
Diversion
Diversion Dam
Doctrine Of Appropriation
Dominant Discharge
Drainage
Drainage Area
Drainage Basin
Drainage Class
Drainage Coefficient
Drainage Divide
Drainage System
Dredging
Drinking Water Standards
Dystrophic Lakes
Earth Dam
Ebb
Ebb Current
Ebb Tide
Effluent Stream
Enrichment
Ephemeral Stream
Epilmnion
Eulittoral
Euphotic
Eutrophication
Eutrophic Lakes
Evaporation
External Drainage
Fecal Coliform
Fecal Coliform Bacteria
Fecal Streptococcus
Flood
Flood Control
Flood Control Pool
Flood Control Project

Flood Duration Curve
Flood Forecasting
Flood Frequency
Flood Frequency Curve
Flood Fringe
Flood Peak
Flood Plain
Flood Plain Information Reports
Flood Plain Management
Flood Plain Regulation
Flood Routing
Flood Stage
Floodway
Floodway Retarding Structure
Flow Duration Curve
Flume
Flushing Rate
Forebay
Forebay Area
Freeboard
Free Groundwater
Fresh Water
Gaging Station
Gravity Dam
Groundwater
Groundwater Basin
Groundwater Flow
Groundwater Runoff
Hardness
Hard Water
Headwater
Headwaters Lake
Heavy Metals
High-Water Line
Hydraulic Gradient
Hydrogeology
Hydrograph
Hydrography
Hydrologic Budget
Hydrologic Cycle
Hydrologic Group
Hydrology
Hypolimnion
Impervious Surfaces
Impoundment
Infiltration
Intermittent Stream
Internal Drainage
Interstate Waters
Intertidal

Intertidal Zone
Jackson Turbidity Units
    (JTU's)
Jetty
Lacustrine
Lagoon
Lake
Lakescape
Lake Turnover
Lake Zonation
Leachate
Leaching
Levee
Limnology
Lotic
Low Flow Frequency Curve
Mass Diagram
Maximum Probable Flood
Mean Depth
Mean High Water (MHW)
Mean High Water Line
Mean Lower Low Water
Mean Low Water (MLW)
Mean Velocity
Meromictic Lake
Movable Dam
Natural Flow
Natural Runoff
Natural System
Nitrogen
Nonpoint Source
Nonuniform Flow
Nutrients
Observation Well
Observed Runoff
Oil Spill
Oligotrophic
Oligotrophic Lakes
One Hundred Year Floodplain
Operational Losses
Outfall
Outlet
Overturn
Perched Water
Perched Water Table
Percolation
Perennial Stream
pH
Phosphates
Photic Zone

Piezometer
Piezometric Surface
Pond
Potable Water
Rainfall Intensity
Rain Shadow
Reasonable-Use Rule
Receiving Waters
Recharge
Recharge Area
Recharge Of An Aquifer
Recharge Well
Regime
Regional Entity
Regulated Flow
Regulatory Floodway
Relief Well
Reservoir
Retention
Retention Dams
Revetment
Riffle
Riffle-Pool Concentration
Riffle Zone
Riparian
Riparian Rights
Riprap
River Basin
River Delta Estuary
Runoff
Runoff Plots
Safe Yield
Saline
Saline Contamination
Saline Water
Salinity
Salinity Wedge
Saltation Load
Salt Water Intrusion
Scenic River
Secchi Disk
Sediment Basin
Sediment Discharge
Sediment Load
Sediment Pool
Seepage
Settling Basin
Sheet Flow
Shoreline
Slough

Sluice
Spillway
Standard Project Flood
Stilling Basin
Stream
Streambanks
Streamflow
Streamflow Depletion
Stream Load
Stream Profile
Subirrigated Land
Surface Profile
Surface Runoff
Surface Water
Suspended Load
Suspended Solids
Tailwater
Thermal Pollution
Thermal Stratification
Thermocline
Tidal Basin
Tidal Flooding
Tidal Influence
Tidal Prism
Tidal Water
Tidal Wave
Tide
Time Of Concentration
Toxic Materials
Tributary
Turbidimeter
Turbidity
Turnover
Unconfined Aquifer
Underground Water
Underground Watercourse
Uniform Flow
Urban Runoff
Water Area
Water Classification
Water Disturbance
Water Penetration
Water Pollution
Water Quality
Water Quality Criteria
Water Quality Standards
Water Resources
Water Rights
Watershed
Watershed Area

Watershed Management
Watershed Planning
Watershed Project
Watershed Protection
Water Table
Water Year
Water Yield
Weir
Wild River Area

Terms

Acidity. Federal--GSA. Capacity of water to neutralize bases
        imparted by dissolved carbon dioxide or mineral acids;
        measure of the corrosiveness of water. General 2: p.
        IX-17.

Annual Flood. Regional--River Basin. The highest peak discharge
        in a water year. Water 5: p. 1015.

Annual Flood Series. Regional--River Basin. A list of annual
        floods. Water 5: p. 1011.

Approved Regional Plan. Federal--Water Resources Council. A com-
        prehensive water resource management plan adopted by a
        river basin commission or other Council-designated region-
        al entity, or by the Council following appropriate review
        and comment under statutory or other requirements of Fed-
        eral agencies, and in consultation with affected States.
        Such plan or revision adopted or approved in this manner
        shall be the "approved regional plan". The plan shall in-
        clude an evaluation of all reasonable alternative means
        of achieving development of water and related land re-
        sources of the region reflective of the region's prefer-
        ence between the two national objectives of National Eco-
        nomic Development and Environmental Quality. Legal Jar-
        gon 36: p. 1.

Aquatic. Federal--GSA. Growing, living in, or frequenting water.
        General 2: p. XI-27.

Aquifer. Federal--EPA. An underground bed or layer of earth,
        gravel, or porous stone that contains water. General 1:
        p. 2.

        Federal--ERDA. A subsurface formation containing suffi-
        cient saturated permeable material to yield significant
        quantities of water. General 7: p. g-1.

        Federal--GSA. Water-bearing stratum of permeable rock,
        sand, or gravel. General 2: p. VII-19.

        Federal--Interior. An underground bed or stratum of earth,
        gravel, or porous stone that contains water. A geological
        rock formation, bed, or zone that may be referred to as a
        water-bearing bed. General 4: Glossary, p. 1.

        Federal--Interior. A porous soil or geological formation
        lying between impermeable strata in which water may move

for long distances; yields ground water to springs and wells. General 3: p. 7.

National--Engineering. Underground water-bearing geologic formation or structure. Water 1: no page number.

Regional--River Basin. A geologic formation that is water-bearing and that transmits water from one point to another. Energy/Utility 7: p. 255.

Regional--River Basin. A rock formation, bed, or zone containing water that is available to wells. An aquifer may be referred to as a water-bearing formation or water-bearing bed. Water 5: p. 1011.

State--Water Resources. A porous, water-bearing geologic formation. Generally restricted to materials capable of yielding an appreciable supply of water. Energy/Utility 2: Appendix E.

Artesian Aquifer. National--Engineering. Aquifer that contains water under artesian pressure. Water 1: no page number.

Artesian Pressure. National--Engineering. Pressure within a groundwater aquifer developed as a result of hydrostatic head. Water 1: no page number. (Revised).

Artesian Water. Regional--River Basin. Ground water under sufficient pressure to rise above the level at which the water-bearing bed is reached in a well. The pressure in such an aquifer commonly is called artesian pressure, and the rock containing artesian water is an artesian aquifer. Water 5: p. 1011.

Artificial Recharge. State--Water Resources. Replenishment of the groundwater supply by means of spreading basins, recharge wells, irrigation, or induced infiltration of surface water. Energy/Utility 2: Appendix E.

Assimilation. Federal--EPA. The ability of a body of water to purify itself of pollutants. General 1: p. 2.

Assimilative Capacity. State--Water Resources. The capacity of a natural body of water to receive: (1) wastewater, without deleterious effects, (2) humans who consume the water, and (3) BOD, within prescribed dissolved oxygen limits. Energy/Utility 2: Appendix E.

Available Oxygen. State--Water Resources. The quantity of dissolved oxygen available for oxidation of organic matter in a water body. Energy/Utility 2: Appendix E.

Average Annual Flow. State--Natural Resources. The mean volume
        of water passing a given point during a one year period.
        General 12: p. 335.

Average Daily Flow. State--Water Resources. The total quantity
        of liquid tributary to a point divided by the number of
        days of flow measurement. Energy/Utility 2: Appendix E.

Average Flow. State--Water Resources. Arithmetic average of
        flows measured at a given point. Energy/Utility 2: Appen-
        dix E.

Bank Storage. National--Engineering. Water entering the banks
        of stream channels during high stages of stream flow,
        most of which returns to stream flow during falling
        stages. Water 1: no page number.

        National--Soil Conservation. Water absorbed by the bed
        and banks of a stream, reservoir, or channel and returned
        in whole or in part as the water level falls. Earth 4:
        p. 8g.

Base Flood. Federal--Water Resources. That flood which has a
        one percent chance of occurrence in any given year--also
        known as a 100-year flood. This term is used in the Na-
        tional Flood Insurance Program (NFIP) to indicate the
        minimum level of flooding to be used by a community in its
        floodplain management regulations. Water 6: p. 5.

Base Flow. Federal--DOT. The amount of flow in a river which is
        maintained by groundwater inflow to the river and is,
        therefore, relatively constant even during dry periods.
        Transportation 3: p. 15.

Base Runoff. Regional--River Basin. Sustained or fair weather
        runoff. In most streams, base runoff is composed largely
        of groundwater effluent. The term base flow is often used
        in the same sense as base runoff. However, the distinc-
        tion is the same as that between streamflow and runoff.
        When the concept in the terms base flow and base runoff
        is that of the natural flow in a stream, base runoff is
        the logical term. Water 5: p. 1011.

Basin. Regional--River Basin. A geographic area drained by a
        major stream. Energy/Utility 8: p. 527.

Bathymetry. Federal--Commerce. The measurement of depths of
        water areas; underwater topography. General 6: p. 310.

Bed Load. <u>National--Engineering</u>. Coarse material moving on or
    near the bed of a flowing stream. Water 1: no page num-
    ber.

    <u>State--Natural Resources</u>. Sediments that move along a
    stream bed. General 12: p. 335.

Benthal Deposit. <u>State--Water Resources</u>. Accumulation on the
    bed of a water-course of deposits containing organic
    matter arising from natural erosion or discharges of
    wastewaters. Energy/Utility 2: Appendix E.

Benthic. <u>Federal--Commerce</u>. Occurring or living on or in the
    bottom of a water body. General 6: p. 310.

Benthic Region. <u>Federal--EPA</u>. The bottom layer of a body of
    water. General 1: p. 2.

    <u>Federal--GSA</u>. Bottom of a body of water. General 2:
    p. IX-17.

Biochemical Oxygen Demand (BOD). <u>Federal--EPA</u>. The dissolved
    oxygen required to decompose organic matter in water. It
    is a measure of pollution since heavy waste loads have a
    high demand for oxygen. General 1: p. 2.

    <u>Federal--GSA</u>. Amount of dissolved oxygen that the bio-
    chemical breakdown of organic matter removes from the
    water. General 2: p. IX-17.

    <u>Federal--Interior</u>. Represents the amount of dissolved
    oxygen that will be required from water during the bacter-
    ial assimilation of organic pollutants. The difference
    in oxygen concentration of a water sample after 5 days of
    incubation. General 3: p. 12.

    <u>Federal--Interior</u>. The amount of oxygen required to de-
    compose a given amount of organic compounds to simple,
    stable substances within a specified time at a specified
    temperature. BOD is an index of the degree of organic
    pollution in water. General 4: Glossary, p. 1.

    <u>National--Resources</u>. A measure of the amount of oxygen
    used by microorganisms to break down organic waste mater-
    ials in water. Energy/Utility 5: p. 3.

    <u>Regional--Commission</u>. The quantity of oxygen utilized in
    the biochemical oxidation of organic matter in a specified
    time and at a specified temperature. It is not related to
    the oxygen requirements in chemical combustion, but is

determined entirely by the availability of the material
as a biological food and by the amount of oxygen utilized
by the micro-organisms during oxidation. General 13: p.
587.

State--Natural Resources. The quantity of oxygen utilized
in the bio-chemical oxidation of organic matter in a specified time and at a specified temperature. General 12:
p. 335.

City--Planning Department. An abbreviation for biochemical oxygen demand; BOD_5 is the quantity of oxygen used in
the biochemical oxidation of organic matter in a five-day
period, at a specified temperature, and under specified
conditions. This is a standard test used to assess water
and wastewater quality. Energy/Utility 9: p. 1.

Biological Oxygen Demand 5 (BOD_5). Federal--EPA. The amount of
dissolved oxygen consumed in 5 days by biological processes
breaking down organic matter in an effluent. General 1:
p. 3.

Bloom. Federal--EPA. A proliferation of algae and/or higher
aquatic plants in a body of water, often related to pollution. General 1: p. 3.

Federal--Interior. A concentrated growth of phytoplankton
(algae). General 3: p. 13.

State--Water Resources. Large masses of microscopic and
macroscopic plant life, such as green algae, occurring in
bodies of water. Energy/Utility 2: Appendix E.

Brackish Water. Federal--Army. Water, salty between the concentrations of fresh water and sea water; usually 5-10 parts
per thousand. General 9: p. 7.

Federal--EPA. A mixture of fresh and salt water. General
1: p. 3.

Regional--River Basin. Water having a mineral content in
the general range between fresh and sea water. Water containing from 500 to 10,000 mg/1 of dissolved solids. Water
2: p. I-21.

Braided. State--Natural Resources. In reference to streams,
having diverging and converging channels separated by
islands and bars. General 12: p. 336.

Braided Stream. <u>Federal--DOT</u>. Composed of a series of anastamos-
      ing (connecting) channels separated by islands. They are
      usually formed in loose, readily shifted materials such as
      glacial debris or moraine gravel. Transportation 3: p. 17.

Canal. <u>National--Engineering</u>. Constructed open channels of an
      irrigation system for transporting water from the source
      of supply to the point of use. Water 1: no page number.

Capillary Soil Moisture. <u>National--Engineering</u>. Moisture in soil
      held by surface tension forces against the force of grav-
      ity. Water 1: no page number.

Catch Basin. <u>National--Engineering</u>. Basin designed to catch sed-
      iment. Water 1: no page number.

Catchment Basin. <u>Federal--Interior</u>. A unit watershed; an area
      from which all the drainage water passes into one stream
      or other body of water. General 3: p. 17.

Channel. <u>Federal--Water Resources</u>. A natural or artificial water-
      course of perceptible extent, with a definite bed and banks
      to confine and conduct continuously or periodically flowing
      water. Water 6: p. 5.

      <u>National--Soil Conservation</u>. A natural stream that con-
      veys water; a ditch or channel excavated for the flow of
      water. Earth 2: p. 12g.

Channel Improvement. <u>National--Soil Conservation</u>. The improve-
      ment of the flow characteristics of a channel by clearing,
      excavation, realignment, lining, or other means in order
      to increase its capacity; sometimes used to connote chan-
      nel stabilization. Earth 2: p. 12g.

Channelization. <u>Federal--EPA</u>. To straighten and deepen streams
      so water will move faster, a flood reduction or marsh
      drainage tactic that can interfere with waste assimilation
      capacity and disturb fish habitat. General 1: p. 3.

      <u>Federal--Interior</u>. The alteration of a natural stream by
      excavation, realignment, lining or other means to acceler-
      ate the flow of water. Recreation 3: p. 6.

Channel Stabilization. <u>National--Soil Conservation</u>. Erosion pre-
      vention and stabilization of velocity distribution in a
      channel using jetties, drops, revetments, vegetation, and
      other measures. Earth 2: p. 12g.

Check Dam. <u>National--Soil Conservation</u>. Small dam constructed in a gully or other small watercourse to decrease the stream-flow velocity, minimize channel scour, and promote deposition of sediment. Earth 2: p. 12g.

Chemical Oxygen Demand (COD). <u>Federal--EPA</u>. A measure of the oxygen required to oxidize all compounds in water, organic and inorganic. General 1: p. 3.

> <u>Federal--ERDA</u>. A measure of the extent to which all chemicals contained in a water sample use dissolved oxygen in a given period of time; therefore, a measure of residual dissolved oxygen in the water available for use by organisms such as fish. General 7: p. g-1.

> <u>Federal--GSA</u>. Total quantity of oxygen required for oxidation of organic matter to carbon dioxide and water regardless of the biological assimilability of the substances. General 2: p. IX-18.

> <u>Federal--Interior</u>. A measure of the oxygen equivalent which is required for the oxidation of an organically polluted water supply. General 3: p. 18.

> <u>Regional--River Basin</u>. The quantity of oxygen utilized in the chemical oxidation of organic matter. It is a measure of the amount of such matter present. Energy/Utility 7: p. 255.

Civil-Law Drainage Rule. <u>National--Engineering</u>. Drainage law stating that the owner of higher land is entitled to the natural advantage which the elevation of his land gives him and that lower lying land must receive surface water flowing to him through natural channels. Water 1: no page number.

Coefficient Of Discharge. <u>National--Engineering</u>. Ratio of observed to theoretical discharge. Water 1: no page number.

Coefficient Of Roughness. <u>National--Engineering</u>. Factor in fluid flow or formulas expressing the character of a channel surface and its frictional resistance to flow. Water 1: no page number.

Cofferdam. <u>Federal--Interior</u>. A barrier constructed in a body of water to form an enclosure from which the water can be pumped to permit free access to the area within. General 3: p. 19.

City--Planning. Watertight enclosure from which water has been pumped to expose the bottom of a body of water to permit construction. General 8: p. 1.

Coliform. Federal--ERDA. A measure of the bacterial content of water; a high coliform count indicates potential contamination of a water supply by human waste. General 7: p. g-2.

Federal--Interior. A general term for the group of bacteria which comprise all of the aerobic and facultavely anaerobic, gram-negative (type of stain related to cell wall composition), nonspore-forming, rod-shaped bacteria which ferment lactose (milksugar) with gas formation within 48 hours at 35 degrees C. Examples: Escherichia, Citrobacter, Klebsiella. General 3: pp. 19-20.

State--Natural Resources. Bacteria found in human and animal feces, indicative of organic pollution. General 12: p. 336.

Coliform Bacteria. Regional--River Basin. A species of genus escherichia bacteria, normal inhabitant of the intestine of man and all vertebrates. Energy/Utility 7: p. 255.

Coliform-Group Bacteria. State--Water Resources. A group of bacteria predominantly inhabiting the intestines of man or animal, but also occasionally found elsewhere. It includes all aerobic and facultative anaerobic, Gram-negative, non-spore-forming bacilli that ferment lactose with production of gas. Also included are all bacteria that produce a dark, purplish-green colony with metallic sheen by the membrane-filter technique used for coliform identification. The two groups are not always identical, but they are generally of equal sanitary significance. Energy/Utility 2: Appendix E.

Coliform Index. Federal--EPA. A rating of the purity of water based on a count of fecal bacteria. General 1: p. 4.

Coliform Organisms. Federal--EPA. Organisms found in the intestinal tract of humans and animals, their presence in water indicates pollution and potentially dangerous bacterial contamination. General 1: p. 4.

Coliforms. Federal--GSA. Group of bacteria whose presence in water indicates recent sanitary pollution or inadequate disinfection of domestic sewage. General 2: p. IX-18.

City--Planning Department. A diverse group of bacteria, some of which normally inhabit human and animal intestinal tracts. Used as an indicator of fecal pollution of water and hence of the probability of presence of organisms causing human disease. Energy/Utility 9: p. 1.

Common Enemy Drainage Rule. National--Engineering. Common-law rule of legal water rights to the effect that surface water is a common enemy and a landowner may lawfully protect his land from water flowing from adjoining higher land. Water 1: no page number.

Cone Of Depression. National--Engineering. Depression, roughly conical in shape, produced in a water table or piezometric surface by the extraction of water from a well. Water 1: no page number.

Confined Aquifer. Federal--ERDA. A subsurface water bearing region having defined relatively impermeable upper and lower boundaries and whose pressure is significantly greater than atmospheric throughout. General 7: p. g-1.

Confined Water. National--Engineering. Groundwater, constrained by an overlying confining bed, which is under sufficient pressure to rise above the bottom of the confining bed. Water 1: no page number.

Regional--River Basin. Water under artesian pressure. Water that is not confined is said to be under water table conditions. Water 5: p. 1011.

Conservation Pool. Federal--Interior. The permanent water storage volume provided in a reservoir. Recreation 3: p. 7.

Consistency. Federal--Water Resources Council. Means that Federal water and related resource activities are carried out in agreement with regional plans to the maximum extent practicable. Legal Jargon 36: p. 1.

Consumptive Use. Regional--River Basin. The quantity of water discharged to the atmosphere or incorporated in the products in the process of vegetative growth, food processing, industrial processes, or other uses. Water 5: p. 1012.

Contamination. State--Water Resources. Any introduction into water of microorganisms, chemicals, wastes, or wastewater in a concentration that makes the water unfit for its intended use. Energy/Utility 2: Appendix E.

Critical Action. Federal--Water Resources. Any activity for which
     even a slight chance of flooding would be too great.
     Water 6: p. 5.

Critical Velocity. National--Soil Conservation. Velocity at which
     a given discharge changes from tranquil to rapid flow;
     that velocity in open channels for which the specific en-
     ergy (sum of the depth and velocity head) is a minimum for
     a given discharge. Earth 2: p. 16g.

Cultural Eutrophication. Federal--EPA. Increasing the rate at
     which water bodies "die" by pollution from human activ-
     ities. General 1: p. 4.

Current Meter. National--Soil Conservation. An instrument used
     for measuring the velocity of flowing water. The velocity
     of the water is proportional to the revolutions per unit
     of time of the propeller, vane, or wheel of the meter.
     Earth 2: p. 17g.

Dam. National--Soil Conservation. A barrier to confine or raise
     water for storage or diversion, to create a hydraulic head,
     to prevent gully erosion, or for retention of soil, rock,
     or other debris. Earth 2: p. 17g.

Deep Percolation. National--Engineering. Water which percolates
     below the root zone and cannot be used by plants. Water
     1: no page number.

Density Stratification. National--Soil Conservation. The arrange-
     ment of water masses into separate, distinct horizontal
     layers as a result of differences in density; may be caused
     by differences in temperature or dissolved and suspended
     solids. Earth 4: p. 18g.

Depletion. Federal--Interior. Total loss of water from the stream
     due to consumptive uses, evaporation, seepage, and evapo-
     transpiration. General 3: p. 24.

     Regional--River Basin. That portion of water supply that
     is consumptively used, beneficially or nonbeneficially.
     Water 4: p. 341.

     State--Water Resources. The continued withdrawal of water
     from a stream or from a surface or groundwater reservoir
     or basin at a rate greater than the rate of replenishment.
     Energy/Utility 2: Appendix E. (Revised).

Depletion Curve. Federal--EPA. A graphical representation of
     water depletion from storage-stream channels, surface soil

and groundwater. A depletion curve can be drawn for base
flow, direct runoff, or total flow. General 1: p. 5.

Desalinization. Federal--EPA. Removing salt from ocean or brackish water. General 1: p. 5.

Detention Dam. Federal--Forest. A dam built to store streamflow
or surface runoff, and to control the release of such
stored water. General 3: p. 24.

Detergent. Federal--EPA. Synthetic washing agent that helps water
to remove dirt and oil. Most contain large amounts of
phosphorus compounds which may kill useful bacteria and
encourage algae growth in the receiving water. General
1: p. 5.

State--Water Resources. Any of a group of synthetic, organic, liquid or water-soluble cleaning agents that are
inactivated by hard water and have wetting-agent and emul-
sifying-agent properties but, unlike soap, are not prepared from fats and oils. Energy/Utility 2: Appendix E.
(Revised).

Digotrophic. Federal--DOT. Deep lakes which have a low supply
of nutrients. Thus they support very little organic production. Dissolved oxygen is present at or near saturation throughout the lake during all seasons. Transportation 3: p. 26.

Dike. State--Water Resources. An embankment constructed to prevent overflow of water from a stream or other body of
water. An embankment constructed to retain water in a
reservoir. The term "dam" is usually used for a structure
constructed across a watercourse or stream channel, and
"dike" for one constructed solely on dry ground. Energy/
Utility 2: Appendix E.

Dike Embankment. Federal--Interior. An embankment of earth for
restraining the waters of a river. General 3: p. 25.

Dilution Ratio. Federal--EPA. The relationship between the volume
of water in a stream and the volume of incoming waste. It
can affect the ability of the stream to assimilate waste.
General 1: p. 5.

Discharge. Regional--River Basin. In its simplest concept, discharge means outflow; therefore, the use of this term is
not restricted as to course or location and it can be used
to describe the flow of water from a pipe or a drainage
basin. Water 5: p. 1013.

State--Water Resources. As applied to a stream or conduit, the rate of flow, or volume of water flowing in the stream or conduit at a given place and within a given period of time. The passing of water or other liquid through an opening or along a conduit or channel. The rate of flow of water, silt, or other mobile substance which emerges from an opening, pulp, or turbine, or passes along a conduit or channel, usually expressed as cubic feet per second, gallons per minute, or million gallons per day. Energy/Utility 2: Appendix E.

Dissolved Oxygen (DO). Federal--Army. An amount of gaseous oxygen dissolved in volume of water. General 9: p. 12.

Federal--EPA. A measure of the amount of oxygen available for biochemical activity in a given amount of water. Adequate levels of DO are needed to support aquatic life. Low dissolved oxygen concentrations can result from inadequate waste treatment. General 1: p. 5.

Federal--GSA. Concentration of oxygen in water available for aquatic life and chemical processes. General 2: p. IX-18.

Federal--Interior. The oxygen dissolved in water, necessary for the life of fish and other aquatic organisms. General 4: Glossary, p. 4.

Federal--Interior. Perhaps the most commonly employed measurement of water quality. Low DO levels adversely affect fish and other aquatic life. The total absence of DO will lead to the development of an anaerobic condition with the eventual development of odor and esthetic problems. Ideal DO for fish life is between 7 to 9 mg/l. Critical levels of DO, for nearly all fish, are between 3 and 6 mg/l. Most fish cannot survive when DO falls below 3 mg/l. General 3: p. 26.

Regional--Commission. The oxygen dissolved in a stream, sewage effluent or other water, usually expressed in milligrams per liter or percent of saturation. General 13: p. 587.

State--Water Resources. The oxygen dissolved in water, wastewater, or other liquid, usually expressed in milligrams per liter, parts per million, or percent of saturation. Energy/Utility 2: Appendix E.

City--Planning Department. Water quality factor which partially determines the water's capability to support life. Energy/Utility 9: p. 1.

Dissolved Solids. <u>Federal--EPA</u>. The total of disintegrated or-
     ganic and inorganic material contained in water. Excesses
     can make water unfit to drink or use in industrial pro-
     cesses. General 1: p. 5.

     <u>Federal--GSA</u>. Inorganic salts, small amounts of organic
     matter, and dissolved gases; amount of dissolved solids
     determines the hardness of water. General 2: p. IX-18.

     <u>Federal--Interior</u>. The total amount of dissolved material,
     organic and inorganic, contained in water and wastes. Ex-
     cessive dissolved solids can make water unsuitable for in-
     dustrial uses, unpalatable for drinking, and even cathar-
     tic. General 4: Glossary, p. 4.

     <u>Regional--River Basin</u>. Solids that are present in water
     in solution--solids that cannot be removed by filtering.
     Water 2: p. I-21.

Distillation. <u>Federal--EPA</u>. Purifying liquids through boiling.
     The steam condenses to pure water and pollutants remain in
     a concentrated residue. General 1: p. 5.

Diversion. <u>Regional--River Basin</u>. The taking of water from a
     stream or other body of surface water into a canal, pipe
     line, or other conduit. Water 2: p. I-21.

Diversion Dam. <u>Federal--Interior</u>. A barrier built across a stream
     to divert all or some of the water. General 3: p. 26.

Doctrine Of Appropriation. <u>National--Engineering</u>. Legal doctrine
     of water rights which asserts that all rights are based on
     use. Water 1: no page number.

Dominant Discharge. <u>State--Natural Resources</u>. The flood flow,
     occurring on the average about two out of three years,
     which transports the most sediment. General 12: p. 336.

Drainage. <u>Federal--GSA</u>. Rapidity and extent of the removal of
     water from soil in terms of surface runoff, permeability,
     and internal drainage. These factors, along with slope,
     influence the wetness of a soil and the depth at which the
     water table will be located for most of the year. Soil
     drainage is described in seven classes: (1) Very poor.
     Water removed slowly; water table at or near ground sur-
     face for most of the year; usually occurs on level or de-
     pressed sites that are frequently ponded; (2) Poor. Gen-
     erally wet; water table at or near surface for considerable

part of the year due to high water table, impermeable
layers; and/or seepage; (3) Moderate. Water removed slowly; wet for significant periods; soil at depth slowly
permeable, has high water table and/or seepage problems;
(4) Moderately good. Water removed fairly slowly due to
slow permeability near surface, relatively high water
table, or seepage; (5) Good. Water readily, but not rapidly removed; (6) Very good. Water removed rapidly, particularly in sand soils; (7) Excessive. Water removed very
rapidly due to steep slope and/or high permeability, danger
of groundwater contamination. General 2: p. VIII-28.

Drainage Area. Regional--River Basin. The drainage area of a
stream, measured in a horizontal plane, which is enclosed
by a drainage divide. Water 5: p. 1013.

Drainage Basin. Regional--River Basin. A part of the surface of
the earth that is occupied by a drainage system, which consists of a surface stream or body of impounded surface
water together with all tributary surface streams and bodies of impounded surface water. Water 5: p. 1013.

State--Water Resources. An area from which surface runoff
is carried away by a single drainage system. Also called
catchment area, watershed, drainage area. The largest
natural drainage area subdivision of a continent. The
United States has been divided at one time or another,
for various administrative purposes, into some 12 to 18
drainage basins. Energy/Utility 2: Appendix E.

Drainage Class. Regional--River Basin. The relative terms used
to describe natural drainage are explained as follows:
(1) Excessive--very porous and rapidly permeable, and have
low water-holding capacity, (2) Somewhat Excessive--very
permeable and are free from mottling throughout their profile, (3) Good--well drained soils that are nearly free
of mottling and are commonly of intermediate texture, (4)
Moderately Good--moderately well drained soils that commonly have a slowly permeable layer in or immediately beneath
the solum. They have uniform color in the surface layers
and upper subsoil, and mottling in the lower subsoils and
substrata, (5) Somewhat Poor--wet for significant periods,
but not all the time. They commonly have a slowly permeability layer in the profile, a high water table, additions through seepage, or a combination of these conditions, (6) Poor--wet for long periods of time. They are
light gray and generally mottled from the surface downward, although mottling may be absent or nearly so in some
soils. Earth 1: p. 373.

Drainage Coefficient. <u>National--Engineering</u>. Design rate at which
      water is to be removed from a drainage area. It may be ex-
      pressed in inches of depth per day or in terms of flow rate
      per unit of area. It may also be expressed in terms of
      flow rate per unit of area, which rate varies with the
      total size of the area. Sometimes called drainage modulus.
      Water 1: no page number.

Drainage Divide. <u>Regional--River Basin</u>. The line of highest el-
      evation which separates adjoining drainage basins. Water
      5: p. 1013.

Drainage System. <u>National--Engineering</u>. Collection of open and/
      or closed drains, together with structures and pumps used
      to collect and dispose of excess surface or subsurface
      water. Water 1: no page number.

Dredging. <u>Federal--EPA</u>. To remove earth from the bottom of
      water bodies using a scooping machine. This disturbs the
      ecosystem and causes silting that can kill aquatic life.
      General 1: p. 5.

Drinking Water Standards. <u>State--Water Resources</u>. Standards pre-
      scribed by the U.S. Public Health Service for the quality
      of drinking water supplied to interstate carriers. Stan-
      dards prescribed by state or local jurisdictions for the
      quality of drinking water supplied from surface-water,
      groundwater, or bottled-water sources. Energy/Utility 2:
      Appendix E.

Dystrophic Lakes. <u>Federal--EPA</u>. Shallow bodies of water that
      contain much humus and organic matter. They contain many
      plants but few fish and are almost eutrophic. General 1:
      p. 5.

      <u>Federal--Interior</u>. Lakes with a very low lime content and
      high humus content resulting in a brown color of the water.
      The lakes are generally nutrient poor. General 3: p. 27.

      <u>National--Soil Conservation</u>. Shallow lakes with brown
      water, high organic matter content, low nutrient avail-
      ability, poor bottom fauna, and high oxygen demand; ox-
      ygen is continually depleted and pH is usually low. In
      lake aging, the age between a eutrophic lake and a swamp.
      Earth 4: p. 20g.

Earth Dam. <u>National--Soil Conservation</u>. Dam constructed of com-
      pacted soil materials. Earth 2: p. 20g.

Ebb. State--Water Resources. The flowing out of the tide, away
      from the shore downstream; the return of the tidal wave to
      the sea. Energy/Utility 2: Appendix E.

Ebb Current. State--Water Resources. A current in a body of water
      affected by the tide that flows seaward or downstream. En-
      ergy/Utility 2: Appendix E.

Ebb Tide. National--Soil Conservation. That period of tide be-
      tween a high water and the succeeding low water; falling
      tide. Earth 2: p. 20g.

Effluent Stream. Regional--River Basin. A stream or reach of
      stream fed by ground water. It is also called a gaining
      stream. Water 5: p. 1021.

      State--Water Resources. A stream or stretch of stream
      which receives water from groundwater in the zone of sat-
      uration. The water surface of such a stream stands at a
      lower level than the water table or piezometric surface
      of the groundwater body from which it receives water.
      Energy/Utility 2: Appendix E.

Enrichment. Federal--EPA. Sewage effluent or agricultural runoff
      adding nutrients (nitrogen, phosphorus, carbon compounds)
      to a water body, greatly increasing the growth potential
      for algae and aquatic plants. General 1: p. 6.

Ephemeral Stream. Federal--Forest Service. In areas where precip-
      itation almost totally consists of rainfall, the term is
      commonly used to refer to the short-lived streams which
      flow for only a very short time (a few days at most) after
      each storm event. General 15: p. 74.

      Regional--River Basin. A stream that flows only in re-
      sponse to precipitation. Water 5: p. 1012.

Epilmnion. Federal--Army. The turbulent superficial layer of a
      lake between the surface and a horizontal plane marked by
      the maximum gradient of temperature and density change.
      General 9: p. 18.

      Federal--GSA. Warmer upper water zone in a stratified
      lake, extending from the surface to the thermocline. Gen-
      eral 2: p. IX-18.

      Federal--Interior. The upper, warmer portion of a lake,
      separated from the hypolimnion by a thermocline. General
      3: p. 31.

Eulittoral. <u>Federal--Interior</u>. That area of the shoreline lying
        between the minimum and maximum yearly lake level fluctu-
        ation. General 3: p. 32.

Euphotic. <u>Federal--Army</u>. Of the upper layers of water in which
        sufficient light penetrates to permit growth of green
        plants. General 9: p. 19.

        <u>Federal--Interior</u>. Relating to the upper, well-illumina-
        ted zone of a lake where photosynthesis occurs. General
        3: p. 32.

Eutrophication. <u>Federal--EPA</u>. The slow aging process of a lake
        evolving into a marsh and eventually disappearing. During
        eutrophication the lake is choked by abundant plant life.
        Human activities that add nutrients to a water body can
        speed up this action. General 1: p. 6.

        <u>Federal--GSA</u>. Natural maturing process in a lake; may be
        accelerated by pollution-contributing nutrients that stim-
        ulate increased production of organic matter. General 2:
        p. IX-18.

        <u>Regional--Commission</u>. The process of production of greater
        amounts of organic matter in a body of water than can be
        consumed through existing biologic oxidization processes.
        This condition may be caused by natural or artificial fert-
        ilization in conjunction with other growth factors. Gen-
        eral 13: p. 588.

        <u>Regional--River Basin</u>. The process of overfertilization
        of a body of water by nutrients which produce more organic
        matter than self-purification processes can overcome.
        Energy/Utility 7: p. 255.

Eutrophic Lakes. <u>Federal--EPA</u>. Shallow murky water bodies that
        have lots of algae and little oxygen. General 1: p. 6.

        <u>State--Water Resources</u>. Lake or other contained water body
        rich in nutrients. Characterized by a large quantity of
        planktonic algae, low water transparency with high dis-
        solved oxygen in upper layer, zero dissolved oxygen in
        deep layers during summer months, and large organic de-
        posits colored brown to black. Hydrogen sulfide often
        present in water and deposits. Energy/Utility 2: Appen-
        dix E.

Evaporation. <u>Regional--River Basin</u>. The process by which water
        passes from a liquid state to vapor, the principal process

by which water is converted to atmospheric vapor, either
naturally from surface streams, moist soil, or other moist
surface, or artificially from cooling devices. Water 2:
p. I-21.

External Drainage. Federal--Interior. The movement of water
across the surface of the land to outlets such as natural
stream channels or waterways. General 3: p. 33.

Fecal Coliform. Federal--Interior. Same description as coliform
except fecal coliforms are grown in a water bath at $44.5^{\pm}$
$0.2C$ for $24^{\pm}2$ hours with the production of gas. Example:
Escherichia coli. General 3: p. 34.

Fecal Coliform Bacteria. Federal--EPA. A group of organisms
found in the intestinal tracts of people and animals.
Their presence in water indicates pollution and possible
dangerous bacterial contamination. General 1: p. 6.

Fecal Streptococcus. Federal--Interior. Bacteria of the intestinal tract characterized by ability to grow at relatively
high pH and temperature. Used as an indicator of recent
fecal pollution by warm blooded animals, including man.
General 3: p. 34.

Flood. Federal--Water Resources. A general and temporary condition of partial or complete inundation of normally dry
land areas from the overflow of inland and/or tidal waters,
and/or the unusual and rapid accumulation or runoff of
surface waters from any source. Water 6: p. 5.

Regional--River Basin. Any relatively high streamflow or
an overflow or inundation that comes from a river or other
body of water and causes or threatens damage. Water 5:
p. 1014.

Regional--River Basin. A great flow along a watercourse
or a flow causing inundation of lands not normally covered
by water. Water 3: p. 393.

Flood Control. National--Soil Conservation. Methods or facilities
for reducing flood flows. Earth 2: p. 23g.

Flood Control Pool. Federal--Interior. Reservoir volume above
the conservation or joint-use pool that is kept empty to
catch flood runoff and then evacuated as soon as possible
to keep it in readiness for the next flood. General 3:
p. 35.

Flood Control Project. <u>National--Soil Conservation</u>. A structural
    system installed for protection of land and improvements
    from floods by the construction of dikes, river embank-
    ments, channels and dams. Earth 2: p. 24g.

Flood Duration Curve. <u>Regional--River Basin</u>. A cumulative fre-
    quency curve that shows the percentage of time that spec-
    ified discharges are equaled or exceeded. Water 5:
    p. 1015.

Flood Forecasting. <u>Regional--River Basin</u>. Flood forecasts are
    primarily the responsibility of the National Weather Ser-
    vice, National Oceanic and Atmospheric Administration, and
    are used to predict flood stages and times and indicate
    areas subject to flooding. Water 3: p. 393.

Flood Frequency. <u>Regional--River Basin</u>. The average interval of
    time between floods equal to or greater than a specified
    discharge or stage. It is generally expressed in years.
    Water 3: p. 393.

Flood Frequency Curve. <u>Regional--River Basin</u>. A graph showing
    the number of times per 100 years, or the average interval
    of times within which a flood of a given magnitude will
    be equaled or exceeded. Water 5: p. 1015.

Flood Fringe. <u>Federal--Water Resources</u>. That portion of the
    floodplain outside of the regulatory floodway--often re-
    ferred to as "floodway fringe". Water 6: p. 5.

Flood Peak. <u>Regional--River Basin</u>. The highest value of the
    stage or discharge attained by a flood; peak stage or peak
    discharge. Flood crest has nearly the same meaning but,
    since it connotes the top of the flood wave, it is prop-
    erly used only in referring to stage. Water 5: p. 1015.

Flood Plain. <u>Federal--Forest Service</u>. The extent of a flood
    plain obviously fluctuates with the size of overbank
    stream flows. Thus, no simple, absolute flood plain
    commonly exists. As a consequence, flood plains are de-
    lineated in terms of some specified flood size (e.g., the
    50-year flood plain, the area that would be flooded by
    the largest stream flow that will, on the average, occur
    once within a 50-year period). Such expected flood-return
    frequencies are estimated from historic records of stream
    flows. The largest, absolute flood plain that is ever
    likely to occur is sometimes referred to as the flood
    basin. General 15: p. 80.

Federal--Water Resources. The lowlands and relatively
flat areas adjoining inland and coastal waters including
flood-prone areas of offshore islands, including at a minimum, that area subject to a one percent or greater chance
of flooding in any given year. The base floodplain shall
be used to designate the 100-year floodplain (one percent
chance floodplain). The critical action floodplain is defined as the 500-year floodplain (0.2 percent chance
floodplain). Water 6: p. 5.

Regional--River Basin. A strip of relatively smooth land
bordering a stream that has been or is subject to flooding.
It is called a "living" flood plain if it is overflowed in
times of high water, but a "fossil" flood plain if it is
beyond the reach of the highest flood. Water 5: p. 1015.

Regional--River Basin. Land bordering a stream and which
receives overbank flow or all lands subject to inundation.
Water 3: p. 394.

City--Planning. Nearly level land, consisting of stream
sediment, that borders a stream and is subject to flooding
unless protected artificially. General 5: p. 55.

Flood Plain Information Reports. Regional--River Basin. Reports
prepared to provide local governmental agencies with basic
technical data to assist in planning for wise use and development of their flood plains. Water 3: p. 394.

Flood Plain Management. Regional--Commission. Comprehensive flood
damage prevention program which requires integration of all
alternative measures (structural and nonstructural) in investigation of flood problems and planning for wise use of
the flood plain. General 13: p. 588.

Flood Plain Regulation. Regional--River Basin. A general term
applied to the full range of codes, ordinances, and other
regulations relating to the use of land, water, and construction within a channel or flood plain area. Water 3:
p. 394.

Flood Routing. Regional--River Basin. The process of determining
progressively downstream the timing and stage of a flood
at successive points along a river. Water 5: p. 1015.

Flood Stage. Regional--River Basin. The stage at which overflow
of the natural banks of a stream begins to cause damage in
the reach in which the stage is observed. Water 5:
p. 1015.

Floodway. <u>Regional--River Basin</u>. The channel of a river or stream
and those parts of the flood plains adjoining the channel
which carry and discharge the floodwater or floodflow of
any river or stream. Water 5: p. 1015.

Floodway Retarding Structure. <u>National--Soil Conservation</u>. A
structure providing for temporary storage of floodwater
and for its controlled release. Earth 2: p. 24g.

Flow Duration Curve. <u>Regional--River Basin</u>. A cumulative frequency curve that shows the percentage of time that specified discharges are equaled or exceeded. Water 5: p.
1015.

Flume. <u>Federal--EPA</u>. A natural or man-made channel that diverts
water. General 1: p. 7.

Flushing Rate. <u>Federal--Commerce</u>. The rate at which water in a
water body is replaced, usually expressed as the time needed for one complete replacement. General 6: p. 311.

Forebay. <u>Regional--River Basin</u>. The impoundment immediately above
a dam or hydroelectric plant intake strucutre. Energy/
Utility 6: p. 195.

Forebay Area. <u>State--Water Resources</u>. (1) Any holding reservoir
supplying water to a powerhouse penstock, (2) In groundwater hydrology, especially in California, a free groundwater basin which serves as recharge area to an artesian
basin. Energy/Utility 2: Appendix E, p. 13.

Freeboard. <u>Federal--Interior</u>. The vertical distance between normal water level and the crest of a dam or the top of a
flume. General 3: p. 36.

Free Groundwater. <u>National--Engineering</u>. Groundwater in aquifers
not bounded or confined by impervious strata. Water 1:
no page number.

Fresh Water. <u>Regional--River Basin</u>. Water having a relatively
low mineral content, generally less than 500 mg/l of dissolved solids. Water 2: p. I-22.

Gaging Station. <u>National--Soil Conservation</u>. A selected section
of a stream channel equipped with a gage, recorder, or
other facilities for determining stream discharge. Earth
2: p. 25g.

Gravity Dam. <u>National--Soil Conservation</u>. Dam that depends on its
weight to resist overturning. Earth 2: p. 25g.

Groundwater. <u>Federal--EPA</u>. The supply of fresh water under the
Earth's surface that forms a natural reservoir. General
1: p. 7.

<u>Federal--GSA</u>. Water stored in the voids of unconsolidated
materials (e.g., soils) or in water-bearing rocks (i.e.,
aquifers). Groundwater sustains the flow and quality of
surface water streams and furnishes water supply for various uses through direct pumping via wells. General 2:
p. VIII-28.

<u>Federal--GSA</u>. The supply of fresh water under the earth's
surface in an aquifer or soil that forms the natural reservoir for man's use. Legal Jargon 18: Appendix B-4.

<u>Federal--Interior</u>. Water which is underground in an
aquifer. General 4: Glossary, p. 5.

<u>National--Engineering</u>. Water occuring in the zone of
saturation in an aquifer or soil. Water 1: no page number.

<u>National--Resources</u>. Water beneath the surface of the
earth that supplies wells and springs. Energy/Utility
5: p. 7.

<u>Regional--River Basin</u>. Water in the ground that is in
the zone of saturation from which wells, springs, and
ground water runoff are supplied. Water 4: p. 342.

<u>Regional--River Basin</u>. All water beneath the surface of
the ground. Water 2: p. I-22.

<u>State--Water Resources</u>. Subsurface water occupying the
saturation zone, from which wells and springs are fed.
In a strict sense, the term applies only to water below
the water table. Also called phreatic water, plerotic
water. Energy/Utility 2: Appendix E.

Groundwater Basin. <u>State--Water Resources</u>. A pervious formation
with sides and bottom of relatively impervious material in
which groundwater is held or retained. Also called subsurface water basin. Energy/Utility 2: Appendix E.

Groundwater Flow. <u>National--Engineering</u>. Flow of water in an
aquifer or soil. That portion of the discharge of a stream
which is derived from groundwater. Water 1: no page number.

Groundwater Runoff. National--Engineering. That part of the
groundwater which is discharged into a stream channel as
spring or seepage water. Water 1: no page number.

Hardness. Federal--GSA. Soap-consuming capacity or scale-produc-
ing characteristics in pipes and boilers caused by calcium,
magnesium, strontium, ferrous iron, and manganous ions.
The hardness of natural waters varies, depending upon geological formations, and is derived from contact with soil
and rock formations. General 2: p. IX-18.

Regional--River Basin. A characteristic of water due to
the presence of cations, chiefly calcium and magnesium,
which causes increased consumption of soap, and deposition
of boiler scale. Water 2: p. I-22.

Regional--River Basin. A characteristic of water. It is
commonly computed from the amounts of calcium and magnesium
in the water and expressed as equivalent calcium carbonate.
Energy/Utility 7: p. 256. (Revised).

Hard Water. Federal--EPA. Alkaline water containing dissolved
mineral salts, that interfere with some industrial processes and prevent soap from lathering. General 1: p. 8.

Headwater. National--Soil Conservation. The source of a stream.
The water upstream from a structure or point on a stream.
Earth 4: p. 27g.

Headwaters Lake. Federal--Army. A water resources impoundment
which is located in the upper reaches of a drainage basin.
Recreation 2: p. A-38.

Heavy Metals. Federal--Interior. Metalic elements generally occurring in trace amounts in water, including iron, manga-
nese, copper, aluminum, zinc, cadmium, chromium, lead,
arsenic, mercury, and vanadium. Usually considered to
have an atomic number above 21. General 3: p. 40.

High-Water Line. State--Water Resources. The line of the shore
of a river, lake or sea which is ordinarily reached at high
water. Along the seashore, the intersection of the plane
of mean high water with the shore. Energy/Utility 2: Appendix E.

Hydraulic Gradient. Regional--River Basin. The gradient or slope
of the water table or piezometric surface in the direction
of the greatest slope, generally expressed in feet per mile.
Water 5: p. 1016.

Hydrogeology. Federal--DOT. The science that deals with subsurface waters and related geologic aspects of surface water.
Transportation 3: p. 37.

Hydrograph. National--Engineering. Graphical or tabular representation of flow rate with respect to time. Water 1:
no page number.

National--Soil Conservation. A graph showing variation
in stage (depth) or discharge of a stream of water over
a period of time. Earth 4: p. 28g.

State--Natural Resources. A graph showing, for a given
point on a stream, the discharge, stage, velocity, or
other property with respect to time. General 12: p. 337.

State--Soil and Water Commission. A graph showing for a
given point on a stream or for a given point in any drainage system the discharge, stage, velocity, or other property of water with respect to time. Earth 3: p. G-13.

City--Planning. A graph to show the level, flow, or velocity of water in a river at all seasons of the year.
General 5: p. 56.

Hydrography. National--Engineering. Science of measuring and
analyzing stream flow, precipitation, evaporation and
other natural occurrences of water. Water 1: no page
number.

Hydrologic Budget. Regional--River Basin. An accounting of the
inflow, outflow, and storage in a hydrologic unit, such as
a drainage basin, aquifer, soil zone, lake, reservoir, or
irrigation project. Water 5: p. 1016.

Hydrologic Cycle. Federal--Interior. The continual exchange of
moisture between the earth and the atmosphere, consisting
of evaporation, condensation, precipitation (rain or snow),
stream runoff, absorption into the soil, and evaporation
in repeating cycles. General 4: Glossary, p. 6.

Federal--Interior. The cycle of water movement from the
atmosphere to the earth by precipitation and its return to
the atmosphere by interception, evaporation, run-off, infiltration, percolation, storage, and transpiration. General 3: p. 42.

Regional--Commission. The circulation of water in various
forms from the ocean and land surfaces to the atmosphere by
evaporation and transportation, from the atmosphere to the

land by precipitation and then back to the ocean. General
10: p. IV-46.

Regional--River Basin. The circulation of water from the
sea, through the atmosphere, to the land; and, thence,
with many delays, back to the sea by overland and subterranean routes, and in part by way of the atmosphere without reaching the sea. Water 4: p. 342.

Hydrologic Group. City--Planning Department. Used to estimate
runoff from rainfall. The groups range from A (low runoff potential) through D (high runoff potential). General
5: p. 58.

Hydrology. Federal--EPA. The science of dealing with the properties, distribution, and circulation of water. General
1: p. 8.

Federal--Interior. The science of dealing with water and
snow, including their properties and distribution. General 3: p. 42.

National--Engineering. Science dealing with the properties, distribution and flow of water on or in the earth.
Water 1: no page number.

State--Transportation. A science dealing with the properties, distribution, and circulation of water on the surface of the land, in the soil and underlying rocks and in
the atmosphere. Transportation 1: p. x.

Hypolimnion. Federal--GSA. Colder bottom zone of a stratified
lake, extending from the thermocline to the bottom. General 2: p. IX-18.

Impervious Surfaces. City--Planning. Surfaces that do not absorb
rainfall, including buildings, parking areas, driveways,
roads, sidewalks, and other areas in non-porous concrete
and asphalt. General 5: p. 56.

Impoundment. Federal--EPA. A body of water confined by a dam,
dike, floodgate, or other barrier. General 1: p. 8.

Federal--Interior. A body of water formed by confining
and storing the water. General 4: Glossary, p. 6.

Infiltration. Federal--EPA. The action of water moving through
small openings in the earth as it seeps down into the
groundwater. General 1: p. 8.

Intermittent Stream. <u>Regional--River Basin</u>. A stream that flows only part of the time or through only part of its reach. Water 5: p. 1021.

> <u>State--Natural Resources</u>. A stream course that carries water only part of the time. General 12: p. 337.

> <u>City--Planning</u>. A natural watercourse which is dry/ceases to flow during some portion of the year; delineated on USGS maps by a dashed line. General 5: p. 56.

Internal Drainage. <u>Federal--Interior</u>. The movement of water down through the soil profile to porous aquifers or to surface outlets at lower elevation. General 3: p. 45.

Interstate Waters. <u>Federal--EPA</u>. Defined by law as: (1) waters that flow across or form a part of State or international boundaries; (2) the Great Lakes; and (3) coastal waters. General 1: p. 9.

Intertidal. <u>Federal--Army</u>. The region of marine shoreline between high-tide mark and low-tide mark; where neap, spring and storm tides are important; usage is flexible. General 9: p. 25.

> <u>Federal--Commerce</u>. The area between high and low tide levels, twice daily exposed and flooded. General 6: p. 312.

Intertidal Zone. <u>Federal--Interior</u>. The area of a shore between the levels of a high and low tide. General 3: p. 45.

Jackson Turbidity Units (JTU's). <u>Federal--GSA</u>. Unit derived from the use of a turbidimeter; a device which compares ordinary candle light transmitted through a finely divided suspension with that transmitted by a standard suspension. 1 JTU = 1 mg of SiO_2 per liter. General 2: p. IX-19.

> <u>Regional--River Basin</u>. The JTU, as the name implies, is a measurement of the turbidity, or lack of transparency, of water. It is measured by lighting a candle under a cylindrical transparent glass tube and then pouring a sample of water into the tube until an observer looking from the top of the tube cannot see the image of the candle flame. The number of JTU's varies inversely with the height of the sample. For example, a sample which measures 2.3 cm has a turbidity of 1,000 JTU's whereas a sample measuring 72.9 cm has a turbidity of 25 JTU's. Energy/Utility 7: p. 256.

City--Planning. A measurement of turbidity in water samples caused by suspended matter such as clay, silt, finely-
divided organic particles, inorganic matter, etc. Energy/
Utility 9: p. 2.

Jetty. National--Soil Conservation. A structure built of piles,
rocks, or other material extending into a stream or into
the sea to induce scouring or bank building, or for protection. Earth 2: p. 30g.

Lacustrine. Federal--Army. Originating in, or inhabiting a lake.
General 9: p. 26.

Federal--Interior. Living in lakes. Pertaining to lake
environment. General 3: p. 46.

Lagoon. Federal--Army. A shallow area of water generally separated from a larger body of water by a partial barrier.
General 9: p. 26.

Lake. Federal--Army. A large body of water contained in a depression of the earth's surface and supplied from drainage of a larger area. Locally may be called a pond. General 9: p. 26.

Lakescape. Federal--Army. The whole or any portion of a lake;
its water surface, islands, shoreline features, and
scenery which can be viewed from a point on or along the
lake. Recreation 2: p. A-40.

Lake Turnover. Federal--Army. The complete top-to-bottom circulation of water in a lake which occurs when the density
of the surface water is the same or slightly greater than
that at the lake bottom; most temperate zone lakes circulate in Spring and again in Fall. General 9: p. 26.

Lake Zonation. Federal--DOT. The division of lakes into horizontal and vertical zones based on depth, light penetration, temperature and resident organisms. Transportation
3: p. 40.

Leachate. Federal--EPA. Materials that pollute water as it seeps
through solid waste. General 1: p. 9.

Leaching. Federal--EPA. The process by which nutrient chemicals
or contaminants are dissolved and carried away by water,
or are moved into a lower layer of soil. General 1: p. 9.

Federal--Interior. The process by which soluble materials
in the soil are washed into a lower layer or are dissolved

and carried away by water. General 4: Glossary, p. 7.

National--Engineering. Removal of soluble material from
soil by the passage of water through it. Water 1: no page
number.

Levee. National--Engineering. Embankment to confine water, especially one built along the banks of a river to prevent
inundation of lowlands by flood water. Water 1: no page
number.

Limnology. Federal--Army. The study of the biological, chemical,
and physical features of inland waters. General 9: p. 28.

Federal--EPA. The study of the physical, chemical, meteorological, and biological aspects of fresh water. General 1: p. 9.

Federal--GSA. Scientific study of physical, chemical,
meteorological, and biological conditions in fresh water.
General 2: p. XI-28.

State--Water Resources. Scientific study of bodies of
fresh water, as lakes or ponds, with reference to their
physical, geographical, biological, and other features.
More recently extended to include streams. Energy/Utility
2: Appendix E.

Lotic. Federal--DOT. Flowing bodies of water. Transportation 3:
p. 43.

Low Flow Frequency Curve. Regional--River Basin. A graph showing
the magnitude and frequency of minimum flows for a period
of given length. Frequency is usually expressed as the
average interval, in years, between recurrences of an
annual minimum flow equal to, or less than, that shown by
the magnitude scale. Water 5: p. 1017.

Mass Diagram. National--Soil Conservation. A graphical representation of cumulative quantities, such as the integral of
a timeflow curve; an integral curve; each point on the
curve is the sum of all preceding quantities considered.
The diagram is used extensively in water storage analyses.
Earth 2: p. 33g.

Maximum Probable Flood. Regional--River Basin. The largest flood
for which there is any reasonable expectancy in the geographical region involved. Water 5: p. 1015.

Mean Depth. <u>National--Engineering</u>. Average depth; cross-sectional
    area of a stream divided by its surface width. Water 1:
    no page number.

Mean High Water (MHW). <u>Federal--Commerce</u>. A tidal datum; the
    arithmetic average of the high water heights observed over
    a specific 18.6 Metonic Cycle (the National Tidal Datum
    Epoch). General 6: p. 312.

Mean High Water Line. <u>Federal--Commerce</u>. The line formed by the
    intersection of the tidal plane of mean high water with
    the shore. General 6: p. 312.

Mean Lower Low Water. <u>City--Planning</u>. The average of the lower
    of the two low tides along coasts where the two daily low
    tides are unequal. General 8: p. 3.

Mean Low Water (MLW). <u>Federal--Commerce</u>. A tidal datum; the
    arithmatic average of the low water heights observed over
    a specific 18.6 water Metonic Cycle (the National Tidal
    Datum Epoch). General 6: p. 312.

Mean Velocity. <u>National--Engineering</u>. Average velocity; velocity
    obtained by dividing the flow rate by a cross-sectional
    area. Water 1: no page number.

Meromictic Lake. <u>Federal--Interior</u>. A lake with incomplete circ-
    ulation (mixing). General 3: p. 50.

    <u>National--Soil Conservation</u>. Lakes in which dissolved
    substances create a gradient of density differences with
    depth; preventing complete mixing or circulation of water
    masses. Earth 4: p. 33g.

Movable Dam. <u>National--Soil Conservation</u>. A movable barrier that
    may be opened in whole or in part, permitting control of
    the flow of water through or over the dam. Earth 2: p.
    34g.

Natural Flow. <u>Regional--River Basin</u>. The flow in a stream as it
    would be if unaltered by activities of man. Water 2: p.
    I-22.

Natural Runoff. <u>Regional--River Basin</u>. Flow of a stream unaltered
    by acts of man. Water 2: p. I-23.

Natural System. <u>National--Engineering</u>. System of drainage in
    which the main drain follows the largest natural depres-
    sion from the outlet to the upper end of the area.

Laterals branch off the main to drain isolated areas.
Water 1: no page number.

Nitrogen. Federal--GSA. Fertilizing element essential to the
growth of algae. The two most common nitrogen compounds,
ammonia and nitrates, are inter-converted by natural processes. Excessive levels of nitrogen compounds in the
water resulting from fertilizer runoff or sewage can lead
to algae blooms. General 2: p. IX-19.

Nonpoint Source. Federal--EPA. A contributing factor to water
pollution that can't be traced to a specific spot; like
agricultural fertilizer runoff, sediment from construction. General 1: p. 10.

Nonuniform Flow. National--Engineering. Flow in which the velocity is not the same at successive channel cross-sections.
If the velocity at a given cross-section is constant with
time, it is referred to as steady nonuniform flow. If
the velocity changes with time at each cross-section, it
is known as unsteady nonuniform flow. Water 1: no page
number.

Nutrients. Federal--GSA. Inorganic compounds (e.g., phosphates,
nitrates) essential for plant growth. Waters with heavy
growths of aquatic weeds and algae generally have high
levels of nutrients. General 2: p. IX-19.

Observation Well. National--Soil Conservation. Hole bored to a
desired depth below the ground surface, used for observing
the water table or piezometric level. Earth 4: p. 36g.

Observed Runoff. Regional--River Basin. Flow of a stream as observed at a specific point. Observed runoff normally reflects upstream regulations and uses by man. Water 2:
p. I-23.

Oil Spill. Federal--EPA. Accidental discharge into bodies of
water, can be controlled by chemical dispersion, combustion, mechanical containment, and absorption. General 1:
p. 10.

Oligotrophic. Federal--Army. Lakes characterized by abundant
oxygen in deep water as a consequence of small nutrient
supply and low productivity of organic material. General
9: p. 32.

Oligotrophic Lakes. Federal--EPA. Deep clear lakes with low
nutrient supplies. They contain little organic matter and
have a high dissolved oxygen level. General 1: p. 10.

Federal--GSA. Deep lakes having a low supply of nutrients
and therefore supporting little organic production. Dissolved oxygen is generally present at or near saturation.
General 2: p. IX-19.

National--Soil Conservation. Deep lakes that have a low
supply of nutrients; thus they support very little organic
production. Dissolved oxygen is present at or near saturation throughout the lake during all seasons of the
year. Earth 4: p. 36g.

One Hundred Year Floodplain. Federal--GSA. Area covering sites
considered to have one chance in 100 years of flooding.
General 2: p. VIII-29.

Operational Losses. Federal--Interior. Losses of water due to
evaporation and seepage. General 3: p. 56.

Outfall. National--Engineering. Point where water flows from a
conduit, stream or drain. Water 1: no page number.

Outlet. National--Soil Conservation. Point of water disposal
from a stream, river, lake, tidewater, or artificial drain.
Earth 2: p. 36g.

Overturn. Federal--Army. The complete circulation or mixing of
the upper and lower waters of a lake when the temperature
(and densities) are similar. General 9: p. 32.

Federal--EPA. The period of mixing (turnover), by top
to bottom circulation, of previously stratified water
masses. This phenomenon may occur in spring and/or fall;
the result is a uniformity of physical and chemical properties of the water at all depths. General 1: p. 11.

Perched Water. Regional--River Basin. Ground water separated
from the underlying water table by a zone of impervious
or relatively impervious material. Water 5: p. 1018.

Perched Water Table. National--Engineering. A water table, usually of limited area, maintained above the normal groundwater supply by the presence of an intervening, relatively
impervious confining stratum. Water 1: no page number.

City--Planning. The top of a zone of saturation that bottoms on an impermeable horizon above the level of the general water table in the area. General 5: p. 56.

Percolation. National--Engineering. A qualitative term applying

to the downward movement of water through soil. Water 1:
no page number.

Perennial Stream. Regional--River Basin. A stream that flows
continuously. Water 5: p. 1021.

City--Planning. A natural watercourse in which water
flows during all seasons of the year; delineated on USGS
maps as a solid line. General 5: p. 56.

pH. Federal--GSA. Intensity of the acid or alkaline condition of
water. Units of measurement for pH range from 0 to 14; a
value of 7 is neutral, values less than 7 are acid, values
greater than 7 are alkaline. General 2: p. IX-19.

Phosphates. Federal--GSA. Inorganic compounds essential to the
growth of algae. Excessive levels of phosphates in the
water resulting from agricultural runoff or sewage can lead
to algae blooms. General 2: p. IX-19.

Photic Zone. Federal--Army. The region of aquatic environments
in which the intensity of light is sufficient for photosynthesis. General 9: p. 34.

Piezometer. Federal--DOT. A pressure-sealed observation tube installed into water-bearing strata which is used to measure
water pressure within the strata. Transportation 3: p. 51.

Piezometric Surface. Regional--River Basin. An imaginary surface
that everywhere coincides with the static level of the
water in the aquifer. Water 5: p. 1018.

Pond. Federal--Army. A small lake. General 9: p. 35.

Potable Water. Regional--River Basin. Water that does not contain objectionable pollution, contamination, minerals, or
infectious agents, and is considered satisfactory for domestic use. Water 2: p. I-22.

Rainfall Intensity. National--Soil Conservation. The rate at
which rain is falling at any given instant, usually expressed in inches per hour. Earth 2: p. 41g.

Rain Shadow. Federal--Interior. Refers to an area in which little
or no rain falls because it is located to the leeward of
mountains which on the opposite side are exposed to mois-
ture-ladden winds. General 3: p. 66.

Reasonable-Use Rule. National--Engineering. A concept of water
law in which a landowner is given the right to the reason-

able use of water. His right, however depends on his degree of need and the damage a neighbor would suffer from
his operations. Water 1: no page number.

Receiving Waters. Federal--DOT. Rivers, lakes, oceans or other
water bodies that receive treated or untreated waste
waters. Transportation 3: p. 57.

Recharge. Federal--EPA. Process by which water is added to the
zone of saturation, as recharge of an aquifer. General
1: p. 12.

Regional--River Basin. The addition of water to the zone
of saturation. Infiltration of precipitation and its movement to the water table is one form of natural recharge;
injection of water into an aquifer through wells is one
form of artificial recharge. Water 5: p. 1018.

State--Water Resources. Addition of water to the zone of
saturation from precipitation, infiltration from surface
streams, and other sources. Energy/Utility 2: Appendix E.

Recharge Area. Federal--DOT. An area in which water is absorbed
that eventually reaches the zone of saturation in one or
more aquifers. Transportation 3: p. 57.

Federal--GSA. Area where underground aquifers meet the
land surface and replenish the groundwater supply. Composed of highly permeable material, recharge areas are
usually located in topographical positions that permit runoff to be channeled or concentrated onto their surfaces
from upslope drainage areas; they are described in terms
of permeability and area (either in acres or square miles).
General 2: p. VIII-29.

Recharge Of An Aquifer. State--Water Resources. The replenishment of an aquifer by either artificial or natural means.
Energy/Utility 2: Appendix E.

Recharge Well. State--Water Resources. A well constructed to conduct surface water or other surplus water into an aquifer
to increase the groundwater supply. Sometimes called diffusion well. Energy/Utility 2: Appendix E.

Regime. National--Engineering. Condition of a stream with respect
to its rate of flow, as measured by the volume of water
passing different cross-sections in a given time. Water
1: no page number.

Regional Entity. Federal--Water Resources Council. A "regional
    entity" is a Title II (P.L. 89-80) river basin commission,
    a Federal-interstate compact commission approved by Con-
    gress and the respective State legislatures, or other re-
    gional organizations designated by the Council in conjunc-
    tion with the affected States, to function similar to a
    Title II river basin commission. The Council may desig-
    nate itself to act as a "regional entity" where no Title
    II river basin commission or Federal-interstate compact
    commission exists. Legal Jargon 36: p. 2.

Regulated Flow. Regional--River Basin. The flow in a stream where
    it is controlled by reservoirs, diversions, exportations,
    importations, and changes in consumptive use associated
    with man's activities. Water 2: p. I-22.

Regulatory Floodway. Federal--Water Resources. The area regulated
    by Federal, State or local requirements; the channel of a
    river or other watercourse and the adjacent land areas that
    must be reserved in an open manner, i.e., unconfined or un-
    obstructed either horizontally or vertically, to provide
    for the discharge of the base flood so the cumulative in-
    crease in water surface elevation is no more than a desig-
    nated amount (not to exceed one foot as set by the NFIP).
    Water 6: pp. 5-6.

Relief Well. National--Soil Conservation. Well, pit, or bore
    penetrating the water table to relieve hydrostatic pres-
    sure by allowing flow from the aquifer. Earth 2: p. 42g.

Reservoir. Federal--Army. An artificially impounded body of
    water; also, the supply of any commodity, as a reservoir
    of infection, etc. General 9: p. 38.

    Federal--EPA. Any holding area, natural or artificial,
    used to store, regulate, or control water. General 1:
    p. 13.

    National--Engineering. Body of water, such as a natural
    or constructed lake, in which water is collected and stored
    for use. Water 1: no page number.

    Regional--River Basin. A pond, lake, or basin, either nat-
    ural or artificial, for the storage, regulation, and con-
    trol of water. Water 5: p. 1019.

    State--Water Resources. A pond, lake, tank, basin, or
    other space, either natural or created in whole or in part
    by the building or engineering structures, which is used
    for storage, regulation, and control of water. Sometimes

called impoundment. Energy/Utility 2: Appendix E.

Retention. National--Soil Conservation. The amount of precipi-
    tation on a drainage area that does not escape as runoff.
    It is the difference between total precipitation and total
    runoff. Earth 2: p. 43g.

Retention Dams. Federal--Interior. Small earthen dams designed
    to retain water for only short periods of time in order to
    prevent excessively rapid runoff and erosion. General 3:
    p. 68.

Revetment. Federal--Interior. A structure or obstacles placed
    along the margins of a stream in order to protect the
    banks from erosion. General 3: p. 68.

Riffle. National--Engineering. Shallow rapids in an open stream,
    where the water surface is broken into waves by obstruc-
    tions wholly or partly submerged. Water 1: no page number.

    State--Natural Resources. A shallow rapid in a stream.
    General 12: p. 338.

Riffle-Pool Concentration. Federal--GSA. Arrangement within a
    stream of alternating riffles (i.e., rapidly-flowing, shal-
    low areas) and pools (i.e., deep, slow-moving areas). Rif-
    fles are areas of high productivity by phytoplankton; pools
    provide resting places for fish and other organisms. The
    specific configuration of riffles and pools is important
    to the potential of a stream as aquatic habitat. General
    2: p. XI-29.

Riffle Zone. Federal--DOT. A shallow rapids in an open stream
    where the water surface is broken into waves by wholly or
    partly submerged obstructions. Transportation 3: p. 58.

Riparian. Regional--River Basin. Pertaining to the banks of
    streams, lakes, or tidewater. Water 5: p. 1019.

Riparian Rights. Federal--EPA. Entitlement of a land owner to
    the water on or bordering his property, including the right
    to prevent diversion or misuse of it upstream. General 1:
    p. 13.

Riprap. State--Natural Resources. Material placed on a stream
    bank and bed for protection from stream or wave action;
    can consist of broken rock or other materials such as car
    bodies or trees. General 12: p. 338.

River Basin. Federal--EPA. The land area drained by a river and
          its tributaries. General 1: p. 13.

River Delta Estuary. Federal--DOT. Found at the mouths of large
          rivers, semi-enclosed bays, channels and brackish marshes
          which are formed by shifting silt deposits. Transportation
          3: p. 58.

Runoff. Federal--Commerce. The portion of precipitation on land
          that flows over the land surface; overland flow. General
          6: p. 313.

          Federal--EPA. Water from rain, snow melt, or irrigation
          that flows over the ground surface and returns to streams.
          It can collect pollutants from air or land and carry them
          to the receiving waters. General 1: p. 13.

          National--Engineering. The portion of precipitation or
          irrigation water which is returned to the stream as sur-
          face flow. Water 1: no page number.

          Regional--Commission. That part of the precipitation that
          appears in surface streams. It is the same as streamflow
          unaffected by artificial diversions, storage or other works
          of man in or on the stream channels. General 13: p. 591.

          Regional--River Basin. That part of rainfall or other
          precipitation that reaches watercourses or drainage sys-
          tems. Energy/Utility 7: p. 257.

          City--Planning. The part of the precipitation upon a
          drainage area that is discharged from the area in stream
          channels. The water that flows off the surface without
          sinking in is called surface runoff; water that enters
          the ground before reaching surface streams is called
          groundwater runoff. General 5: p. 57.

Runoff Plots. National--Soil Conservation. Areas of land, usu-
          ally small, arranged so the portion of rainfall or other
          precipitation flowing off and perhaps carrying soluble
          materials and soil may be measured. Usually, the flow
          from runoff plots includes only surface flow. Earth 4:
          p. 43g.

Safe Yield. Federal--GSA. Rate of diversion of extraction for
          consumptive use that can be indefinitely extracted from
          surface and groundwater systems within the limits of eco-
          nomic feasibility and under specific conditions of water
          supply development. General 2: p. VII-19.

Saline. <u>City--Planning Department</u>. Water containing dissolved
        salts, usually at concentrations from 10,000 to 33,000
        milligrams per liter. Energy/Utility 9: p. 2.

Saline Contamination. <u>State--Water Resources</u>. Contamination of
        water by intrusion of salt water. Energy/Utility 2: Apen-
        dix E.

Saline Water. <u>Regional--River Basin</u>. Water containing more than
        250 mg/1 of chlorides or more than 500 mg/1 of dissolved
        solids. Water 2: p. I-23.

Salinity. <u>Federal--Commerce</u>. A measure of the quantity of dis-
        solved salts in water expressed in parts per thousand of
        water (ppt). General 6: p. 313.

        <u>Regional--River Basin</u>. The relative concentration of
        salts, usually sodium chloride, in a given water sample.
        It is usually expressed in terms of the number of parts
        per thousand of chlorine (Cl). Energy/Utility 7: p. 257.

        <u>State--Water Resources</u>. The relative concentration of
        salts, usually sodium chloride, in a given water. It is
        usually expressed in terms of the number of parts per mil-
        lion of chlorine (Cl). A measure of the concentration of
        of dissolved mineral substances in water. Energy/Utility
        2: Appendix E.

Salinity Wedge. <u>Federal--Army</u>. The movement of subsurface saline
        water into an aquifer, or, in an estuary. Of a body of
        saline (sea) water under fresh water. General 9: p. 39.

Saltation Load. <u>National--Engineering</u>. Material bouncing along
        the bed, or moved, directly or indirectly, by the impact
        of the bouncing particles. Water 1: no page number.

Salt Water Intrusion. <u>Federal--Commerce</u>. The movement of salt
        water inland into subterranean aquifers. General 6: p.
        314.

        <u>Federal--EPA</u>. The invasion of fresh surface or ground
        water by salt water. If the salt water comes from the
        ocean it's called sea water intrusion. General 1: p. 13.

Scenic River. <u>Federal--Forest Service</u>. Wild and Scenic River Act
        usage. Those rivers or sections of rivers that are free
        of impoundments, with shorelines or watersheds still large-
        ly primitive and shorelines largely undeveloped, but ac-
        cessible in places by roads. General 15: p. 186.

Secchi Disk. <u>City--Planning Department</u>. A device used to determine the clarity of water. Usually used to make on-site
determinations as opposed to laboratory methods, such as
the Jackson Turbidity Test. Energy/Utility 9: p. 2.

Sediment Basin. <u>State--Soil and Water Commission</u>. A depression
formed from the construction of a barrier or dam built at
a suitable location to retain sediment and debris. Earth
3: p. G-19.

Sediment Discharge. <u>National--Soil Conservation</u>. The quantity
of sediment, measured in dry weight or by volume, transported through a stream cross-section in a given time.
Sediment discharge consists of both suspended load and
bedload. Earth 4: p. 45g.

> <u>Regional--River Basin</u>. The rate at which dry weight of
> sediment passes a section of a stream of the quantity of
> sediment, as measured by dry weight or by volume, that is
> discharged in a given time. Water 5: p. 1019.

Sediment Load. <u>National--Engineering</u>. Amount of sediment carried
by running water. Water 1: no page number.

Sediment Pool. <u>National--Soil Conservation</u>. The reservoir space
alloted to the accumulation of submerged sediment during
the life of the structure. Earth 2: p. 45g.

Seepage. <u>Federal--EPA</u>. Water that flows through the soil. General 1: p. 13.

Settling Basin. <u>State--Department of Agriculture</u>. Enlargement
in the channel of a drain to permit settling of sediment
carried in suspension. Energy/Utility 4: p. 2.

Sheet Flow. <u>National--Engineering</u>. Water, usually storm runoff,
flowing in a thin layer over the soil or other smooth surface. Water 1: no page number.

Shoreline. <u>State--Water Resources</u>. The line along which the land
surface meets the water surface of a lake, sea, or ocean.
Strictly speaking, it is not a line, but a narrow strip
or area embracing that part of the land surface which comes
in contact with wave action both above and below the surface of the water. The term does not apply on tidal flats
or marshes which are inundated by the tides, but essentially to strips where the land surface has an appreciable
slope toward the water. Energy/Utility 2: Appendix E.

Slough. State--Water Resources. (1) A small muddy marshland or
    tidal waterway, which usually connects other tidal areas,
    (2) A tideland or bottom land creek. A side channel or
    inlet, as from a river or a bayou, may be connected at
    both ends to a parent body of water. Energy/Utility 2:
    Appendix E.

Sluice. National--Soil Conservation. Channel serving to drain
    off surplus water from behind a flood gage; conduit for
    carrying water at high velocity; an opening in a struc-
    ture for passing debris. Also, to cause water to flow at
    high velocities for ejecting debris. Earth 2: p. 47g.

Spillway. Federal--Interior. Overflow channel of a dam. General
    3: p. 72.

Standard Project Flood. Regional--Commission. A hypothetical
    flood that might result from the most severe combination
    of meteorological and hydrological conditions that are rea-
    sonably characteristic of the geographic region involved.
    The SPF is the usual basis for design of flood control
    structures. General 13: p. 591.

Stilling Basin. Federal--Interior. An area or basin used to dis-
    sipate the energy of the flow. General 3: p. 73.

Stream. National--Engineering. Body of water usually flowing in
    a natural surface channel, also open or closed conduit; a
    jet of water. Water 1: no page number.

    Regional--River Basin. A general term for a body of flow-
    ing water. In hydrology, the term is generally applied to
    the water flowing in a natural channel as distinct from a
    cannal. More generally, as in the term stream gaging, it
    is applied to the water flowing in any channel, natural or
    artificial. Water 5: p. 1021.

Streambanks. National--Soil Conservation. The usual boundaries,
    not the flood boundaries, of a stream channel. Right and
    left banks are named facing downstream. Earth 4: p. 54g.

Streamflow. Regional--Commission. The rate of discharge of a
    stream as defined by the unit used to measure its flow.
    A stream flow of cubic foot per second would be the rate
    of discharge of a stream whose channel is one square foot
    in cross-sectional area and whose average velocity is one
    foot per second. General 10: p. IV-46.

    Regional--River Basin. The discharge that occurs in a
    natural channel. Although the term discharge can be

applied to the flow of a canal, the word streamflow uniquely describes the discharge in a surface stream course.
Streamflow is a more general term than runoff, as streamflow may be applied to discharge whether or not it is affected by diversion or regulation. Water 5: p. 1021.

Streamflow Depletion. Regional--River Basin. The amount of water
that flows into a valley, or onto a particular land area,
minus the water that flows out of the valley or off from
the particular land area. Water 5: p. 1013.

Stream Load. National--Soil Conservation. Quantity of solid and
dissolved material carried by a stream. Earth 2: p. 54g.

Stream Profile. Regional--Commission. Stream elevation plotted
against its horizontal distance, representing the slope or
gradient of the stream channel from upper to lower end.
General 10: p. IV-46.

Subirrigated Land. Regional--River Basin. Land with a high water
table condition, either natural or artifically controlled,
that normally supplies a crop irrigation requirement.
Earth 1: p. 381.

Surface Profile. National--Soil Conservation. The longitudinal
profile assumed by the surface of a stream flowing in an
open channel; the hydraulic grade line. Earth 4: p. 55g.

Surface Runoff. Federal--GSA. Rate at which water is removed by
flowing over the soil surface. This rate is determined by
the texture of the soil, slope, climate, and land use cover
(e.g., paved surface, grass, forest, bare soil). General
2: p. VIII-30.

Federal--Interior. That part of the precipitation which
travels over the soil surface to the nearest stream channel. General 4: Glossary, p. 10.

Surface Water. Federal--Forest Service. Water which remains on
top of the land, such as a river or lake. General 15:
p. 213.

Federal--GSA. Water found in streams, rivers, ponds, lakes
and marshes or wetlands. Streams and rivers are measured
in terms of width, depth, and flow rate (usually in cubic
feet per second of discharge (CFS), and lakes and other
standing water bodies in depth and acres or square miles.
General 2: p. VIII-30.

Suspended Load. <u>National--Soil Conservation</u>. Solids or sediments
        suspended in a fluid by the upward components of turbulent
        currents or by colloidal suspension. Earth 2: p. 55g.

Suspended Solids. <u>Federal--GSA</u>. Particles in water that do not
        pass through a filter. Suspended solids may enter the
        water naturally by the turbulent action of water over a
        muddy or sandy bottom, or as a result of human activity
        that causes runoff and sewage effluents. General 2: p.
        IX-20.

        <u>Regional--River Basin</u>. Solids that either float on the
        surface of, or are in suspension in, water or waste water,
        that can be removed by filtering. Water 2: p. I-23.

Tailwater. <u>National--Soil Conservation</u>. In hydraulics, water,
        in a river or channel, immediately downstream from a
        structure. Earth 2: p. 55g. (Revised).

        <u>Regional--River Basin</u>. The water surface immediately
        downstream from a dam or hydroelectric powerplant. En-
        ergy/Utility 6: p. 199.

Thermal Pollution. <u>Federal--Army</u>. The excessive raising or low-
        ering of water temperatures above or below normal seasonal
        ranges in streams, lakes, or estuaries or oceans as the
        result of discharge of hot or cold effluents into such
        waters. General 9: p. 46.

        <u>Federal--EPA</u>. Discharge of heated water from industrial
        processes that can affect the life processes of aquatic
        plants and animals. General 1: p. 15.

        <u>Federal--Forest Service</u>. Altering the amount of energy
        available in the environment by the addition of heat or
        cold. Most frequently this occurs in the form of heat
        added by returning water which has been used for cooling
        purposes to a river, lake, estuary, bay or ocean. However
        "thermal pollution" may also occur with the release of
        large volumes of the cold, deep water in reservoirs, into
        warmer bodies of water. General 15: p. 217.

        <u>Federal--Interior</u>. The warming of the environment, es-
        pecially streams and other bodies of water, by waste heat
        from power plants and factories. Drastic thermal pollu-
        tion endangers many species of aquatic life. General 4:
        Glossary, p. 11.

Thermal Stratification. <u>Federal--Army</u>. The seasonal formation of
        horizontal layers of water in lakes and oceans (warm

surface, cool bottom) of markedly varying temperatures,
separated by a zone with a steep temperature gradient.
General 9: p. 46.

Thermocline. Federal--Army. A narrow (horizontal) zone of water
in lakes and oceans with a steep temperature gradient, separating a warmer surface layer (epilimnion, epithalassa)
from a cooler bottom layer (hypolimnion, hypothalassa);
as a thermocline is a plane, but a zone is observed, the
preference or usual term is metalimnion. General 9: p.
46.

Federal--GSA. Transition zone between the warm epilimnion
and cold hypolimnion of stratified lakes. General 2: p.
IX-20.

Federal--Interior. A place, in relation to a lake's depth,
where there is an abrupt temperature change. An obvious
temperature change between the upper warm portion of a
lake and the lower cold portion. General 3: p. 76.

Tidal Basin. State--Water Resources. A dock, basin, or bay connected with an ocean or tidal estuary, in which the water
level changes with fluctuations of the tides. Energy/
Utility 2: Appendix E.

Tidal Flooding. State--Environmental Protection. Inundation of
land caused by an abnormally high tidal water having an
average frequency of once in 100 years, although the event
may occur in any year. General 6: p. 314.

Tidal Influence. State--Environmental Protection. Waters which
measurably rise and fall with twice-daily tides. General
6: p. 314.

Tidal Prism. City--Planning. The total amount of water that flows
into a tidal basin or estuary, such as San Francisco Bay,
and out again with movement of the tide, excluding any
fresh-water flow. General 8: p. 4.

Tidal Water. State--Water Resources. A river, estuary, bay, lake,
or sound where tides can be measured. Energy/Utility 2:
Appendix E.

Tidal Wave. State--Water Resources. (1) The general rise and
fall of the surface of the ocean that is generated by the
sun and the moon and that travels around the earth each
day, causing the water level of the ocean to rise and fall,
(2) An extreamely large wave caused by a seismic disturbance or a great storm, which often causes overflow of low

lying lands not usually inundated by ordinary wave or
tidal action. Energy/Utility 2: Appendix E.

Tide. <u>Federal--Commerce</u>. The periodic rise and fall of the water
resulting from gravitational interaction between the sun,
moon, and earth. The vertical component of the particulate motion of a tidal wave. In each lunar day of 24
hours and 49 minutes there are two high tides and two low
tides. General 6: p. 314.

Time Of Concentration. <u>National--Soil Conservation</u>. Time required
for water to flow from the most remote point of a watershed, in a hydraulic sense, to the outlet. Earth 4: p.
56g.

Toxic Materials. <u>Federal--GSA</u>. Any water contaminants that harm
or kill living matter. Toxic materials generally include
heavy metals, pesticides, phenol, and cyanides. The low
levels at which toxic materials are present often render
detection impossible without sophisticated laboratory
analyses and bioassays. General 2: p. IX-20.

Tributary. <u>National--Engineering</u>. Branch of a stream, ditch,
drain or other channel which contributes flow to the
main channel. Water 1: no page number.

<u>National--Soil Conservation</u>. Secondary or branch of a
stream, drain, or other channel that contributes flow to
the primary or main channel. Earth 4: p. 57g.

Turbidimeter. <u>Federal--EPA</u>. A device that measures the amount
of suspended solids in a liquid. General 1: p. 15.

Turbidity. <u>Federal--Army</u>. Condition of water resulting from suspended matter; water is turbid when its load of suspended
material is conspicuous. General 9: p. 48.

<u>Federal--Commerce</u>. Reduced water clarity resulting from
presence of suspended matter. General 6: p. 315.

<u>Federal--GSA</u>. Measure of the interference with the pas-
sage of light through water. General 2: p. IX-20.

<u>Federal--Interior</u>. A measure of the extent to which light
passing through water is reduced due to suspended materials. Excessive turbidity may interfere with light penetration and minimize photosynthesis, thereby causing a
decrease in primary productivity. It may alter water
temperature and interfere directly with essential physiological functions of fish and other aquatic organisms,

making it difficult for fish to locate a food source. General 3: p. 79.

Regional--River Basin. A measure of fine suspended matter, usually colloidal in liquids. Energy/Utility 7: p. 257.

Turnover. Federal--Interior. A process whereby the water on the bottom of a deep lake exchanges with the water on the top. The process is generally caused by warming of weather with concomitant seasonal wind patterns. The same process occurs as the weather cools into winter season. General 3: p. 80.

Unconfined Aquifer. Federal--ERDA. An aquifer that has a water table or surface at atmospheric pressure. General 7: p. g-10.

Underground Water. State--Water Resources. Water that occurs in the lithosphere. It may be in liquid, solid, or gaseous state. It comprises suspended water and groundwater. Energy/Utility 2: Appendix E.

Underground Watercourse. State--Water Resources. A known and defined subterranean channel, created by natural conditions, that contains flowing water. Energy/Utility 2: Appendix E.

Uniform Flow. National--Engineering. Flow in which the velocity and depth is the same at each cross-section. Water 1: no page number.

Urban Runoff. Federal--EPA. Storm water from city streets, usually carrying litter and organic wastes. General 1: p. 15.

Water Area. Regional--River Basin. Water areas of more than 40 acres and water courses more than one-eighth mile wide. Earth 1: p. 382.

Water Classification. National--Soil Conservation. Separation of water in an area into classes according to usage, such as domestic consumption, fisheries, recreation, industrial, agricultural, navigation, waste disposal, etc. Earth 4: p. 59g.

Water Disturbance. State--Environmental Protection. Measurable change in biological, chemical or physical water quality. General 6: p. 315.

Water Penetration. <u>National--Soil Conservation</u>. The depth to
    which irrigation water or rain penetrates soil before the
    rate of downward movement becomes negligible. Earth 2:
    p. 59g.

Water Pollution. <u>Federal--EPA</u>. The addition of enough harmful
    or objectionable material to damage water quality. Gen-
    eral 1: p. 16.

    <u>State--Environmental Control</u>. The manmade or man-induced
    alteration of the chemical, physical, biological, and ra-
    diological integrity of water. Legal Jargon 25: p. 3.

Water Quality. <u>Federal--Army</u>. A graded value of the components
    (organic and inorganic, chemical or physical) which com-
    prise the nature of water. Established criteria determine
    the upper and/or lower limits of those values which are
    suitable for particular uses. Recreation 2: p. A-51.

Water Quality Criteria. <u>Federal--EPA</u>. The levels of pollutants
    that affect use of water for drinking, swimming, raising
    fish, farming or industrial use. General 1: p. 16.

    <u>National--Soil Conservation</u>. A scientific requirement on
    which a decision or judgment may be based concerning the
    suitability of water quality to support a designated use.
    Earth 4: p. 59g.

Water Quality Standards. <u>Federal--EPA</u>. A management plan that
    considers: (1) what water will be used for; (2) setting
    levels to protect those uses; (3) implementing and en-
    forcing the water treatment plans; and (4) protecting ex-
    isting high quality waters. General 1: p. 16.

    <u>National--Soil Conservation</u>. Minimum requirements of
    purity of water for various uses; for example, water for
    agricultural use in irrigation systems should not exceed
    specific levels of sodium bicarbonates, pH, total dissolved
    salts, etc. Earth 4: p. 59g.

Water Resources. <u>Federal--Interior</u>. Water in any of its forms,
    wherever located--atmosphere, surface or ground--which is
    or can be of value to man. Recreation 3: p. 27.

    <u>National--Engineering</u>. Supply of water in a given area
    or watershed, usually interpreted in terms of availability
    of surface or underground water. Water 1: no page number.

Water Rights. <u>National--Engineering</u>. Legal rights to water sup-
    plies derived from common law, court decisions or statutory

enactments. Water 1: no page number.

Watershed. Federal--Army. The total surface drainage area that
    contributes water runoff to an impoundment. Recreation
    2: p. A-51.

    Federal--Army. An entire drainage basin including all
    living and non-living components of the system. General
    9: p. 49.

    Federal--EPA. The land area that drains into a stream.
    General 1: p. 16.

    Federal--Interior. The area from which water drains to
    a single point. In a natural basin, the area contributing
    flow to a given place on a stream. General 4: Glossary,
    p. 11.

    National--Engineering. Total land area above a given
    point on a stream or waterway that contributes runoff to
    that point. Water 1: no page number.

    Regional--River Basin. A term to signify drainage basin
    or catchment area. Water 5: p. 1022.

Watershed Area. National--Soil Conservation. All land and water
    within the confines of a drainage divide or a water prob-
    lem area consisting in whole or in part of land needing
    drainage or irrigation. Earth 4: p. 59g.

Watershed Management. National--Soil Conservation. Use, regu-
    lation, and treatment of water and land resources of a
    watershed to accomplish stated objectives. Earth 4: p.
    59g.

Watershed Planning. National--Soil Conservation. Formulation of
    a plan to use and treat water and land resources. Earth
    4: p. 59g.

Watershed Project. Regional--Commission. Comprehensive program
    of structural and nonstructural measures to preserve or
    restore a watershed to good hydrologic conditions. These
    measures may include detention reservoirs, dikes, channels,
    contour trenches, terraces, furrows, gully plugs, revege-
    tation, and possibly other practices to reduce flood peaks
    and sediment production. General 13: p. 593.

Watershed Protection. Regional--River Basin. The treatment of
    watershed lands in accordance with such predetermined ob-
    jectives as the control of erosion, streamflow, silting

floods, and water, forage, or timber yield. Earth 1: p.
383.

Water Table. Federal--Army. The upper limit of that part of the
ground which is saturated with water. General 9: p. 49.

Federal--ERDA. Upper boundary of an unconfined aquifer
surface below which saturated groundwater occurs; defined
by the levels at which water stands in wells that barely
penetrate the aquifer. General 7: p. g-11.

Federal--GSA. Upper surface of a zone of saturation except where that surface is formed by an impermeable body.
General 2: p. VIII-30.

Federal--Interior. The upper surface of free ground water.
General 3: p. 82.

National--Engineering. Soil water surface at which the
pressure is equal to the atmospheric pressure. Water 1:
no page number.

Regional--River Basin. The upper zone of saturation. No
water table exists where that surface is formed by an impermeable body. Water 5: p. 1022.

Water Year. Regional--River Basin. The 12-month period, October
1 through September 30. The water year is designated by
the calendar year in which it ends. Water 5: p. 1022.

Water Yield. Federal--Forest Service. (1) The runoff from a
watershed, including groundwater outflow, (2) Water yield
is the precipitation minus the evapotransportation. General 15: p. 230.

Regional--River Basin. Runoff, including ground-water
outflow that appears in the stream, plus ground-water outflow that leaves the basin underground. Water yield is
the precipitation minus the evapotranspiration. Water 5:
p. 1022.

Weir. National--Engineering. Dam across a stream to stop, raise
or divert the flow; device for measuring the flow of water.
Water 1: no page number.

City--Planning Department. A structure used for control
and/or measurement of flow. Energy/Utility 9: p. 2.

Wild River Area. <u>Federal--Forest Service</u>. Wild and Scenic River
    Act usage. Those rivers or sections of rivers that are
    free of impoundments and generally inaccessible except by
    trail, with watersheds or shorelines essentially primitive
    and waters unpolluted. These represent vestiges of prim-
    itive America. General 15: p. 234.

7.   PLANT/ANIMAL

Index

Cleaning Cut
Clearance
Clear Cut
Climax
Climax Community
Climax Species
Climax State
Climax Vegetation
Clutch
Cold-Water Fishery
Community
Conifer
Contact Herbicide
Cover
Cover Crop
Covert
Crown
Cruising Radius
Deciduous
Deciduous Forest
Density
Density-Dependent Factor
Disclimax
Dispersion
Distribution
Diversity
Dominance
Dominant
Dominant Species
Dredge
Dredge Sampler
Dripline
Ecological Balance
Ecological Climax
Ecological Factor
Ecological Impact
Ecological Stress
Ecological Study Area
Ecological Succession
Ecology
Economic Poisons
Ecosystem
Ecotone
Ecotype
Edaphic
Edge
Edge Effect
Emigration
Endangered Species
Endemic

Endemic Species
Energy Budget
Evapotranspiration
Even-Aged
Evergreen
Exotic
Exotic Species
Exotic Vegetation
Eyrie
Family
Fauna
Feral Species
Fish
Fish And Wildlife Enhancement
Fish Attractors
Fishery Management Plan (FMP)
Fishery Resources
Fish Stocking
Fishway
Flora
Floristic
Fluvial
Flyway
Food Chain
Food Cycle
Food Web
Forage
Forest Cover
Forest Cover Type
Forest Game
Forest Lands
Fungi
Game Fish
Game Range
Genus
Grab Samplers
Grazing Food Chain
Groundcover
Group Cutting
Habitat
Habitat Structure
Harvest
Hedgerow
Herbicide
Herbivore
Home Range
Homing
Homogeneous Response Unit
Hoop, Fyke, Trap or Pound Nets
Hunter Success

Hybrid
Immigration
Improvement Cutting
Indigenous
Indigenous Species
Integrated Pest Management
Intensive Forest Management
Interspecific
Intraspecific
Introduction
Invader Plant Species
Irruption
Key Management Species
Keystone Species
Key Utilization Species
Key Winter Range
Lentic Habitat
Life Cycle
Limiting Factor
Lotic Habitats
Medium-Stocked Stand
Migration
Mutualism
National Wildlife Refuges
Native
Native Species
Native Vegetation
Natural Selection
Nature Preserves
Nest Counts
Niche
Non-Game Species
Order
Organophosphates
Overpopulation
Overstory
Paleoecology
Patch Cutting
Peripheral Species
Persistent Pesticide
Pesticide
Pesticide Tolerance
Photosynthesis
Phylum
Phytoplankton
Phytotoxic
Pioneer Plant
Plankton
Plant Litter
Plant Succession

Poletimber Stand
Poorly Stocked Stand
Population
Population Density
Population Dynamics
Population Index
Population Irruption
Population Pressure
Potential Natural Vegetation
Predator
Prey
Prey Organism
Primary Succession
Productivity
Protozoa
Radioecology
Range
Range Capacity
Rare Species
Recruitment
Reforestation
Removal Cutting
Removal Regeneration Cutting
Riparian
Risk Cutting
Roadside Car Census
Rookery
Root Zone
Rotation
Rough Fish
Salvage Cutting
Sanctuary
Sapling And Seeding Stand
Sawtimber Stand
Scat
Scavenger
Secondary Succession
Seed Tree Cutting
Selective Pesticide
Sere
Shelter Belt
Shelterwood Cutting
Silviculture
Special Cutting
Species
Species Diversity
Species Diversity Indices
Stand
Stand Cutting
Status Undetermined Species

Strip Census
Strip Cutting
Strip Survey
Sub-Climax
Succession
Successional
Symbosis
Sympatric
Synecology
Systemic Pesticide
Systems Ecology
System Stability
Taxonomy
Temporal Census
Territoriality
Territory
Threatened Species
Threshold
Timber Management Compartment
Timber Resource System
Time-Area Counts
Tolerance
Tolerance Limit (TL)
Track Counts
Transect
Transient Species
Type Conversion
Types
Undergrowth
Understory
Uneven-Aged
Unique Species
Virgin Forest
Warm-Water Fishery
Waterfowl
Water Pumps
Weed Cutting
Well-Stocked Stand
Wildlife
Wildlife And Fish Habitat Sys-
      tem
Wildlife Enhancement
Wildlife Habitat
Wildlife Management
Wildlife Refuge
Wildlife Stocking
Windbreak
Xerophyte
Year Class
Zone

Zoogeography
Zoological Area
Zooplankton
Zymogenous Flora

Terms

Abiotic Productivity. <u>Federal--Forest Service</u>. The amount of
        material yielded by an inorganic, renewable resource pro-
        cess (e.g., abiotic products include water, sand, gravel).
        General 15: p. 164.

Acclimation. <u>Federal--EPA</u>. The physiological and behaviorial ad-
        justments of an organism to changes in the environment.
        General 1: p. 1.

        <u>Federal--Interior</u>. The adjustment of an organism to a
        new habitat or environment. General 3: p. 1.

Acclimatization. <u>Federal--EPA</u>. The adaptation over several gen-
        erations of a species to a marked change in the environ-
        ment. General 1: p. 1.

Adaptation. <u>Federal--EPA</u>. A change in structure or habit of an
        organism that produces better adjustment to its surround-
        ings. General 1: p. 1.

Aeration. <u>Federal--GSA</u>. Process of adding oxygen to water. May
        be physical (i.e., atmosphere oxygen directly dissolving
        in water) or biological (i.e., plants releasing oxygen
        during photosynthesis). General 2: p. IX-17.

Aerial Counts. <u>Regional--Commission</u>. This method is done during
        fall, winter, and spring to obtain counts of large game
        present within a specific area by doing systematic counts
        of the animals along predetermined routes or in sample
        areas. The same routes or same areas are used each sea-
        son and each year for comparing results. General 10: p.
        IV-26.

Aerial Cover. <u>Federal--Interior</u>. The ground area circumscribed
        by the perimeter of the branches and leaves of a given
        plant or group of plants (generally used as a measure of
        relative density). General 3: p. 2.

Aerial Nesting Inventories. <u>Regional--Commission</u>. This is a
        sampling census in which sample strips are 1/8 mile wide
        and up to 18 miles long. They are flown at 100 to 200
        feet above the ground at speeds of 75 to 100 miles per
        hour. General 10: p. IV-27.

Aerial Photography. <u>Regional--Commission</u>. This technique of sur-
        veying waterfowl is most useful during the winter when the
        waterfowl are wintering in large concentrations, but it is

not that suitable during the breeding season of the spring
or summer because the birds are either too spread out or
hidden by vegetation cover. General 10: p. IV-27.

Age Group. Federal--Interior. Animals of the same age in a population. General 3: p. 3.

Algae. Federal--EPA. Simple rootless plants that grow in bodies
of water in relative proportion to the amounts of nutrients
available. Algal blooms, or sudden growth spurts can affect water quality adversely. General 1: p. 1.

Federal--Interior. Simple plants containing chlorophyll.
Most live submerged in water. General 3: p. 4.

Algal Bloom. Federal--Army. Rapid and flourishing growth of
algae. General 9: p. 2.

Allochthonous. Federal--Interior. The exotic species of a given
area. Also refers to deposits of material that originated
elsewhere; e.g., drifted plant material on the bottom of
a lake. General 3: p. 4.

Alpine. Federal--Interior. The zone in a mountain system which
lies above the timberline. General 3: p. 5.

Anadromous. Federal--Interior. Fish which live in salt water but
migrate to fresh water to spawn. General 3: p. 6.

Anadromous Fish. Regional--River Basin. A marine species of fish
that ascends a river to spawn in fresh water. The young
remain in the river for a short period of time then go to
the sea. Water 2: p. I-21.

Animal Community. Federal--Interior. The combination of animal
species occupying and interacting in one area or habitat.
Recreation 3: p. 3.

Animal Unit Month. Regional--River Basin. The amount of food or
forage required by an animal unit for one month. An animal
is one mature cow with calf under 6 months of age or equivalent. Animal unit equivalent varies by agencies for
horse, sheep, goats, or wildlife. Earth 1: p. 371.

Antagonism. Federal--Interior. Growth inhibition of one organism
caused by unfavorable conditions created by the presence
of another organism. General 3: p. 6.

Aquatic Growth. State--Water Resources. The aggregate of passively floating or drifting or attached organisms in a

body of water; plankton. Energy/Utility 2: Appendix E.

Aquiculture. Federal--Interior. The use of artificial means to
increase the production of aquatic organisms in fresh or
salt water. General 3: p. 7.

Association. Federal--Army. A definite or characteristic assemblage of plants living together in an area essentially
uniform in environmental conditions; any ecological unit
of more than one species. General 9: p. 4.

Federal--Interior. All organisms living together in any
given combination of environmental conditions. General
3: p. 8.

Autecology. Federal--Interior. That part of ecology which deals
with individual species. General 3: p. 9.

Autotrophic. General--EPA. An organism that produces food from
inorganic substances. General 1: p. 2.

Available Water. Federal--Interior. That portion of water in
the soil that can be readily absorbed by plant roots.
General 3: p. 10.

Barrier. Federal--Interior. The blockage of migratory or daily
movement of biota. General 3: p. 11.

Basal Application. Federal--EPA. In pesticides, the spreading of
a chemical on stems or trunks just above the soil line.
General 1: p. 2.

Basal Area. Federal--Interior. The area enclosed by the circum-
ference of a tree trunk, measured at a level 4 feet (1.21
meters) above the ground. General 3: p. 11.

Benthic Organism. Federal--GSA. Organism that spends all or part
of its life cycle at the bottom of a body of water. General 2: p. XI-27.

Benthos. State--Water Resources. The aggregate of organisms
living on or at the bottom of a body of water. Energy/
Utility 2: Appendix E.

Biocoenosis. Federal--Army. An ecological unit comprising both
the plant and animal populations of a habitat; a biological or biotic community. General 9: p. 6.

National--Soil Conservation. The plants and animals comprising a community. Earth 2: p. 9g.

Biological Diversity. Federal--Army. The number of kinds of or-
      ganisms per unit area or volume; the richness of species
      in a given area. General 9: p. 6.

      Federal--Forest Service. In common usage "biological di-
      versity" is usually equivalent to species diversity--i.e.,
      the number of different species occurring in some location
      or under some condition such as pollution. "Biological
      diversity" may also be used in a more general sense to re-
      fer to the number of higher taxonomic levels or types and
      amounts of organismal relationships in some location or
      under some condition--e.g., number of genera, families,
      orders or phyla present, or the number of biotic commun-
      ities present; or the number of energy, nutrient, or food
      chain pathways present. General 15: p. 34.

Biological Handling. Federal--HEW. How a living organism pro-
      cesses a substance; as the uptake, storage, metabolism,
      and exretion of that substance. Health 1: p. 34.

Biological Magnification. Federal--DOT. The tendency for some
      substances (such as DDT) to become concentrated instead of
      dispersed with each link in the food chain. Transporta-
      tion 3: p. 16.

      Federal--EPA. The concentration of certain substances up
      a food chain. A very important mechanism in concentrating
      pesticides and heavy metals in organisms such as fish.
      General 1: p. 2.

Biological Management Unit. Federal--Forest Service. U.S. Forest
      Service usage. This term includes a big-game management
      unit as recognized by cooperating states even though it
      may not be strictly a herd unit. It may include a drain-
      age system in the case of fishery management, any unit for
      species management, or any unit of intensive or special
      management. General 15: p. 34.

Biological Mechanisms. Federal--HEW. The ways used by the living
      organism to process substances. Health 1: p. 34.

Biological Systems. Federal--HEW. Living organisms and their
      vital processes. Health 1: p. 34.

Biomass. Federal--EPA. The amount of living matter in a given
      unit of the environment. General 1: p. 3.

      Federal--ERDA. The total mass of living and dead organisms
      present in an area, volume, or ecological system. General
      7: p. g-1.

Federal--GSA. Amount of living matter (as in a unit area or volume of habitat). General 2: p. XI-27.

Federal--Interior. The total mass or amount of living organisms in a particular area or volume. General 3: p. 13.

State--Natural Resources. Mass of life forms, often applied to one or more species in a particular area. General 12: p. 335.

Biome. Federal--DOT. A large community of living organisms having a peculiar form of dominant vegetation and associated characteristic animals. Transportation 3: p. 16.

Federal--Interior. A complex community of all living organisms; e.g., tundra biome, grassland biome, desert biome. General 3: p. 13.

Biometry. Federal--Interior. The application of statistical methods in the study of biological problems. General 3: p. 13.

Biophysics. Federal--HEW. A branch of knowlege concerned with applying physical properties and methods to biological problems. Health 1: p. 34.

Biota. Federal--Commerce. The plant and animal assemblage of a biological community. General 6: p. 310.

Federal--EPA. All living organisms that exist in an area. General 1: p. 3.

Federal--ERDA. The plant and animal life of a region. General 7: p. g-1.

Federal--Interior. Living organisms; the plant and animal life of a region. General 4: Glossary, p. 2.

State--Water Resources. Animal and plant life, or fauna and flora, of a stream or other water body. Energy/Utility 2: Appendix E.

Biotic. Federal--Forest Service. All the natural living organisms in a planning area and their life processes. The term "biotic" in land use planning contexts is most commonly used as a resource classification category which subdivides the natural resources and properties into either the biota and living characteristics or the abiotic (nonliving) entities and characteristics. General 15: p. 36.

(Revised).

Federal--GSA. Caused or produced by living beings. General 2: p. X-27.

Biotic Factors. National--Soil Conservation. In ecology, those environmental factors which are the result of living organisms and their activities, distinct from physical and chemical factors, such as competition and predation. Earth 2: p. 9g.

Biotic Potential. Federal--Army. The inherent ability of members of a population to grow in numbers within a given time and under stated environmental conditions. General 9: p. 7.

Biotic Productivity. Federal--Forest Service. (1) The amount of living matter actually produced by the unit being discussed, (2) The growth products and by-products of living organisms (e.g., wood, meat, other plant fibers). General 15: p. 164.

Biotic Pyramid. Federal--Army. The set of all food chains or hierarchic arrangements of organisms as eaters and eaten in a prescribed area when tabulated by numbers or by biomasses, usually takes a pyramidal form. General 9: p. 7.

Biotype. National--Soil Conservation. A group of individuals occurring in nature, all with essentially the same genetic constitution. A species usually consists of many biotypes. Earth 2: p. 9g.

Bosque. Federal--Interior. A grove or community of trees in a given area. General 3: p. 14.

Botanical Area. Federal--Forest Service. U.S. Forest Service usage. An area which has been designated by the Forest Service as containing specimens or group exhibits of plants, plant groups, and plant communities which are significant because of form, color, occurrences, habitat, location, life history, arrangement, ecology, environment, rarity, and/or other features. General 15: p. 37.

Breeding Density. Federal--Interior. The density of sexually mature organisms in a given area during the breeding period. General 3: p. 14.

Breeding Potential. Federal--Interior. The maximum rate of increase in numbers of individuals of a species or population under optimum conditions. General 3: p. 14.

Breeding Rate. Federal--Interior. The actual rate of increase of
     new individuals in a given population; the breeding poten-
     tial minus limiting factors. General 3: p. 15.

Browser. Federal--Interior. An animal which feeds on leaves,
     twigs, and young shoots of trees or shrubs; i.e., deer.
     General 3: p. 15.

Buffer Species. Federal--Interior. Alternate prey species ex-
     ploited by predators when a more preferred prey is in rel-
     atively short supply; i.e., if rabbits are scarce, foxes
     will exploit more abundant rodent populations. General 3:
     p. 15.

     National--Soil Conservation. A nongame species that serves
     as food for predators and thus relieves predation on a game
     species. Earth 4: p. 10g.

Canopy. Federal--Army. The leafy cover of vegetation, e.g., the
     uppermost leafy layer in forests. General 9: p. 8.

     Federal--GSA. Uppermost layer of spreading branches in a
     forest. General 2: p. XI-27.

     Federal--Interior. The uppermost layer of a forest or
     woodland, consisting of crowns of trees or shrubs. Gen-
     eral 3: p. 16.

Canopy Cover. City--Planning. Areal coverage of ground by foliage
     as determined from air photographic interpretation; usually
     expressed as a percent. General 5: p. 53.

Carnivore. Federal--GSA. Organism that subsists or feeds on other
     animals. General 2: p. XI-27.

Carrying Capacity. Federal--Army. The maximum population size of
     a given species in an area beyond which no significant in-
     crease can occur without damage occurring to the area.
     General 9: p. 8.

     Federal--EPA. The maximum number of animals an area can
     support during a given period of the year. General 1: p.
     3.

     Federal--Forest Service. Strictly speaking, any level of
     use greater than zero will always result in some altera-
     tion of natural environments, and therefore, it is not use-
     ful to define ecological carrying capacity in "no change"
     terms. Some decision about what is an acceptable amount of
     alteration of natural values must first be made before

ecological carrying capacity can be operational for setting
levels of use. General 15: p. 40.

Federal--Interior. The maximum number of a wildlife species that a given area will support through the most critical period of the year. General 3: p. 16.

Federal--Interior. The maximum population size of a given
species in an area or concentration of pollutants in a
stream, beyond which any increase would cause degradation
of the resource. General 4: Glossary, p. 2.

Regional--River Basin. Ability of a given amount of habitat to support a certain population of animals at a given
time. Plant/Animal 1: p. 451.

Casual Species. Federal--Interior. Species which occur rarely or
without regularity in a given community. General 3: p. 16.

Catabolism. Federal--Interior. The metabolic process whereby complex or simple organic compounds are broken down into sim-
pler substances. General 3: p. 16.

Chemosterilant. National--Soil Conservation. A pesticide chemical that controls pests by destroying their ability to
reproduce. Earth 2: p. 12g.

Chlorinated Hydrocarbons. Federal--EPA. A class of persistent,
broad-spectrum insecticides, notably DDT, that linger in
the environment and accumulate in the food chain. Other
examples are aldrin, diedrin, heptachlor, chlorodane, lin-
dane, endrin, mirex, benzene, hexachloride, and toxaphene.
General 1: p. 3.

Chlorosis. Federal--EPA. Discoloration of normally green plant
parts that can be caused by disease, lack of nutrients or
various air pollutants. General 1: p. 4.

Chorology. Federal--Interior. The scientific study of the geographic distribution of organisms. General 3: p. 18.

Clarke-Bumpus Sampler. Regional--Commission. This instrument,
slowly towed behind a boat, samples a horizontal strata.
It collects primarily larger planktonic organisms and has
a revolving flowmeter that registers water sample volumes.
Sampling should be done at various depths at the same time
of day during the four seasons to establish an accurate
plankton census. The plankton sample is taken to the lab
to be identified, counted, and the number present per acre
of volume calculated. General 10: p. IV-23.

Cleaning Cut. Federal--Forest Service. Timber management usage.
    Cleaning is applied to stands not past the sampling stage.
    Usually, if otherwise practical, this is best done as soon
    as the stand has passed browse height, but not until the
    removal cut has been completed. At this stage, genetic
    differences are likely to be easily recognized; opportun-
    ity to improve species composition is at a maximum; and
    not much production loss to unusable stems will have to
    be accumulated--with the possible exception of areas where
    rainfall exceeds vegetative needs. General 15: p. 43.

Clearance. City--Planning Department. Removal of existing vege-
    tative cover; should be described by the extent to which
    vegetation is removed (e.g., clearcutting, removal of un-
    derstory, thinning of trees, stands, etc.). General 5:
    p. 53.

Clear Cut. Federal--EPA. A forest management technique that in-
    volves harvesting all the trees in one area at one time.
    Under certain soil and slope conditions it can contribute
    sediment to water pollution. General 1: p. 4.

Climax. Federal--Army. The final, stable community in an ecolog-
    ical succession which is able to reproduce itself indef-
    initely under existing conditions. General 9: p. 9.

Climax Community. Federal--Interior. Final or stable community
    in a successional series. General 3: p. 19.

Climax Species. Federal--Forest Service. The members (or a member
    of) those plant and animal species present in a climax com-
    munity. Frequently used in the sense of species usually
    present (or at least common) only as members of a climax
    community. General 15: p. 44.

Climax State. State--Natural Resources. For an ecosystem, the
    condition of being capable of perpetuation under the pre-
    vailing climatic and soil conditions. General 12: p. 336.

Climax Vegetation. Federal--Forest Service. The group of plant
    species which is the culminating stage in plant succession
    for a given set of environmental conditions. General 15:
    p. 44.

Clutch. Federal--Interior. All the eggs of birds, reptiles, or
    amphibians of a given nest. General 3: p. 19.

Cold-Water Fishery. Federal--Forest Service. Stream and lake
    waters which support predominantly cold-water species of
    game or food fishes, which have maximum, sustained water

temperature tolerances of about 70 degrees Fahrenheit in
the summer. Salmon, trout, graying, and northern pike are
examples. General 15: p. 46.

Community. <u>Federal--Army</u>. All of the plants and animals in an
area or volume; a complex association usually containing
both animals and plants. General 9: p. 10.

> <u>Federal--GSA</u>. Any assemblage of populations of organisms
> in a prescribed area of physical habitat. General 2: p.
> XI-27.

> <u>Federal--Interior</u>. A group of plants and animals which
> occupy a given locale. General 3: p. 20.

> <u>Federal--Interior</u>. All of the plants and animals in an
> area with defined boundaries; a complex association usu-
> ally containing both animals and plants. General 4: Glos-
> sary, p. 2.

Conifer. <u>Federal--Army</u>. Pines, cedars, hemlocks, etc.; any of
a type of (mostly) evergreen trees and shrubs with (botan-
ically) true cones. General 9: p. 10.

> <u>Federal--Interior</u>. Cone-bearing trees and shrubs. Mostly
> evergreens such as pine, cedar, spruce, etc. General 3:
> p. 20.

Contact Herbicide. <u>National--Soil Conservation</u>. A herbicide that
kills primarily by contact with plant tissue rather than
as a result of translocation. Earth 2: p. 15g.

Cover. <u>Federal--GSA</u>. Any form of environmental protection that
helps an animal stay alive (primarily shelter from weather
and concealment from predators). General 2: p. XI-27.

Cover Crop. <u>State--Soil and Water Commission</u>. A close-growing
crop grown primarily for the purpose of protecting and
improving soil between periods of permanent vegetation.
Earth 3: p. G-6.

Covert. <u>Federal--Interior</u>. A geographical unit of wildlife cover.
General 3: p. 21.

Crown. <u>National--Soil Conservation</u>. The upper part of a tree,
including the branches and foliage. Earth 2: p. 16g.

Cruising Radius. <u>Federal--Interior</u>. Distance between geographic locations at which an individual animal can be found at
various hours of the day, season, or various years.

General 3: p. 22.

Deciduous. <u>Federal--Army</u>. Falling off or actively shed at maturity or at certain seasons. General 9: p. 11.

<u>Federal--Interior</u>. Falling off at an end of a growing
period (season) or at maturity, as some leaves, antlers,
insect wings, etc. Commonly used term to distinguish trees
which shed their leaves as opposed to evergreens which retain their leaves. Examples: cottonwood, oak, elm, aspen.
General 3: p. 23.

<u>State--Natural Resources</u>. Having leaves which are lost
seasonally. General 12: p. 336.

Deciduous Forest. <u>Federal--Interior</u>. Forests composed of trees
which shed their leaves at maturity or at certain seasons.
General 4: Glossary, p. 3.

Density. <u>Federal--Interior</u>. The number per unit of individuals
of any given species at any given time. General 3: p. 24.

<u>Regional--River Basin</u>. The number of animals present on
a given sized area. Animal 1: p. 451.

Density-Dependent Factor. <u>Federal--Interior</u>. Any environmental
factor that is dependent upon population density to be
fully effective. General 3: p. 24.

Disclimax. <u>Federal--Army</u>. A climax which is the consequence of
repeated or continuous disturbance by man, domesticated
animals of natural events. General 9: p. 12.

<u>Federal--DOT</u>. An ecological community, normally stable
under certain climatic conditions, that has been altered
by man or other influences. Transportation 3: p. 26.

<u>Federal--Interior</u>. An enduring climax community altered
by man's disturbance. General 3: p. 26.

Dispersion. <u>Federal--Interior</u>. The distribution of animals in
a given locale. General 3: p. 26.

Distribution. <u>Federal--Army</u>. The geographic range of a species.
General 9: p. 12.

Diversity. <u>Federal--Commerce</u>. The variety of species present in
a habitat or ecosystem. High diversity indicates environmental health. General 6: p. 311.

National--Soil Conservation. The variety of species within a given association of organisms. Areas of high diversity are characterized by a great variety of species; usually relatively few individuals represent any one species. Areas with low diversity are characterized by a few species; often relatively large numbers of individuals represent each species. Earth 4: p. 20g.

Dominance. Federal--Army. The degree of influence (usually inferred from the amount of area covered) that a species exerts over a community. General 9: p. 13.

Federal--Interior. The condition in communities in which one or more species, by means of their numbers, coverage, or size, have considerable influence upon the conditions of existence of associated species. General 3: p. 27.

Dominant. State--Natural Resources. Referring to species, the one having the most influence in an ecosystem. General 12: p. 336.

Dominant Species. Federal--DOT. In the ecological sense, an organism or several organisms in a community which so behave, either passively or actively, as to dominate or control the whole habitat. Transportation 3: p. 26.

Dredge. Regional--Commission. For freshwater sampling work, there are two main types of dredges to sample shellfish; they are the sledge dredge and the closing dredge. These dredges are towed by a power boat across the bottom to collect the shellfish. The specimens are collected in nets, and the closing dredge has a calculating device for determining the distance towed across the bottom. The specimens are taken from the collection nets, and then identified, their age determined, their species diversity determined, and their number per area sampled is calculated. General 10: p. IV-25.

Dredge Sampler. Regional--Commission. There are two types of dredge samplers used for freshwater bodies; they are the bottom skimmer and the sledge dredge. These samplers are dragged across the bottom of a stream for sampling. These instruments are mainly used for exploratory studies since they do not provide quantitative data. They are very adaptable to different habitat types and can be used under adverse weather conditions. Organisms caught are classified, counted and the data recorded. General 10: p. IV-23.

Dripline. Federal--GSA. Approximately circular area directly
       beneath the leaf canopy of a tree, where most of the feed-
       ing roots are located. General 2: p. XI-27.

Ecological Balance. Federal--Interior. The stability of an eco-
       system resulting from interacting processes of its compon-
       ents. Recreation 3: p. 9.

Ecological Climax. Federal--Interior. The final, stable commun-
       ity in an ecological succession which is able to reproduce
       itself indefinitely under existing conditions. General 4:
       Glossary, p. 4.

Ecological Factor. National--Soil Conservation. Any part or con-
       dition of the environment that influences the life of one
       or more organisms. Earth 4: p. 20g.

Ecological Impact. Federal--EPA. The total effect of an environ-
       mental change, natural or man-made, on the community of
       living things. General 1: p. 6.

Ecological Stress. Federal--Interior. The result or consequent
       state of a physical or chemical, or social stimulus on an
       organism or system; perturbations likely to cause observ-
       able changes in ecosystems; usually departures from normal
       or optimum. General 4: Glossary, p. 4.

Ecological Study Area. Federal--Army. A portion of land whose
       unique or significant natural resources lend themselves to
       scrutiny by institutions for scientific, educational, or
       aesthetic purposes. Public use is generally restricted.
       Recreation 2: p. A-34.

Ecological Succession. Federal--Interior. The replacement of one
       assemblage of plants and animals by another. General 4:
       Glossary, p. 4.

       Federal--Interior. The transition of species of a given
       area through a definite ecological stage; i.e., through
       succession of species composition, grasslands become tree-
       bearing forests. General 3: p. 28.

Ecology. Federal--Army. The study of the interrelationships of
       organisms with and within their environment. General 9:
       p. 13.

       Federal--EPA. The relationships of living things to one
       another and to their environment, or the study of such re-
       lationships. General 1: p. 6.

Federal--ERDA. That branch of biological science which deals with the relationships between organisms and their environment. General 7: p. g-4.

National--Resources. A branch of science concerned with the interrelationships of all animals, plants, insects and other organisms and their environment. Energy/Utility 5: p. 5.

Economic Poisons. Federal--EPA. Chemicals used to control pests and to defoliate cash crops such as cotton. General 1: p. 6.

Ecosystem. Federal--EPA. The interacting system of a biological community and its nonliving surroundings. General 1: p. 6.

Federal--ERDA. An assemblage of biota (community) and habitat. General 7: p. g-1.

Federal--GSA. Complex of a community and its environment, which functions as an interrelated system in nature, linked by the cycling of materials and the flow of energy. General 2: p. XI-27.

Federal--Interior. A community and its living and non-living environment considered collectively, the fundamental unit in ecology. General 4: Glossary, p. 4.

Federal--Interior. A complex system composed of a community of fauna and flora taking into account the chemical and physical environment with which the system is inter-related. General 3: p. 28.

Regional--Commission. An interacting system of one to many living organisms and their nonliving environment. General 13: p. 587.

State--Natural Resources. A life community and its physical environment. General 12: p. 336.

Ecotone. Federal--Commerce. An edge or border zone between different habitats usually with high species diversity. General 6: p. 311.

Federal--Interior. A transition area between plant communities which has some of the characteristics of each. General 3: p. 28.

National--Soil Conservation. A transition line or strip
of vegetation between two communities, having characteristics of both kinds of neighboring vegetation as well as
characteristics of its own. Earth 4: p. 21g.

Ecotype. National--Soil Conservation. A locally adapted population of a species which has a distinctive limit of tolerance to environmental factors. Earth 4: p. 21g.

Edaphic. Federal--Interior. The chemical, physical, or biological characteristics of a given water and soil environment that influence organisms. General 3: p. 28.

Edge. Regional--River Basin. The border between vegetative types
favorable to wildlife. Animal 1: p. 451.

Edge Effect. Federal--Interior. The influence of two communities
upon their adjoining margins or fringes, affecting the
composition and density of the populations in these bordering areas. General 3: p. 28.

Emigration. Federal--Interior. The movement of an animal out of
a given area, generally without returning. General 3: p.
29.

Endangered Species. Federal--Forest Service. An endangered species, or sub-species, of animal or plant is one whose pros-
pects of survival and reproduction are in immediate jeop-
ardy. Its peril may result from one or many causes--loss
of habitat or change of habitat, overexploitation, predation, competition, disease or even unknown reasons. An
endangered species must have help, or extinction may follow. It must be designated in the Federal Register by the
appropriate Secretary as an "endangered species". General
15: p. 204.

Federal--GSA. Species in danger of extinction throughout
all or a significant part of their range. General 2: p.
XI-27.

Federal--Interior. Any species which, as determined by
the Fish and Wildlife Service, is in danger of extinction
throughout all or a significant portion of its range other
than a species of the class Insecta determined to constitute a pest whose protection would present an overwhelming
and overriding risk to man. General 4: Glossary, p. 5.

Federal--Interior. Generally taken to mean any species or
subspecies whose survival is threatened with extinction.
General 3: p. 29.

Endemic. Federal--Interior. A species restricted to a given geo-
    graphical location. Native species to a given locale.
    General 3: p. 29.

Endemic Species. National--Soil Conservation. An organism or
    species that is restricted to a relatively small geograph-
    ical area or to an unusual or rare type of habitat. Earth
    4: p. 21g.

Energy Budget. Federal--Army. A quantitative account sheet of
    inputs, transformations, and outputs of energy in an eco-
    system. May apply to the long-wave radiation (heat) of
    an organism or lake, or to the food taken in and subse-
    quently reduced to heat by an individual or a population.
    General 9: p. 16.

Evapotranspiration. Federal--DOT. Loss of water from a land area
    through transpiration of plants and evaporation from the
    soil. Also, the volumes of water lost through evapotrans-
    piration. Transportation 3: p. 30.

Even-Aged. Federal--Interior. Refers to a stand of trees in which
    only small differences in age occur between the individ-
    uals. General 3: p. 32.

Evergreen. Federal--Interior. A tree or shrub which has green
    leaves throughout the year. General 3: p. 33.

Exotic. Regional--River Basin. Refers to an animal which has
    been introduced into a region, but which is not native
    to that region. Animal 1: p. 451.

Exotic Species. Federal--Executive Order. All species of plants
    and animals not naturally occurring, either presently or
    historically, in any ecosystem of the United States. Le-
    gal Jargon 38: p. 363.

    Federal--Forest Service. U.S. Forest Service usage. Any
    species of wildlife not native to the continental United
    States. General 15: p. 204.

    Federal--Interior. Introduced species. Not indigenous
    to a given area. General 3: p. 33.

Exotic Vegetation. Federal--Army. Vegetation which is introduced
    into a region or site where those species do not normally
    flourish. Survival of exotics is dependent upon extreme
    climatic changes, disease, insect pests and competition
    with native species. Recreation 2: p. A-51.

Eyrie. Federal--Interior. The nest or brood of a bird of prey,
such as an eagle. General 3: p. 33.

Family. Federal--Interior. In taxonomy, a category containing
one or more genera which have similar characteristics.
General 3: p. 34.

Fauna. Federal--Army. The animals of a given region taken collectively; as in the taxonomic sense, the species, or
kinds, of animals in a region. General 9: p. 20.

Federal--Forest Service. The animal life of an area,
"animal" being used in the broad sense to include birds,
fish, reptiles, insects, mollusks, crustaceans, etc.,
in addition to mammals. General 15: p. 78.

Feral Species. Federal--Forest Service. Non-native species, or
their progeny, which were once domesticated but have since
escaped from captivity and are now living as wild animals,
such as wild horses, burrors, hogs, cats and dogs. General
15: p. 204.

Fish. Federal--Forest Service. U.S. Forest Service usage. Includes all species of fresh or salt-water fishes, as well
as crustaceans, mollusks, and other underwater organisms
which are considered part of the fishery resource. General 15: p. 79.

Fish And Wildlife Enhancement. Federal--Army. The addition, deletion or rearrangement of natural or man-made elements on
project lands which strengthens and/or intensifies and
exist within the project. Generally applies to habitat
improvement. Recreation 2: p. A-36.

Fish Attractors. Federal--Army. Natural or man-made features
which are introduced into a lake, river, or stream which
increase the concentration of fish in that area by providing food, cover or spawning habitats. Recreation 2: p.
A-35.

Fishery Management Plan (FMP). Federal--Commerce. A plan developed by a Regional Fishery Management Council to manage a
fishery resource pursuant to the FCMA. Plant/Animal 2:
p. 106.

Fishery Resources. Federal--Army. Those physical elements which
contribute to the reproduction of aquatic species for the
purpose of observation, sport fishing or commercial fishing
such as water quality, watershed management, pool level
control, bank and bottom conditions, etc. Recreation 2:

p. A-36.

Fish Stocking. Federal--Army. The introduction of native fish species into a stream, river, or lake for the purpose of restoring certain fish populations which have been depleted due to overuse, poor water quality, poor habitat or managed reductions in species type. Recreation 2: p. A-49.

Fishway. National--Soil Conservation. A passageway designed to enable fish to ascend a dam; cataract, or velocity barrier. Also called a fish ladder. Earth 2: p. 23g.

Flora. Federal--Army. Plants, organisms of the plant kingdom; specifically, the plants growing in a geographic area; as the Flora of Illinois. General 9: p. 21.

Floristic. Federal--GSA. Of or pertaining to the plant species present in an area. General 2: p. XI-27.

Fluvial. Federal--Army. Applied to plants growing on streams. General 9: p. 21.

Flyway. Federal--Army. A specific corridor of both land and air space within which migratory birds travel and feed during seasonal migrations. Recreation 2: p. A-36.

Federal--Army. The routes taken by migratory birds usually waterfowl during migration. General 9: p. 21.

Federal--Interior. Any one of several established migratory routes of birds. General 3: p. 35.

Food Chain. Federal--Commerce. The step-by-step transfer of food energy and materials, by consumption, from the primary source in plants through increasingly higher forms of animals. General 6: p. 312.

Food Cycle. National--Soil Conservation. All the interconnecting food chains in a community; also called the food web. Earth 4: p. 24g.

Food Web. Federal--Commerce. The network of feeding (trophic) relationships in and between (a) biological community (ies). General 6: p. 312.

Forage. State--Environmental Protection. Food source. General 6: p. 312.

Forest Cover. Federal--Interior. Trees and associated vegetation within a forest. Recreation 3: p. 10.

Forest Cover Type. <u>Federal--Army</u>. A specific forest community
        which is composed of trees, shrubs, herbs, and ground lit-
        ter and inhabits a defined area. Recreation 2: p. A-36.

        <u>Federal--Army</u>. All trees and other woody plants (under-
        brush) covering the ground in a forest. Includes trees,
        herbs, and shrubs, litter and the rich humus of partly
        decayed vegetable matter at the surface of the soil. Gen-
        eral 9: p. 21.

Forest Game. <u>Federal--Army</u>. Any forest fauna upon which hunting
        is regulated by regulations. Recreation 2: p. A-36.

Forest Lands. <u>Federal--Army</u>. Lands which the dominant flora are
        woody species over 10 m. in height. Recreation 2: p.
        A-36.

        <u>Federal--Forest Service</u>. Land at least 10 percent occupied
        by forest trees of any size or formerly having had such
        tree cover and not currently developed for nonforest use.
        General 15: p. 83.

        <u>Regional--Commission</u>. Land which is at least 10 percent
        stocked by trees of any size and land from which the trees
        have been removed to less than 10 percent stocking, but
        which has not been developed for other uses. General 13:
        p. 588.

Fungi. <u>Federal--EPA</u>. Tiny plants that lack chlorophyll. Some
        cause disease, others stabilize sewage and break down solid
        wastes for compost. General 1: p. 7.

Game Fish. <u>Federal--EPA</u>. Species like trout, salmon, bass, etc.,
        caught for sport. They show more sensitivity to environ-
        mental changes than "rough" fish. General 1: p. 7.

        <u>Federal--GSA</u>. Species of fish valued for sporting qual-
        ities on fishing tackle (e.g., salmon, trout, striped
        bass); usually considered to be more sensitive to environ-
        mental change than other fish. General 2: p. IX-18.

Game Range. <u>Federal--Interior</u>. A wildlife area for game animals.
        Recreation 3: p. 11.

Genus. <u>Federal--GSA</u>. Category of biological classification com-
        prising a group of similar species. General 2: p. XI-28.

Grab Samplers. <u>Regional--Commission</u>. Examples include the Ekman
        Grab, Peterson Grab, van Veen Grab, and the Ponar Grab.
        These samplers are of various sizes and shapes, but all

essentially have open jaws that close upon impact with the
bottom or shut by messenger sent down the retrieving cord.
The sampler grabs a measured volume of bottom substrate
and the organisms within that substrate. The organisms
present are identified and counted. In addition, the number of organisms present per area of the sample is determined. General 10: p. IV-23.

Grazing Food Chain. Federal--DOT. Starting from a green plant
base, goes to grazing herbivores (i.e., organisms eating
living plants) and on to carnivores (i.e., animal eaters).
Transportation 3: p. 34.

Groundcover. Federal--DOT. (1) Vegetative growth at the ground
surface: the lowest layer in a multi-layered plant community, (2) Grasses or other plants grown to keep soil
from being blown or washed away. Transportation 3: p. 34.

Group Cutting. Federal--Forest Service. U.S. Forest Service
usage. A clearcutting system variation in which the logging operation removes all merchantable timber from an
area smaller than that normally recognized in timber types
and condition class mapping. Group cuts should not exceed ten acres, nor usually be less than one-fifth acre.
General 15: p. 90. (Revised).

Habitat. Federal--Army. The environment, the natural environment
in which a population of plants or animals occurs. General
9: p. 23.

Federal--Commerce. Place of residence of plants and animals; community of species. General 6: p. 312.

Federal--EPA. The sum of environmental conditions in a
specific place that is occupied by an organism, population,
or community. General 1: p. 8.

Federal--ERDA. The abiotic characteristics of the place
where biota live. General 7: p. g-5.

Federal--GSA. Place or type of site where a plant or animal naturally or normally lives and grows. General 2: p.
XI-28.

Federal--Interior. An area where a plant or animal lives.
(Sum total of environmental conditions in the area.) General 3: p. 39.

Federal--Interior. The natural abode of a plant or animal,
including all biotic, climatic, and soil conditions, or

other environmental influences affecting life. General
4: Glossary, p. 5.

Regional--River Basin. Area which supplies food, water,
shelter, and space necessary for a particular animal's
existence. Plant/Animal 1: p. 452.

Habitat Structure. Federal--Army. The physical structure of a
habitat; e.g., the layering of vegetation in a forest, or
the grain of a coral reef. General 9: p. 23.

Harvest. Regional--River Basin. Number of animals or fish taken
from a population by man for sport or commercial purposes,
usually refers to an annual period, and often expressed
as a percentage of the total population. Plant/Animal 1:
p. 452.

Hedgerow. Federal--Army. A dense thicket of trees or shrubs
which both physically and visually separate two parcels
of land. Also utilized to provide habitat for certain
wildlife species. Recreation 2: p. A-38.

Herbicide. Federal--EPA. A chemical that controls or destroys
undesirable plants. General 1: p. 8.

Herbivore. Federal--GSA. Plant-eating animal. General 2: p.
XI-28.

Home Range. Federal--Army. The area or space of normal activity
of an individual animal; sometimes, but not necessarily
defended against intrusion by other individuals. General
9: p. 24.

Homing. Federal--Interior. The return of animal species to their
spawning or breeding grounds. General 3: p. 41.

Homogeneous Response Unit. Federal--Forest Service. A particular
land (or water) unit which is delineated on the assumption
that it is sufficiently uniform within its boundaries to
respond in a homogeneous manner throughout the area to any
set of inputs. That is, any treatment, stimulus, or set
of conditions will cause the same reaction, response, or
set of outputs when applied to each location within the
unit. General 15: p. 94.

Hoop, Fyke, Trap or Pound Nets. Regional--Commission. Hoop and
Fyke nets are very useful for determining the population
size and distribution of many species in lake or river environments. Trap nets are more efficient than the Hoop
or Fyke nets in deeper waters. The effectiveness of these

nets will vary with fishing conditions and the efficiency
can be increased by baiting. Catches should be studied
in order to establish optimal fishing times. General 10:
p. IV-25.

Hunter Success. Federal--Interior. An expression of individual
hunter satisfaction usually in terms of daily or seasonal
take of game. Recreation 3: p. 11.

Hybrid. Federal--Interior. An organism resulting from a cross
breeding between parents of different genotypes. General
3: p. 42.

Immigration. Federal--Interior. The movement of an animal into
an area not previously occupied by that animal. General
3: p. 43.

Improvement Cutting. Federal--Forest Service. U.S. Forest Service usage. Cutting made in an immature stand to improve
the stand composition by removing trees of poor form and/
or genetic qualities. General 15: p. 96.

Indigenous. Federal--Commerce. Having originated in and being
produced, growing, or living naturally in a particular
region or environment; native species. General 6: p. 312.

Federal--Interior. A species which is native to a given
area. General 3: p. 44.

Indigenous Species. Federal--Forest Service. U.S. Forest Service
usage. Any species of wildlife native to a given land or
water area by natural occurrence. For planning purposes,
indigenous species will include introduced or exotic species which have established a niche in the area's ecology
and are compatible with national forest management objectives. Hungarian and chukar partridges are examples.
General 15: p. 205.

Integrated Pest Management. Federal--EPA. Combining the best of
all useful techniques--biological, chemical, cultural,
physical, and mechanical--into a custom-made pest control
system. General 1: p. 8.

Intensive Forest Management. Federal--Army. Management of forest
resources which provides a sustained yield of merchantible
timber. Recreation 2: p. A-39.

Interspecific. Federal--Interior. Relations between species.
General 3: p. 45.

Intraspecific. Federal--Interior. Relations between individuals
     within a species. General 3: p. 45.

Introduction. Federal--Executive Order. The release, escape, or
     establishment of an exotic species into a natural eco-
     system. Legal Jargon 38: p. 363.

Invader Plant Species. Federal--Interior. Species, often annuals,
     which are not a part of the climax vegetation. General 3:
     p. 45.

Irruption. Federal--Interior. A sudden or dramatic increase in
     a given wildlife population. General 3: p. 46.

Key Management Species. National--Soil Conservation. Forage
     species whose use serves as an indicator of the degree of
     use of associated species. Species on which management
     of a specific unit is based. Earth 2: p. 30g.

Keystone Species. Federal--Army. A species whose removal causes
     marked changes to the community. General 9: p. 26.

Key Utilization Species. National--Soil Conservation. Species
     whose use indicates the degree of use of key grazing areas.
     Earth 2: p. 30g.

Key Winter Range. Federal--Forest Service. The smaller portion
     of the total year's range where big game animals find food
     and/or cover during severe winter weather. "Key winter"
     areas limit the number of animals the range can support.
     General 15: p. 170.

Lentic Habitats. Federal--DOT. Plant and animal habitats found
     in standing bodies of water like ponds and lakes. Trans-
     portation 3: p. 42.

Life Cycle. Federal--Interior. The various stages an animal pass-
     es through from egg fertilization to death. General 3:
     p. 47.

Limiting Factor. Regional--River Basin. That which holds a pop-
     ulation at a certain level at a specific time. Plant/
     Animal 1: p. 452.

Lotic Habitats. Federal--DOT. Plant and animal habitats found
     in flowing waters like rivers and streams. Transportation
     3: p. 43.

Medium-Stocked Stand. Regional--River Basin. A stand that is 40
     to 69 percent stocked with present or potential growing

stock trees. Earth 1: p. 381.

Migration. Federal--Army. A regular movement from one region to
    another. General 9: p. 29.

Mutualism. Federal--Interior. An association of two or more
    species where each species derives some benefit from the
    other. General 3: p. 53.

National Wildlife Refuges. Federal--Interior. Nationally signif-
    icant areas established to enhance wild animal and bird
    populations and public enjoyment. Recreation 3: p. 16.

Native. City--Planning Department. Trees and shrubs and ground-
    cover presently existing or specified to have existed with-
    in the escarpment area in similar soils and topographic
    position. General 5: p. 58.

Native Species. Federal--Executive Order. All species of plants
    and animals naturally occurring, either presently or his-
    torically, in any ecosystem of the United States. Legal
    Jargon 38: p. 363.

Native Vegetation. Federal--Army. Vegetation which naturally
    flourishes in a region and is a part of an existing plant
    association. Recreation 2: p. A-51.

Natural Selection. Federal--EPA. The process of survival of the
    fittest, by which organisms that adapt to their environ-
    ment survive and those that don't disappear. General 1:
    p. 10.

Nature Preserves. Federal--Interior. Natural areas, often limited
    in size, established to preserve distinctive natural com-
    munities of plants and animals of scientific and esthetic
    interest. Recreation 3: p. 17.

Nest Counts. Regional--Commission. This method involves the use
    of a helicopter to count the raptor nest and employs three
    persons: one pilot and two observers. Counting should be
    done over the site area two to five times during the winter
    and spring, prior to the appearance of new leaves on trees.
    Locations of nests and types of raptors should be noted on
    form sheets and plotted on maps. General 10: p. IV-29.

Niche. Federal--Army. The range of sets of environmental condi-
    tions which an organism's behavioral morphological and
    physiological adaptations enable it to occupy; the role an
    organism plays in the functioning of a natural system, in
    contrast to habitat. General 9: p. 31.

Federal--Interior. The specific part or smallest unit of a habitat occupied by an organism. General 3: p. 55.

Non-Game Species. Federal--Army. Those fish and/or wildlife species which are excluded from hunting or fishing activities by law. Utilized primarily for observation, education, research, photography, etc. Recreation 2: p. A-41.

Order. Federal--Interior. In taxonomy, a group of organisms allied between family and class. General 3: p. 56.

Organophosphates. Federal--EPA. Pesticide chemicals that contain phosphorus, used to control insects. They are short-lived but some can be toxic when first applied. General 1: p. 10.

Overpopulation. National--Soil Conservation. A population density that exceeds the capacity of the environment to supply the health requirements of the individual organism. Earth 2: p. 37g.

Overstory. Federal--Interior. The layer of foliage in a forest canopy. General 3: p. 58.

Paleoecology. Federal--Interior. Ecological studies of the past based on evidence collected from fossil remains. General 3: p. 59.

Patch Cutting. Federal--Forest Service. U.S. Forest Service usage. "Patch cuts" are logging operations of the size generally mapped for timber type and condition and for control, but which do not include the entire stand of which they are a part. The minimum size will be 10 acres and the maximum size, 100 acres. General 15: p. 149.

Peripheral Species. Federal--Forest Service. U.S. Forest Service usage. A peripheral species, or subspecies, is one whose occurrence in the United States is at the edge of its natural range, and which is rare or endangered within the United States, although not in its range as a whole. Special attention is necessary to ensure retention in the nation's fauna. General 15: p. 205.

Federal--Interior. A species or subspecies whose geographical distribution is at the margin of its range. General 3: p. 60.

Persistent Pesticide. Federal--EPA. Pesticides that do not break down chemically and remain in the environment after a growing season. General 1: p. 11.

Pesticide. <u>Federal--EPA</u>. Any substance used to control pests
        ranging from rats, weeds, and insects to algae and fungi.
        Pesticides can accumulate in the food chain and can con-
        taminate the environment if misused. General 1: p. 11.

Pesticide Tolerance. <u>Federal--EPA</u>. The amount of pesticide res-
        idue allowed by law to remain in or on a harvested crop.
        By using various safety factors, EPA sets these levels
        well below the point where the chemicals might be harm-
        ful to consumers. General 1: p. 11.

Photosynthesis. <u>Federal--EPA</u>. The manufacture by plants of car-
        bohydrates and oxygen from carbon dioxide and water in the
        presence of chlorophyll, using sunlight as an energy
        source. General 1: p. 11.

Phylum. <u>Federal--Interior</u>. In taxonomy, a primary division of
        the plant and animal kingdom. General 3: p. 62.

Phytoplankton. <u>Federal--GSA</u>. Plankton plant life (e.g., algae).
        General 2: p. XI-28.

Phytotoxic. <u>Federal--EPA</u>. Something that harms plants. General
        1: p. 11.

        <u>State--Board</u>. Toxic to plants. Air 2: no page number.

Pioneer Plant. <u>Federal--DOT</u>. One of the first plants to grow on
        a site which has never before been occupied by vegetation,
        or on which the vegetation has been disturbed, as in clear-
        ing. Transportation 3: p. 52.

Plankton. <u>Federal--Army</u>. Small organisms (animals, plants, or
        microbes) passively floating in water; macroplankton are
        relatively large (1.0 mm to 1.0 cm); mesoplankton of in-
        termediate size; microplankton are small. General 9: p.
        34.

Plant Litter. <u>Federal--DOT</u>. The amount of organic matter incorp-
        orated in all plant elements of the above-ground and under-
        ground parts of the community that die annually, and in
        plants or parts of plants that die in the course of aging
        or natural thinning. Transportation 3: p. 53.

Plant Succession. <u>Regional--River Basin</u>. The series of plant com-
        munities, each replacing another until the climax commun-
        ity stage finally is reached. Plant/Animal 1: p. 452.

Poletimber Stand. <u>Regional--River Basin</u>. Stand at least 10 per-
        cent stocked with growing-stock trees, with half or more

of this stocking in sawtimber and poletimber trees, and
with poletimber stocking exceeding sawtimber stocking.
Earth 1: p. 381.

Poorly Stocked Stand. Regional--River Basin. A stand that is 10
to 39 percent stocked with present or potential growing-
stock trees. Earth 1: p. 381.

Population. Federal--Army. A group of organisms of the same species. General 9: p. 35.

Federal--Interior. An interbreeding group of plants or
animals. The entire group of organisms of one species.
General 3: p. 64.

Population Density. Federal--Army. The number of individuals of
a population per unit area, or volume. General 9: p. 35.

Population Dynamics. Federal--Interior. The process of numerical
and structural change within populations resulting from
births, deaths, and movements. General 3: p. 64.

Population Index. Federal--Army. An estimate of size or other
characteristics of a population, obtained by indirect
means. General 9: p. 35.

Population Irruption. Federal--Army. A sudden, large increase in
population density, resulting in emigration or immigration.
General 9: p. 35.

Population Pressure. Federal--Army. A metaphor implying the magnitude of demand of a population on space or other resources. General 9: p. 35.

Potential Natural Vegetation. Federal--GSA. Vegetation that would
naturally exist in an area by virtue of its climatic, topographic, and geological characteristics, assuming no human
interference or alteration. General 2: p. XI-28.

Predator. Federal--GSA. Animal that survives by eating other
animals. General 2: p. XI-28.

Federal--Interior. An animal which gains nutrients by
capturing and feeding upon other animals. General 3: p.
64.

Prey. Federal--Interior. An animal hunted or killed and used as
food source by another animal. General 3: p. 64.

Prey Organism. Regional--River Basin. Living form that serves
      as sustenance for higher and usually larger forms in a
      food chain. Plant/Animal 1: p. 452.

Primary Succession. Federal--DOT. Colonization, by plants, of a
      site that was not previously occupied by a plant community
      (as on a rock outcrop, for example). Transportation 3:
      p. 55.

Productivity. Federal--GSA. Rate of production of biomass by an
      organism. General 2: p. XI-28.

Protozoa. Federal--Interior. Single-celled animals which inhabit
      fresh water, salt water, and soils. General 3: p. 65.

Radioecology. Federal--EPA. The study of the effects of radia-
      tion on plants and animals in natural communities. Gen-
      eral 1: p. 12.

Range. Federal--Army. The geographic area of occurrence of a
      species; the region over which a given form occurs, nat-
      urally or after introduction. General 9: p. 37.

Range Capacity. Regional--River Basin. This is the maximum stock-
      ing rate of the range possible without inducing damage to
      vegetation and related resources or without preventing re-
      habilitation of previous damage by overgrazing. Capacity
      is discussed in terms of animal unit months or acres per
      animal unit month. Earth 1: p. 378.

Rare Species. Federal--Forest Service. A rare species, or sub-
      species, is one that, although not presently threatened
      with extinction, is in such small numbers, throughout its
      range, that it may be endangered if its environment wors-
      ens. General 15: p. 206.

Recruitment. Federal--Interior. The increase in population caused
      by natural reproduction or immigration. General 3: p. 67.

Reforestation. Federal--Army. The revegetation of a land area
      utilizing primarily woody tree species for the purposes
      of erosion control, wildlife enhancement, visual enhance-
      ment, recreation improvement, etc. Recreation 2: p. A-47.

      Federal--Forest Service. (1) The natural or artificial re-
      stocking of an area with forest trees; most commonly used
      in reference to the latter, (2) "Reforestation" includes
      measures to obtain natural regeneration as well as tree
      planting and seeding. Reforestation is done to produce
      timber and other forest products, protect watersheds,

prevent erosion and improve other social and economic
values of the forests, such as wildlife, recreation and
natural benefits. General 15: p. 174.

Removal Cutting. Federal--Forest Service. The removal of the
seed source, typically the seed trees, after adequate regeneration has taken place. General 15: p. 178.

Removal Regeneration Cutting. Federal--Forest Service. Cutting
a stand which has an advanced manageable understory of
1-20 year old age class reproduction. The seed-tree cutting is bypassed and all the overstory competition is removed in one cutting. The stand is reduced to the single
age class for future management. General 15: p. 178.

Riparian. Federal--Interior. Living one or adjacent to a water
supply such as a riverbank, lake, or pond. General 3:
p. 69.

Risk Cutting. Federal--Forest Service. Cutting to remove trees
that are likely to die before the next periodic cut. General 15: p. 182.

Roadside Car Census. Regional--Commission. This method is used
to survey wintering hawks and other raptor species. A
40-mile route is chosen that is surveyed in a 0.5 mile
wide strip. Two people are needed: one as the driver,
the other as the observer. The survey should be done for
3 or 4 days in the fall and winter between the hours of
1 and 4 P.M. The observer should record the types and numbers on form sheets and maps. General 10: p. IV-29.

Rookery. Federal--Commerce. A communal breeding site for certain
species of aquatic birds. General 6: p. 313.

Root Zone. National--Soil Conservation. The part of the soil
that is penetrated or can be penetrated by plant roots.
Earth 2: p. 43g.

Rotation. Federal--Forest Service. The period of years between
the initial establishment of a stand of timber and the time
when it is considered ready for cutting and regeneration.
General 15: p. 184.

Rough Fish. Federal--Army. A non-sport fish, usually omnivorous
in eating habits but not prized due to poor flavor. Usually not covered under fishing regulations, e.g., carp,
suckers, etc. Recreation 2: p. A-48.

Salvage Cutting. <u>Federal--Forest Service</u>. Cutting primarily to
      utilize dead and downed material and scattered poor-risk
      trees that will not be merchantable if left in the stand
      until the next scheduled cut. General 15: p. 184.

Sanctuary. <u>Federal--Army</u>. A land or water area which is set aside
      by legislation or easements for the preservation and pro-
      tection of flora or fauna. Recreation 2: p. A-48.

      <u>Federal--Army</u>. An area, usually set aside by legislation
      or deed restrictions, for the preservation and protection
      of organisms. General 9: p. 39.

Sapling And Seeding Stand. <u>Regional--River Basin</u>. Stand at least
      10 percent stocked with growing-stock trees, with more than
      half of this stocking in saplings and/or seedlings. Earth
      1: p. 381.

Sawtimber Stand. <u>Regional--River Basin</u>. Stand at least 10 per-
      cent stocked with growing-stock trees, with half or more
      of this stocking in sawtimber and poletimber trees, and
      with sawtimber stockings at least equal to poletimber
      stocking. Earth 1: p. 381.

Scat. <u>Federal--GSA</u>. Animal fecal dropping. General 2: p. XI-29.

Scavenger. <u>Federal--Interior</u>. An animal which eats the decom-
      posing corpses of other animals not killed by itself.
      General 3: p. 70.

Secondary Succession. <u>Federal--DOT</u>. Colonization, by plants, of
      a site where an established vegetative community has been
      eliminated by a natural disaster or through human action.
      Transportation 3: p. 60.

Seed Tree Cutting. <u>Federal--Forest Service</u>. A regeneration cut-
      ting where the planned source of the new stand is from
      seed existing on, or to be produced by, trees standing on
      the cutover area. The cutting removes all the mature
      timber except for the number of seed trees which are needed
      to provide seed for reproducing the stand. General 15: p.
      187.

Selective Pesticide. <u>Federal--EPA</u>. A chemical designed to affect
      only certain types of pests leaving other plants and ani-
      mals unharmed. General 1: p. 13.

Sere. <u>Federal--DOT</u>. The sequence of communities that replace one
      another in a given area is called a sere. Transportation
      3: p. 60.

Shelter Belt. Federal--Interior. A long windbreak of living
      trees and shrubs extending over an area larger than a
      single farm. General 3: p. 71.

Shelterwood Cutting. Federal--Forest Service. Cutting which
      leaves enough trees to provide about half shade or more
      on the ground. In some places, more trees than needed
      to provide shade for reproduction must be left in order
      to prevent windthrow. General 15: p. 188.

Silviculture. Federal--EPA. Management of forest land for timber.
      Sometimes contributes to water pollution, as in clear-cut-
      ting. General 1: p. 14.

Special Cutting. Federal--Forest Service. U.S. Forest Service
      usage. The term usually applies to logging activities in
      special areas, such as recreation areas and administrative
      sites, where other uses or other values override timber
      production. General 15: pp. 202-203.

Species. Federal--GSA. Category of biological organization, com-
      prising related organisms potentially capable of inter-
      breeding. General 2: p. XI-29.

      Federal--Interior. The basic category of biological clas-
      sification intended to designate a single kind of animal
      or plant. Any variation among the individuals may be re-
      garded as not affecting the essential sameness which dis-
      tinguishes them from all other organisms. General 3: p.
      72.

Species Diversity. Federal--DOT. The number of different species
      of plants and animals present in an area of study. Trans-
      portation 3: p. 63.

Species Diversity Indices. Federal--DOT. Mathematical formulae
      used to measure the characteristics of plant or animal
      populations; often used to differentiate areas containing
      many species of equal abundance from areas containing one
      dominant and several subordinate species. Transportation
      3: p. 63.

Stand. Federal--Forest Service. U.S. Forest Service usage. A
      growth of trees on a minimum of 1 acre of forest land
      that is at least 10 percent stocked by forest trees of
      any size. General 15: p. 207.

Stand Cutting. Federal--Forest Service. U.S. Forest Service
      usage. A clearcutting system variation in which the log-
      ging operation removes all merchantable timber from areas

that are large enough to be practical for future management as uniform even-aged stands. The minimum size will be 100 acres and the maximum size, 500 acres. General 15: p. 208.

Status Undetermined Species. <u>Federal--Forest Service</u>. U.S. Forest Service usage. A species, or subspecies, that has been suggested as possibly being endangered but about which there is not enough information to determine its true status. General 15: p. 206.

Strip Census. <u>Regional--Commission</u>. Used during all seasons and involves a person or small crew walking predetermined lines or routes and recording animals seen and the distance away. General 10: p. IV-26.

Strip Cutting. <u>Federal--Forest Service</u>. U.S. Forest Service usage. A clear cutting system variation in which the logging operation removes all merchantable timber from areas that run through a stand and are usually of a width equal to one or two times the general stand height. General 15: p. 210.

Strip Survey. <u>National--Soil Conservation</u>. A survey of one or more sample strips in a forest, these commonly being based on regularly-spaced, open traverses (whence termed a linear survey) along which recording of data is continuous. A survey employing continuous narrow strips as sampling units. Strips $\frac{1}{2}$, 1, or 2 chains wide are run across the area to be surveyed; also called strip cruise. Earth 2: p. 54g.

Sub-Climax. <u>Federal--DOT</u>. Either a climax which has not reached a stable level by reason of factors other than climate or a stage which precedes a true climax but which persists for an unusually long time. Transportation 3: p. 65.

Succession. <u>Federal--Army</u>. The replacement of one community by another; the definition includes the (controversial or hypothetical) possibility of "retrograde" succession. General 9: p. 44.

Successional. <u>State--Environmental Protection</u>. Plant species or vegetative community which will be successively replaced by more stable communities. A sub-climax vegetation type. General 6: p. 314.

Symbosis. <u>Federal--Interior</u>. Two or more species living together for their mutual benefit. General 3: p. 75.

Sympatric. <u>Federal--Interior</u>. Pertaining to two or more closely
    related species occupying identical or overlapping ter-
    ritories. General 3: p. 75.

Synecology. <u>Federal--Interior</u>. That part of ecology that deals
    with groups or organisms. General 3: p. 75.

Systemic Pesticide. <u>Federal--EPA</u>. A chemical that is taken up
    from the ground or absorbed through the surface and car-
    ried through the systems of the organism being protected,
    making it toxic to pests. General 1: p. 15.

Systems Ecology. <u>Federal--Army</u>. That branch of ecology which
    incorporates the viewpoints and techniques of systems
    analysis and engineering especially those having to do
    with the simulation of systems using computers and math-
    ematical models. General 9: p. 45.

System Stability. <u>Federal--Army</u>. The degree to which a system
    continues to function relatively unchanged when stressed
    (perturbed). General 9: p. 45.

Taxonomy. <u>Federal--Interior</u>. The science of classification ac-
    cording to relationships of organisms. General 3: p. 76.

Temporal Census. <u>Regional--Commission</u>. This method is an eco-
    logically stratified ground census that covers 10-15 sites
    per person-day. Good census coverage should be 10-15 days
    per season; mid-winter and the fall and spring migrations.
    Counts for dabbling ducks should be made early in the morn-
    ing (near sunrise) or late in the evening (near sunset),
    but diving ducks can normally be counted throughout the
    day. General 10: p. IV-28.

Territoriality. <u>Federal--GSA</u>. Tendency of an animal species to
    utilize and defend a particular habitat area. General
    2: p. XI-29.

Territory. <u>Federal--Interior</u>. An area over which an animal or
    group of animals establishes jurisdiction. General 3:
    p. 76.

Threatened Species. <u>Federal--Forest Service</u>. Any species which
    is likely to become an endangered species within the fore-
    seeable future throughout all or a significant portion of
    its range and which has been designated in the Federal
    Register by the Secretary of Interior as a threatened
    species. General 15: p. 207.

Federal--Interior. Those species, as determined by the
Fish and Wildlife Service, which are likely to become endangered within the foreseeable future throughout all or
a significant portion of their range. General 4: Glossary,
p. 11.

Threshold. National--Soil Conservation. The maximum or minimum
duration or intensity of a stimulus that is required to
produce a response in an organism; also called the critical level. Earth 4: p. 56g.

Timber Mangement Compartment. Federal--Forest Service. U.S. Forest Service usage. A compartment is defined as an organization unit or small subdivision of forest area for purposes of orientation, administration, and silvicultural
operations, and defined by permanent boundaries (either
of natural features or artificially marked) which are not
neccessarily coincident with stand boundaries. General
15: p. 218.

Timber Resource System. Federal--Forest Service. U.S. Forest
Service usage. The role of the Timber Resource System
is to grow and make available wood for the nation on a
continuing basis. Thus, the system includes those activities necessary to (1) protect, improve, grow, and harvest
timber from forest land and (2) protect, process, and utilize wood and wood-related products. In addition to wood,
the system produces other goods and services, either by
design or incidentally. General 15: p. 218.

Time-Area Counts. Regional--Commission. This process is done by
recording the number of game species observed during the
various seasons over a definite time period on plots of
known size in each ecological type of area. This will require one or more persons to cover each plot in each of
the ecological types. Detailed notes of the time of day
observed, location within plot area, along with detailed
notes on all vegetation present within the plot that is
being used for food, cover, or nesting purposes must be
recorded. General 10: p. IV-26.

Tolerance. Federal--EPA. The ability of an organism to cope with
the changes in its environment. Also the safe level of
any chemical applied to crops that will be used as food or
feed. General 1: p. 15.

Tolerance Limit (TL). National--Soil Conservation. The concentration of a substance that some specified portion of an
experimental population can endure for a specified period
of time with reference to a specified type of response;

for example, TL_{100} means that all test organisms endured
the stress for the specified time; TL_{10} means only 10 percent of the test organisms could tolerate the imposed
stress for the specified time. Earth 2: p. 56g.

Track Counts. Regional--Commission. This method is done during
the fall and spring using a small crew to count and identify tracks of large game species. General 10: p. IV-26.

Transect. Federal--Army. A line (or belt) through a community
on which are indicated the important characteristics of
the individuals of the species observed; sampling along
a transect may be plotless or refer to specific plots.
General 9: p. 47.

Transient Species. Federal--Interior. A species that migrates
through a locality without breeding or overwintering.
General 3: p. 78.

Type Conversion. Federal--Forest Service. The conversion of one
type of vegetation cover to another, e.g., the conversion
of brush or forest covered lands to grass as for grazing
purposes or the conversion of a white fir forest to a
ponderosa pine forest. General 15: p. 220.

Types. Regional--River Basin. A classification of forest land
based upon the predominant species in the present tree
cover. Types are determined on the basis of majority of
stocking by all live trees of various species, considering
both size and spacing. Earth 1: p. 382.

Undergrowth. Federal--Interior. Collectively, the shrubs,
sprouts, seeding and sapling trees, and all herbacious
plants in a forest. General 3: p. 80.

Understory. Federal--Army. Vegetation zone lying between the
forest canopy (overstory) layer and the vegetation covering the ground (ground cover). General 9: p. 48.

Federal--Interior. A layer of foliage below the level of
the main tree canopy. General 3: p. 80.

Uneven-Aged. Federal--Interior. Refers to a forest in which considerable variation occurs in the age of trees. General
3: p. 80.

Unique Species. Federal--Forest Service. U.S. Forest Service
usage. Species which are not endangered, but have considerable scientific, local, or national interest. General 15: p. 207.

Virgin Forest. <u>National--Soil Conservation</u>. A mature or over-
    mature forest essentially uninfluenced by human activity.
    Earth 2: p. 58g.

Warm-Water Fishery. <u>Federal--Forest Service</u>. Stream and lake
    waters which support fishes with a maximum summer water
    temperature tolerance of about 80 degrees Fahrenheit.
    Bluegills, perch and largemouthed bass are examples.
    General 15: p. 229.

Waterfowl. <u>Federal--Army</u>. Birds frequenting water; ordinarily
    referring to game birds such as ducks and geese. General
    9: p. 49.

Water Pumps. <u>Regional--Commission</u>. Water pumps are useful in ob-
    taining plankton from large volumes of water at a given
    strata at depths up to 20-30 meters. Water is pumped
    through a base and filtered through one or more nets to
    capture the plankton. Samples on nets are collected,
    taken back to the lab, counted, identified, and the number
    per liter is calculated to determine the amount of primary
    productivity. General 10: p. IV-23.

Weed Cutting. <u>Federal--Forest Service</u>. Cutting made in a young
    stand, not past the sapling stage, to free the small trees
    from weeds, brush, chaparral, vines, sod-forming grasses,
    or other competing vegetation. General 15: p. 230.

Well-Stocked Stand. <u>Regional--River Basin</u>. A stand that is 70
    percent or more stocked with present or potential growing-
    stock trees. Earth 1: p. 381.

Wildlife. <u>Federal--Commerce</u>. A collective term used for living
    organisms neither human or domesticated. General 6: p.
    315.

Wildlife And Fish Habitat System. <u>Federal--Forest Service</u>. U.S.
    Forest Service usage. This system protects and improves
    wildlife and fish habitat with special emphasis on threat-
    ened and endangered species. Management of wildlife and
    fish habitats is closely coordinated with the states be-
    cause they control wildlife and fish populations. This
    coordination includes (1) close working relations among
    National Forest, State, and private land managers; (2)
    cooperative forestry programs designed to assist non-Fed-
    eral land managers; and (3) research programs that define
    environmental requirements of fish and wildlife and provide
    management alternatives through which these requirements
    can be attained. One of the six "systems" established by
    the U.S. Forest Service to have a systematic, orderly way

to view and evaluate its many diverse but interrelated
activities. The Forest Service has developed this approach
to better respond to the mandates of the Forest and Rangeland Renewable Resources Planning Act of 1974. It has
grouped its various programs into these six "systems",
each of which incorporates all the activities concerned
with developing and managing a specific resource. General 15: pp. 233-234.

Wildlife Enhancement. Federal--Army. Manipulation of wildlife
regions to promote increases in the amount or quality of
living animals. General 9: p. 50.

Wildlife Habitat. Federal--Army. Suitable upland or wetland areas
promoting survival of wildlife. General 9: p. 50.

Wildlife Management. Federal--Interior. The application of techniques for maintaining or modifying wild animal populations. Recreation 3: p. 28.

Wildlife Refuge. Federal--EPA. An area designated for the protection of wild animals, within which hunting and fishing
is either prohibited or strictly controlled. General 1:
p. 12.

Wildlife Stocking. Federal--Army. The introduction of wildlife
species into a region for the purpose of restoring species
which have been extinct in that area, providing more diversity for consumptive uses or for re-establishing a once
balanced ecosystem. Recreation 2: p. A-50.

Windbreak. Federal--Interior. A planting of trees, shrubs, or
other vegetation to protect soil, crops, etc., against the
effects of winds. General 3: p. 82.

Xerophyte. Federal--Interior. Plants which are structurally
adapted to growing in dry or desert conditions. The plants
often have a greatly reduced leaf surface area to prevent
water loss; thick, fleshy parts for water storage; and
many possess hairs, spines, or thorns. Examples: cacti,
Joshua tree, yucca. General 3: p. 83.

Year Class. Federal--Interior. Animals born in a given year.
General 3: p. 83.

Zone. National--Soil Conservation. An area characterized by similar flora and fauna; a belt or area to which certain species are limited. Earth 2: p. 61g.

Zoogeography. Federal--Interior. The science of the geographical
        distribution of animals. General 3: p. 84.

Zoological Area. Federal--Forest Service. U.S. Forest Service
        usage. An area which has been designated by the Forest
        Service as containing authentic, significant and inter-
        esting evidence of our American national heritage as it
        pertains to fauna. The areas are meaningful because they
        embrace animals, animal groups or animal communities that
        are natural and important because of occurrence, habitat,
        location, life history, ecology, environment, rarity or
        other features. General 15: p. 238.

Zooplankton. Federal--ERDA. Microscopic animals that live drift-
        ing in a body of water. General 7: p. g-11.

        Federal--GSA. Animal plankton life. General 2: p. XI-29.

Zymogenous Flora. Federal--Interior. Organisms found in large
        numbers immediately following the addition of redily de-
        composable organic materials. General 3: p. 86.

8. NOISE

Index

Rayleigh Wave
Reflection Barrier
Reverberation
Risk
Sleep Interference
Slow C-Weighted Sound Level
    (L_{CS})
Slow Sound Level (L_{AS})
Sound
Sound Exposure (E)
Sound Exposure Level (L_{AE})
Sound Exposure Level Contour
Sound Level (L_A, L)
Sound Level Meter

Sound Pressure (P)
Sound Pressure Level (L_p)
Speech Interference
Stationary Equipment Noise
Structural Resonance
Task Interference
Vehicular Noise
Vibratory Acceleration (a)
Vibratory Acceleration Level
    (La)
Vibrational Climate
Vibrational Damping
Yearly Day-Night Average Sound
    Level (L_{dny})

Terms

Acoustic Barrier. Federal--DOT. An obstacle composed of sound-
    absorbing material or oriented such that it reflects sound
    weakly or towards the sky. Transportation 3: p. 11.

Ambient Noise. Federal--GSA. Noise of a measurable intensity
    normally present in the background of a given environment.
    General 2: p. V-21.

Ambient Noise Level. Federal--Interior. The background noise
    level within a given area for certain period of time dur-
    ing the 24-hour day. General 4: Glossary, p. 1.

A-Scale Sound Level. Federal--EPA. A measurement of sound approx-
    imating the sensitivity of the human ear, used to note the
    intensity or annoyance of sounds. General 1: p. 2.

    National--Soil Conservation. The measurement of sound
    approximating the auditory sensitivity of the human ear;
    used to measure the relative noisiness or annoyance of
    common sounds. Earth 4: p. 7g.

Audiometer. Federal--EPA. An instrument that measures hearing
    sensitivity. General 1: p. 2.

Average Sound Level (L_T). Federal--Academy of Science. A sound
    level typical of the sound levels at a certain place in
    a stated time period. Technically, average sound level
    in decibels is the level of the mean-sqaure A-weighted
    sound pressure during the stated time period, with ref-
    erence to the square of the standard reference sound pres-
    sure of 20 micropascals. Average sound level differs
    from sound level in that for average sound level, equal
    emphasis is given to all sounds within the stated averaging
    period, whereas for sound level an exponential time weight-
    ing puts much more emphasis on sounds that have just oc-
    curred than those which occurred earlier. Noise 2: p. A-2.

A-Weighted Decibel Scale. Federal--DOT. The measurement of sound
    approximating the auditory sensitivity of the human ear.
    The A-Scale sound level is used to measure the relative
    noisiness or annoyance of common sound. Transportation 3:
    p. 11.

Background Noise. Federal--GSA. Total of all noise in a system
    or situation independent of the presence of the desired
    signal. In accoustical measurements, the term "background
    noise" means electrical noise in the measurement system.

However, in popular usage, the term is also used with the same meaning as "residual noise". General 2: p. V-21.

Community Noise Equivalent Level (CNEL). Federal--GSA. Cumulative measure of community noise. CNEL uses the A-weighted sound level and applies weighting factors that place greater importance upon noise events occurring during the evening (7:00 p.m. to 10:00 p.m.) and night (10:00 p.m. to 6:00 a.m.) hours. General 2: p. V-21.

Community Reactions. Federal--GSA. Community reactions to excessive noise may take the form of complaints to authorities, political action against noisy activities, reduction in land values, high property turnover rates, or changes in family recreational patterns. General 2: p. V-21.

Composite Noise Rating (CNR). Federal--GSA. Noise exposure used for evaluating land use around airports. CNR is in wide use by the Department of Defense in predicting noise environments around military airfields. General 2: p. V-21.

Construction Equipment Noise. Regional--Commission. Construction activities result in noise being generated from mobile or stationary construction equipment. This construction noise is treated separately due to the short-term nature of the noise compared to project operational noise. General 10: p. IV-60.

C-Weighted Sound Exposure Level (L_{CE}). Federal--Academy of Science. In decibels, the level of the time integral of C-Weighted, squared sound pressure, with reference to the square of 20 micropascals and to one second. Noise 2: p. A-3.

Day Average Sound Level (Ld). Federal--Academy of Science. Average sound level over the 15-hour time period from 7 a.m. to 10 p.m. (0700 up to 2200 hours). Noise 2: p. A-2.

Day-Night Average Sound Level (Ldn). Federal--Academy of Science. The 24-hour average sound level, in decibels, from midnight to midnight, obtained after addition of 10 decibels to sound levels in the night from midnight up to 7 a.m. and from 10 p.m. to midnight (0000 up to 0700 and 2200 up to 2400 hours). Noise 2: p. A-2.

Federal--GSA. Day-night average sound level (i.e., 24-hour weighted equivalent sound level, with 10 dB penalty applied to nighttime levels). General 2: p. V-23.

Day-Night Average Sound Level Contour. <u>Federal--Academy of Science</u>. A curved line connecting places on a map where the day-night average sound level is the same. If only one kind of contour is shown on the map the fact may be made known by a single legend, "Contours of day-night average sound level in decibels". In this case only the number of decibels need be marked on a contour. Noise 2: p. A-2.

dB(A). <u>Federal--GSA</u>. Unit of sound level with A-weighted characteristics. General 2: p. V-21.

Decibel (dB). <u>Federal--Academy of Science</u>. A unit measure of sound level and other kinds of levels. Noise 2: p. A-1.

<u>Federal--GSA</u>. Logarithmic measure of the magnitude of a particular quantity (e.g., sound pressure, sound power, intensity) with respect to a standard reference value (e.g., 20 micropascals for sound pressure). General 2: p. V-21.

<u>Federal--Interior</u>. A unit which measures the relative intensity of sound; one decibel being approximately the least change detectable by the average human ear. General 4: Glossary, p. 3.

Dosimeter. <u>Federal--GSA</u>. Instrument that registers the occurrence and cumulative duration of noise exceeding a predetermined level at a chosen point in the environment or on a person. General 2: p. V-21.

Eight Hour Average C-Weighted Sound Level (L_{C8h}). <u>Federal--Academy</u>. Average sound level, in decibels, over a given 8 hour time period, measured with the C-frequency weighting. Noise 2: p. A-3.

Eight Hour Average Sound Level (L_{8h}). <u>Federal--Academy of Science</u>. Average sound level, in decibels, over an 8-hour period. Noise 2: p. A-2.

Equivalent Continuous Sound Level (LeqT). <u>Federal--Academy of Science</u>. Same as average sound level. The pertinent time period must be stated. Noise 2: p. A-2.

Equivalent Sound Level. <u>Federal--GSA</u>. Level of a constant sound which, in a given situation and time period, has the same sound energy as does a time-varying sound. Technically, equivalent sound level is the level of the time-weighted, mean square, A-weighted sound pressure. The time interval over which the measurement is taken should always be specified. General 2: p. V-21.

Fast Sound Level (L$_{AF}$). Federal--Academy of Science. In decibels,
     the exponential-time-averaging sound level measured with
     the squared-pressure time constant of 125ms. The A-fre-
     quency weighting is understood. Noise 2: p. A-1.

Frequency. Federal--GSA. Number of times per second that a per-
     iodic sound repeats itself; expressed in Hertz (Hz), form-
     erly in cycles per second (cps). General 2: p. V-22.

Hearing Impairment. Federal--GSA. Hearing loss exceeding a des-
     ignated criterion (e.g., 25 dB hearing threshold level,
     averaged from the threshold levels at 500, 1000, and 2000
     Hz). General 2: p. V-22.

Hearing Level. Federal--GSA. Difference in sound pressure level
     between the threshold sound for a person (or the median
     value or the average for a group) and the reference sound
     pressure level defining the ASA standard audiometric
     threshold (ASA: 1951), expressed in decibels. The term
     is now commonly used to mean hearing threshold (qv).
     General 2: p. V-22.

Hearing Loss. Federal--GSA. Impairment of auditory sensitivity
     or elevation of a hearing threshold level. Hearing loss
     is known to result from exposures to continuous noise in
     industrial settings, impulsive sound, gunfire, and loud
     music for extended periods. The effect of fluctuating,
     intermittent, or shorter-term exposures has not been fully
     determined. General 2: p. V-22.

Hearing Threshold Level. Federal--GSA. Amount (in decibels) by
     which the threshold of hearing for an ear (or the average
     for a group) exceeds the standard audiometric reference
     zero (ISO 1964; ANSI 1969). General 2: p. V-22.

Hertz (Hz). City--Commission. The abbreviation for Hertz, and is
     equivalent to cycles per second. Noise 1: p. 1.

Hourly Average Sound Level (Lh). Federal--Academy of Science.
     Average sound level, in decibels, over a one-hour time
     period, usually reckoned between integral hours. It may
     be identified by the begining and ending times, or by the
     ending time only. Noise 2: p. A-2.

Impedance. Federal--EPA. The rate at which a substance absorbs
     and transmits sound. General 1: p. 8.

Impulse Noise. <u>Federal--GSA</u>. Noise of short duration (typically
    less than one second) and high intensity, with abrupt onset
    and rapid decay and often rapidly changing spectral comp-
    osition. Impulse noise is characteristically associated
    with such sources as explosions, impacts, discharge of
    firearms, passage of supersonic aircraft (sonic boom),
    and many industrial processes. General 2: p. V-22.

Impulse Sound Level (L_{A1}). <u>Federal--Academy of Science</u>. In deci-
    bels, the exponential-time-average sound level obtained
    with a squared-pressure time constant of 35 milliseconds.
    The A-frequency weighting is understood. Noise 2: p. A-1.

Instantaneous Sound Pressure, Overpressure (P_i). <u>Federal--Academy
    of Science</u>. Pressure at a place and instant considered,
    minus the static pressure there. Noise 2: p. A-3.

L_{10} Level. <u>Federal--GSA</u>. Sound level exceeded 10 percent of the
    time period during which measurement was made. General
    2: p. V-22.

L_{50} Level. <u>Federal--GSA</u>. Sound level exceeded 50 percent of the
    time period during which measurement was made. General 2:
    p. V-23.

L_{90} Level. <u>Federal--GSA</u>. Sound level exceeded 90 percent of the
    time period during which measurement was made. General 2:
    p. V-23.

L_{eq}. <u>Federal--GSA</u>. "Energy averaged" weighted equivalent sound
    level over a given time interval. General 2: p. V-23.

Level. <u>Federal--GSA</u>. Logarithm of the ratio of a quantity of a
    reference quantity of the same kind. The base of the log-
    arithm, the reference quantity, and the kind of level must
    be specified. General 2: p. V-22.

Loudness. <u>Federal--GSA</u>. Attribute of an auditory sensation in
    terms of which sounds may be ordered on a scale extending
    from soft to loud, measured in tones. Loudness is chiefly
    a function of intensity but also depends upon the fre-
    quency and wave form of the stimulus. General 2: p. V-23.

Maximum Sound Level (Lmax). <u>Federal--Academy of Science</u>. The
    greatest sound level during a designated time interval
    or event. More specifically, it is the greatest FAST
    A-weighted sound level of the event. Noise 2: p. A-1.

Night Average Sound Level (Ln). <u>Federal--Academy of Science</u>.
    Average sound level, in decibels, over the split nine-hour

period from midnight up to 7 a.m. and from 10 p.m. to midnight (0000 up to 0700 and 2200 up to 2400 hours). Noise 2: p. A-2.

Noise.   Federal--EPA. Any undesired sound. General 1: p. 10.

Federal--GSA. Disturbing, harmful, or unwanted sound; erratic, intermittent or statistically random oscillation. General 2: p. V-23.

Federal--Interior. Any unwanted extraneous electrical quantity or sound which modifies the transmission, indication, or recording of desired data. General 4: Glossary, p. 8.

Noise Exposure. Federal--GSA. Integrated effect over a given period of time of a number of different events of equal or different noise levels and durations. The integration may include weighting factor for the number of events during certain time periods. General 2: p. V-23.

Noise Exposure Forecast. Federal--GSA. Method currently used for making noise exposure forecasts that uses a perceived noise level scale with additional corrections for the presence of pure tones. General 2: p. V-23.

Noise Hazard. Federal--GSA. Acoustical stimulation of the ear, likely to produce noise-induced permanent threshold shift in some fraction of the population. General 2: p. V-23.

Noise Level (L_A, L). Federal--Academy of Science. Same as sound level, for sound in air. Some people use "noise" only for sound that is undesirable. A sound level meter does not, however, measure people's desires. Hence there is less likelihood of misunderstanding, if what is measured by a sound level meter is called sound level, rather than noise level. Noise 2: p. A-1.

Federal--GSA. Averaged sound level (or weighted sound pressure level), with weighting specified. General 2: p. V-23.

Noise Limit. Federal--GSA. Graphical, tabular, or other numerical expression of the permissible amount of noise that may be produced by a practical source (e.g., a vehicle or an appliance) or that may invade a specified point in a living or working environment (e.g., in a work place or residence) under prescribed conditions of measurements. General 2: p. V-23.

Noise Pollution. <u>Federal--Forest Service</u>. The addition of energy
    in the form of sound to the environment beyond what would
    naturally occur or of a type which would not naturally
    occur. The degree of pollution is measured in terms of
    intensity, duration, frequency of occurrence, and sound
    frequency (wavelength or pitch). General 15: p. 141.

    <u>National--Soil Conservation</u>. The persistent intrusion of
    noise into the environment at a level that may be injurious
    to human health. Earth 4: p. 35g.

    <u>City--Commission</u>. The presence of that amount of acoustic
    energy for that amount of time necessary to: (1) cause
    temporary or permanent hearing loss in persons exposed;
    (2) otherwise be injurious, or tend to be, on the basis
    of current information, injurious, to the public health
    or welfare; (3) cause a nuisance; (4) interfere with the
    comfortable enjoyment of life and property or the conduct
    of business; or (5) exceed standards or restrictions es-
    tablished herein or pursuant to the granting of any permit
    by the Commission. Noise 1: p. 1.

Noise Pollution Level. <u>Federal--DOT</u>. A noise rating descriptor
    or procedure representative of the level of annoyance
    caused by the fluctuating aspect of the noise. Transpor-
    tation 3: p. 48.

Outdoor-Indoor Sound Level Difference (D_A). <u>Federal--Academy of
    Science</u>. Difference in decibels, between the average
    sound level outside a building, at a position two or more
    meters from the facade or roof as appropriate, and the
    space-time average sound level in a designated room due
    to the outdoor sound. When the outdoor sound is caused by
    a moving vehicle it often suffices to measure the indoor
    sound only near the middle of the room. Noise 2: p. A-3.

Peak Sound Level (L_{Apk}). <u>Federal--Academy of Science</u>. The great-
    est instantaneous A-weighted sound level, during a desig-
    nated time interval or event. Noise 1: p. A-1.

Peak Sound Pressure (P_{pk}). <u>Federal--Academy of Science</u>. Greatest
    absolute instantaneous sound pressure in a stated frequency
    band, during a given time interval. Noise 2: p. A-3.

Peak Sound Pressure Level (L_{pk}). <u>Federal--Academy of Science</u>.
    In decibels, twenty times the common logarithm of the
    ratio of a greatest absolute instantaneous sound pressure
    to the reference sound pressure of twenty micropascals
    (0.0002 microbar). Noise 2: p. A-3.

Physiological Stress. <u>Federal--GSA</u>. Stress on the functions and activities of a living cell, tissue, or organism. Noise of excessive intensity and duration has proved to induce the same physiological reactions as other stressors (e.g., emotional stress, and pain): alter the functions of the endocrine, cardiovascular and neurological systems; affect equilibrium; and change the constriction of blood vessels, cetebral vasodilation, blood pressure, heart rhythm, and rate of stomach acid secretions. General 2: p. V-24.

Pitch. <u>Federal--GSA</u>. Attribute of auditory sensation in terms of which sounds may be ordered on a scale extending from low to high. Pitch is primarily a function of the frequency of the sound stimulus, but also depends upon the sound pressure and wave form of the stimulus. General 2: p. V-24.

Rayleigh Wave. <u>Federal--DOT</u>. A type of surface vibration having a retrograde, elliptical motion at the ground. Transportation 3: p. 57.

Reflection Barrier. <u>Federal--DOT</u>. A rigid plane oriented perpendicularly to the ground which reflects or alters the direction of sound waves. Transportation 3: p. 58.

Reverberation. <u>Federal--EPA</u>. The echoes of a sound that persists in an enclosed space after the sound source has stopped. General 1: p. 13.

Risk. <u>Federal--GSA</u>. Percentage of a population whose hearing level, as a result of a given influence, exceeds the specified value, minus that percentage whose hearing level would have exceeded the specified value in the absence of that influence, other factors remaining the same. General 2: p. V-24.

Sleep Interference. <u>Federal--GSA</u>. Noise can prevent sleep, cause awakening, and change the level or pattern of sleep. These interferences may have both temporary and long-term health effects on behavior and performance during waking hours. Survey data indicate that sleep disturbance is often the principal reason reported for annoyance, and some experts consider sleep disturbance to be one of the most severe effects of noise on health. General 2: p. V-24.

Slow C-Weighted Sound Level (L_{CS}). <u>Federal--Academy of Science</u>. In decibels, the exponential time average sound level measured with the squared-pressure time constant of one second and the C-frequency weighting of the sound level meter. Noise 2: p. A-3.

Slow Sound Level (L_{AS}). <u>Federal--Academy of Science</u>. In decibels, the exponential-time-average sound level measured with the squared-pressure time constant of one second. The A-frequency weighting is understood. Noise 2: p. A-1.

Sound. <u>Federal--GSA</u>. Oscillation in pressure, stress, particle displacement, particle velocity, etc., in a medium with internal forces (e.g., elastic, viscous), or the superposition of such propagated alterations. General 2: p. V-24. (Revised).

Sound Exposure (E). <u>Federal--Academy of Science</u>. Time integral of squared, A-frequency-weighted sound pressure over a stated time interval or event. The exponent of sound pressure and the frequency weighting may be otherwise if clearly so specified. Noise 2: p. A-2.

Sound Exposure Level (L_{AE}). <u>Federal--Academy of Science</u>. The level of sound accumulated over a given time period or event. It is particularly appropriate for a discrete event such as the passage of an airplane, a railroad train, or a truck. Sound exposure level is not an average, but a kind of sum. In contrast with average sound level which may tend to stay relatively constant even though the sound fluctuates, sound exposure level increases continuously with the passing of time. Technically, sound exposure level in decibels is the level of the time integral of A-weighted squared sound pressure over a stated time interval or event, with reference to the square of the standard reference pressure of 20 micropascals (0.0002 microbar) and reference duration of one second. Noise 2: pp. A-2, A-3.

Sound Exposure Level Contour. <u>Federal--Academy of Science</u>. A curved line connecting places on a map where the sound exposure level of a discrete event is the same. Noise 2: p. A-3.

Sound Level (L_A, L). <u>Federal--Academy of Science</u>. The quantity in decibels measured by an instrument satisfying requirements of American National Standard Specification for Sound Level Meters S1.4-1971. Fast time-averaging A-frequency weighting are understood, unless others are specified. The sound level meter with the A-weighting is progressively less sensitive to sounds of frequency below 1000 hertz (cycles per second), somewhat as in the ear. With FAST time averaging the sound level meter responds particularly to recent sounds almost as quickly as does the ear in judging the loudness of sound. Noise 2: p. A-1.

Federal--GSA. Weighted sound pressure level, obtained
through metering techniques and the weightings A, B, or
C as specified in the American National Standard Specification for Sound Level Meters, ANSI-S1. 4-1971. The weighting employed must be stated. General 2: p. V-24.

Sound Level Meter. Federal--GSA. An instrument, comprising a
microphone, an amplifier, an output meter, and frequency-
weighting networks, that is used for the measurement of
noise and sound levels in a specified manner. General 2:
p. V-25.

Sound Pressure (P). Federal--Academy of Science. Root-mean-square
of instantaneous sound pressures over a given time interval. The frequency bandwidth must be identified. Noise
2: p. A-3.

Sound Pressure Level (Lp). Federal--Academy of Science. In decibels, twenty times the common logarithm of the ratio of
a sound pressure to the reference sound pressure of twenty
micropascals (0.0002 microbar). The frequency bandwidth
must be identified. Noise 2: p. A-3.

Speech Interference. Federal--GSA. In addition to the reduced
understanding of speech resulting from noise-induced hearing loss, noise can interfere directly with the accuracy,
frequency, and quality of verbal exchange, all of which
are important in formal education, occupational efficiency,
family life patterns, and quality of relaxation. General
2: p. V-25.

Stationary Equipment Noise. Regional--Commission. The operation
of various projects require the use of equipment which generate noise that may be annoying for persons in the vicinity of the project or for employees. General 10: p. IV-60.

Structural Resonance. Federal--DOT. There are for most structures
situations in which the ground vibrations are such that the
largest possible amplitude in the steady state occurs. The
frequency of the vibration for which this takes place is
said to be the resonant frequency for the maximum displacement of the structure. Transportation 3: p. 64.

Task Interference. Federal--GSA. Noise interfers with the performance of tasks that require response to auditory signals. Both high-level continuous and intermittent, unexpected noise exposures have disruptive effect on human performance. Noise may increase the variability of work rate
and affect the accuracy of work requiring mental concentration. Individual thresholds vary widely as a result of

many psychological and social factors. General 2: p.
V-25.

Vehicular Noise. Regional--Commission. Vehicular noise is that
noise generated by cars and trucks on streets and highways. Projects which require new highway construction,
the building of access roads, or generate or attract
traffic will generate vehicular noise. General 10: p.
IV-60.

Vibratory Acceleration (a). Federal--Academy of Science. The
rate of change of speed and direction of a vibration, in
a specified direction. The frequency bandwidth must be
identified. Noise 2: p. A-3.

Vibratory Acceleration Level (La). Federal--Academy of Science.
In decibels, twenty times the common logarithm of the
ratio of a vibratory acceleration to the reference acceleration of ten micrometers per second squared (nearly one-
millionth of the standard acceleration of free fall).
The frequency bandwidth must be identified. Noise 2: p.
A-3.

Vibrational Climate. Federal--DOT. The vibrational characteristics of the area of interest. Transportation 3: p. 71.

Vibrational Damping. Federal--DOT. Attenuation of vibrational
energy due to losses or leakages in a system. Transportation 3: p. 71.

Yearly Day-Night Average Sound Level (L_{dny}). Federal--Academy of
Science. The day-night average sound level, in decibels,
averaged over an entire calendar year. Noise 2: p. A-2.

Cultural Glossaries

9. TRANSPORTATION

Index

Terms

Access Characteristics. <u>Federal--DOT</u>. Refers to the type and
volume of vehicles allowed to access a transportation
facility as well as the frequency and location of access.
Transportation 3: p. 11.

Accessibility And Convenience. <u>Regional--Commission</u>. Accessibility is a purpose of transportation, namely, to move goods
and people as efficiently as possible between locations.
Accessibility and convenience can be measured in terms of
number of alternate modes available, time costs, travel
costs, and ease of access to these modes. General 10:
p. IV-93.

Access Road. <u>Federal--Army</u>. Roadways which are the route of
vehicular travel between an existing public thoroughfare
and a public use area. Recreation 2: p. A-29.

<u>Federal--DOT</u>. A road giving direct access to the land and
premises on one or both sides. Transportation 3: p. 11.

Alternative Analysis. <u>Federal--DOT</u>. A process of studying a
variety of transportation alternatives to any fixed guide-
way project. Its purpose is to assess different transit
modes and technologies appropriate to the service requirements of specific corridors to ensure that the needs of
localities are met. Legal Jargon 35: p. A-1.

Alternatives. <u>State--Transportation</u>. Optional courses of action,
including various transportational modes and facilities,
as well as, non-transportation and non-construction decisions. Transportation 1: p. vi.

Arterial. <u>Federal--DOT</u>. The second rank in the classification of
thoroughfares. A restricted access road with limited vehicular access to adjoining properties, but no direct access from individual driveways. Ideally, traffic flow is
interrupted only at street intersections where it is controlled by traffic signals. Transportation 3: p. 14.

Arterial Highways. <u>Federal--GSA</u>. Highway designed primarily for
through traffic, usually on a continuous route. General
2: p. III-23.

Average Annual Daily Traffic. <u>Federal--GSA</u>. Total yearly volume
divided by the number of days in the year, commonly abbreviated as AADT. General 2: p. III-23.

Average Daily Traffic (ADT). <u>Federal--DOT</u>. The total volume during a given time period in whole days greater than one day and less than one year divided by the number of days in that period, commonly abbreviated as ADT. Transportation 3: p. 14.

    <u>Federal--GSA</u>. Total number during a given time period divided by the number of days in that time period. General 2: p. III-23.

Average Overall Travel Speed. <u>Federal--GSA</u>. Summation of distance traveled by all vehicles or a specified class of vehicles over a given section of roadway during a specific period of time, divided by the summation of overall travel time; travel speed used for estimating level of service on urban and suburban roads. General 2: p. III-24.

Balanced Transportation System. <u>Federal--DOT</u>. All facilities and services for intrametropolitan travel will be treated as part of a single system, each component to be planned in a manner most effectively utilizing its special characteristics in combination with other elements. Transportation 3: p. 15.

Bicycle Lane. <u>Federal--DOT</u>. A portion of a roadway or right-of-way which has been designated for preferential or exclusive use by bicycles. It is distinguished from the portion of the roadway for motor vehicular traffic by a paint stripe, curb or other similar device. Transportation 3: p. 16.

Bicycle Route. <u>Federal--DOT</u>. Any road, street, path or way which in some manner is specifically designated as being open to bicycle travel, regardless of whether such facilities are designated for the exclusive use of bicycles or are to be shared with other transportation modes and/or pedestrians. Transportation 3: p. 16.

Bike Lane. <u>Federal--Interior</u>. A restricted right-of-way, designated for exclusive or semi-exclusive use of bicycles, usually a designated portion of a roadway. Recreation 3: p. 5.

Bypass Road. <u>Federal--DOT</u>. A road which takes through traffic around a congested area and thereby facilitates through movement and relieves local congestion. Transportation 3: p. 18.

Capacity. <u>Federal--DOT</u>. The capability of a roadway to accomodate traffic, usually expressed in the number of vehicles

per lane per hour. The maximum number of vehicles which
has a reasonable expectation of passing over a given section of a lane or a roadway in one direction (or in both
directions) during a given time period under prevailing
roadway and traffic conditions. Transportation 3: p. 18.

Channels Of Movement. Federal--DOT. The collection and distribution routes along which people and goods pass. These include various kinds of roads and rail lines, terminals,
and facilities for parking, loading, unloading, and transferring people or goods. Transportation 3: p. 19.

Collector. Federal--DOT. The third rank in the classification of
thoroughfares. A free access road which provides the link
between arterials and local access streets. It has few
restrictions on entrances and serves both to move vehicles
and, to a lesser degree, to serve adjoining properties.
Traffic control is usually by stop signs. Transportation
3: pp. 20-21.

Commuter. Federal--DOT. One who travels back and forth between a
city and an outside residence, or from a city to an outside
work location, to earn a livelihood. Transportation 3: p.
21.

Congestion. Federal--GSA. Occurring when normal level of service
on a given lane or roadway is restricted, placing limitations on speed, travel time, maneuverability, driving com-
fort and convenience. General 2: p. III-23.

Continuing Urban Transportation Planning. State--Transportation.
The updating and evaluating of all elements leading to the
transportation plan including researching new methods and
additional elements to be used in modifying the plan from
forecasted conditions. Transportation 1: p. vii.

Convenient Commuting Range. Federal--DOT. The distance which can
be traversed easily from home to school or work by public
or private transportation. Transportation 3: p. 23.

Corridor. State--Transportation. That defined area between two
points through which a transportation facility may be proposed. Transportation 1: p. vii.

Crossover. Federal--DOT. Refers to pedestrain or vehicular links
(at-grade or grade-separated) across a transportation facility. Transportation 3: p. 23.

Curb Cut. Federal--DOT. A break in the curb along the edge of a
pavement which defines the edge to vehicle operators. The

break or cut allows vehicular egress and ingress between a highway and abutting land. Transportation 3: p. 24.

Design Alternatives. Federal--DOT. Highway plans incorporating optional combinations of design features; e.g., degree of access control, number of lanes, grade separation, elevated, at-grade or depressed, etc. Transportation 3: p. 25.

Design Study Report. State--Transportation. A report that describes essential elements, such as design standards, number of traffic lanes, access control features, general horizontal and vertical alignment, right-of-way requirements and location of bridges, interchanges, and other structures. Transportation 1: p. viii.

Desire Line. Federal--DOT. Denotes a straight line drawn between the origin and destination of a trip. It is not necessarily the actual route of a journey (which has to follow the existing road system) but symbolizes the need or desire to make the trip. When put on a graph, desire lines summarize the data collected by an origin-destination survey. Transportation 3: p. 25.

Destination Attractiveness. Federal--DOT. Refers to the relative desire of individuals to reach a specific opportunity in an area. Transportation 3: p. 25.

Expressway. Federal--GSA. Divided arterial highway for through traffic with full or partial control of access, generally with grade separations at major intersections. General 2: p. III-23.

Freeway. Federal--GSA. Expressway with full control of access. General 2: p. III-23.

Freeways And Expressways. Federal--DOT. The first rank in the classification of thoroughfares. Divided multi-lane limited access roads with grade-separated interchanges, total control of access and buffers along the sides. Designed to move large volumes of through vehicular traffic. They provide for no vehicular or pedestrian access to adjoining properties. Typified by the interstate highway system. Transportation 3: p. 32.

Functional Class Or Type. Federal--DOT. The division of highways into various categories according to physical characteristics and the function or type of travel needs to be served; e.g., through traffic, local traffic, access to property. Typical classes or types include: arterial highway, expressway, parkway, local street or road. Transportation

3: p. 33.

Headway. Federal--DOT. The time span between the arrival of successive vehicles or units of mass transit. The scheduled
frequency of service, dependent, of course, on the time of
day or night. Transportation 3: p. 34.

Heavy Duty Vehicle (HDV). A motor vehicle either designated primarily for transportation of goods and rated at more than
6,000 pounds, or designated primarily for transportation
of persons and hauling a capacity of more than 12 persons.
Transportation 3: p. 34.

High-Type Pavement. Federal--DOT. Highway pavements having
smooth, hard, well-maintained, wearing, or running surface that minimize friction between pavement and vehicle
tires and impedance due to roughness. Transportation 3:
p. 35.

Horizontal Curvature. Federal--DOT. Curvature in a roadway as
seen in plan view (from above). Transportation 3: p. 35.

Isochrons. Federal--Forest Service. A concentric set of lines
each of which joins distances with equal travel times from
some central reference point of concern. General 15: p.
104.

Jitney. Federal--DOT. A small bus that carries passengers over
a regular route according to a flexible schedule. Transportation 3: p. 39.

Lane. Federal--DOT. A narrow way intended to carry a single line
of moving persons or vehicles. A six-lane highway, for example, is one with 3 lanes in either direction. A rever-
sible lane is one which may be used by vehicles operating
in different directions at different times, switching at
specific times to accommodate peak flows. Transportation
3: p. 41.

Level Of Service. Federal--DOT. A qualitative measure of operating conditions that may occur on a given lane or roadway
when it is accommodating various traffic volumes. Transportation 3: p. 42.

Level Terrain. Federal--GSA. Any combination of gradients, length
of grade, or horizontal or vertical alignment that permits
trucks to maintain speeds that equal or approach the speeds
of passenger cars. General 2: p. III-24.

Light Duty Vehicle (LDV). Federal--DOT. Any motor vehicle either
    designed primarily for transportation of goods and rated
    at 6,000 pounds gross vehicle weight (GVW) or less, or
    designated primarily for transportation of persons and
    hauling a capacity of 12 persons or less. Transportation
    3: p. 42.

Link. Federal--DOT. In traffic assignment, a section of the high-
    way network between two nodes defined for the purpose of
    analysis. Transportation 3: p. 42.

Local Access Street. Federal--DOT. The fourth rank in the class-
    ification of thoroughfares. A free access road which pro-
    vides access for pedestrians and vehicles to properties
    that front it. It is not intended for through traffic.
    Transportation 3: p. 43.

Localized Movement Patterns. Federal--DOT. Observed, recurrent
    trip characteristics within a zone. Transportation 3:
    p. 43.

Local Street Or Road. Federal--GSA. Street or road primarily for
    access to a residence, business, or other abutting prop-
    erty. General 2: p. III-23.

Location Team. Federal--DOT. An organization of professionals
    which is assigned the task of conducting studies of al-
    ternative highway locations and designs. This team may
    have as few as two or three professionals or as many as
    100. It may be an element of a state department of trans-
    portation or highways, some other state or local agency,
    a metropolitan planning council, or a consulting firm hired
    by such agencies. Transportation 3: p. 43.

Major Street Or Highway. Federal--GSA. Arterial highway with in-
    tersections at grade and direct access to abutting prop-
    erty. General 2: p. III-23.

Major Thoroughfare. City--Planning. A cross-town street whose
    primary function is to link districts within the City and
    to distribute traffic from and to the freeways; a route
    generally of citywide significance; as identified in Thor-
    oughfare Plan of the Transportation Element of the Compre-
    hensive Plan. General 8: p. 3.

Market Linkages. Federal--DOT. The series or system of ties be-
    tween an economic region or specific establishment pro-
    ducing goods or services and the geographic areas in which
    they are sold; the route and mode by which they are trans-
    ported. Transportation 3: p. 44.

Market Reach. Federal--DOT. The farthest point from an economic
        enterprise that its goods or services can be offered for
        sale at competitive prices. This concept is particularly
        sensitive to transportation costs. Transportation 3: p.
        44.

Mass Transit. Federal--DOT. The act or means of conveying masses
        of people from place to place along a given right-of-way
        system. Routes are prearranged, and service is operated
        according to prescribed schedules. Transportation 3: p.
        44.

Median Barrier. Federal--DOT. A protective wall or guard-rail
        erected in that portion of a divided highway separating
        the traveled ways for traffic in opposite directions.
        Transportation 3: p. 44.

Modal Alternatives. Federal--DOT. Optional means of transporta-
        tion; e.g., automobile, bus, railroad, rapid transit, bi-
        cycles, etc. Transportation 3: p. 45.

Modal Split. Federal--DOT. The term applied to the division of
        person trips between public and private transportation;
        the process of separating person trips by the mode of
        travel. For example, in a given city, the modal split
        might show that 60 percent of school children walk to
        school, 30 percent use a school bus and 10 percent are
        driven by parents. Transportation 3: p. 45.

        City--Planning. Percentage of the total number of trips
        made by each mode of travel (bus, private car, walking,
        etc.). General 8: p. 3.

Modal-Split Model. State--Transportation. Evaluation of effi-
        ciency of highway and mass transportation facilities and
        forecasting future travel demands by mode. Transportation
        1: p. xi.

Mode. State--Transportation. A means of movement within a trans-
        portation system. The automobile is an example of a trans-
        portation mode. Transportation 1: p. xi.

Mode-Split Percentage. Federal--GSA. Percent of all travelers
        in a given area (city or region) who will use the avail-
        able or proposed transportation modes. Trends can usually
        be observed from historical data or results of mode split
        models. General 2: p. III-24.

Multi-Modal. State--Transportation. As it relates to transpor-
        tation, it is the consideration of alternate modes of

transportation, e.g., highway, mass transit, during all
phases of development. Transportation 1: p. xi.

Multiple Use. Federal--DOT. Widely practiced concept of devoting
portions of a highway right-of-way to non-highway purposes,
e.g., parks, public buildings, apartments, etc. These
other uses make use of space over, under or within the
highway right-of-way. Transportation 3: p. 46.

Network Models. Federal--DOT. A simulated highway system for a
given area composed only of connections (links) between
zone centroids (nodes) without respect to physical street
layout. Used in conjunction with mathematical formulae
that express the actions and interactions of the elements
of a system in such a manner that traffic volumes generated in each node and utilizing each link may be predicted
and evaluated under many given sets of conditions. Transportation 3: p. 48.

No-Build Option. Federal--DOT. An alternative solution frequently
used in highway planning studies that assumes proposed improvements are not made. Is often used as an evaluation
device to compare the costs and benefits of making a highway improvement to those incurred by continuing to use existing facilities. Transportation 3: p. 48.

Origin-Destination Survey. Federal--DOT. A traffic study technique that systematically samples the movement of people,
vehicles, and goods in a given area with a view to determining where they begin and end their journeys, the mode
of travel, the elapsed time and the land use at origin and
destination. Transportation 3: p. 50.

State--Transportation. That process that gathers data on
the number and type of trips including movement of vehicles
and passengers from various zones of origin to various
zones of destination. Transportation 1: p. xii.

Parking Area. Federal--Army. An area designed and constructed
for the parking of more than one vehicle and may include
parking for trailers, bicycles, buses and traffic islands,
interior circulation roads and landscaping. Recreation 2:
p. A-43.

Parking Demand. Federal--GSA. Number of cars requiring parking
spaces and the duration of their stay. General 2: p. III-
24.

Parking Space. Federal--Army. A particular site which is designed
and constructed for the parking of a single vehicle.

Several parking spaces equal a parking area. Recreation
2: p. A-43.

Parkway. Federal--GSA. Arterial highway for noncommercial traffic with full or partial control of access, located within
a park or other green space. General 2: p. III-23.

Peak-Hour Traffic. Federal--GSA. Highest number of vehicles that
pass over a section of a roadway or lane during 60 consecutive minutes. General 2: p. III-24.

Primary Facility. Federal--Forest Service. Transportation planning usage. Primary facilities provide access and service
to large land areas encompassed and served by the public
transportation system. They include public transportation
facilities and form the basic framework around which the
transportation network is designed. A substantial percentage of the traffic served by these facilities is public service traffic, and emphasis is given to travel speed
and efficiency. General 15: p. 162.

Profile. Federal--DOT. An engineering drawing of a roadway
showing a side or sectional elevation. Transportation
3: p. 55.

Project. State--Transportation. A specific length of roadway
located between logical termini or a specific facility.
Transportation 1: p. xii.

Public Transit. Federal--DOT. Transportation services available
to the general public. As public transport, it may also
be regulated as to its operations, charges and profits.
Transportation 3: p. 56.

Public Transportation Routes. Federal--DOT. The precise paths
(sequence of streets) and scheduled stops of public transportation vehicles. Transportation 3: p. 56.

Rapid Transit. Federal--DOT. Mass transit with an exclusive
right-of-way. Intended to convey people as quickly as
possible from point to point. Transportation 3: p. 57.

Ring Road. Federal--DOT. A road, or circumferential highway,
that avoids the core of an urban place, permitting through
traffic to bypass the center and local traffic to distribute itself to various points around the center. Transportation 3: p. 58.

Road. Federal--DOT. An improved line of communication for passage
or travel between different places that is wide enough for

vehicles and is of reasonable length. Transportation 3:
p. 59.

Road Hierarchy. Federal--DOT. The classification of throughfares
by their varying sizes and purposes. The generally accept-
ed basic categories, according to purpose, type of access
control, and other features are freeways and expressways,
arterials, collectors and local access streets. Transportation 3: p. 59.

Roadway. Federal--GSA. Public way for purposes of vehicular and
pedestrian travel. General 2: p. III-24.

Roadway Capacity. Federal--GSA. Maximum number of vehicles that
can reasonably be expected to pass over a given section of
a roadway or lane during a given period of time under prevailing roadway and traffic conditions. General 2: p.
III-24.

Roadway Conditions. Federal--GSA. Factors that affect roadway
capacity, including lane width, lateral clearance, and
other physical features of the roadway. General 2: p.
III-24.

Rolling Terrain. Federal--GSA. Any combination of gradients,
length of grade, or horizontal or vertical alignment that
causes trucks to reduce their speeds substantially below
that of passenger cars on some sections of the road, but
which does not involve sustained crawl speed by trucks for
any substantial distance. General 2: p. III-25.

Safety. Regional--Commission. Safety is a critical consideration
in both the design and operation of a transportation facility. The degree of safety may be expressed in terms of
accidents per vehicle or passenger miles for comparison
with accident rates on similar facilities. A comparison
of relative safety of alternate transportation modes may
also be made. General 10: p. IV-94.

Secondary Facility. Federal--Forest Service. Transportation planning usage. Secondary facilities provide access to smaller
land areas than primary facilities. They are multi-re-
source oriented, providing access and mobility for the
utilization of a variety of resource systems, and are
usually developed and operated for long-term service. Emphasis is given to achieving a balance between travel efficiency and resource service. General 15: p. 186.

Section. Federal--DOT. Longitudinal profile view of a roadway
seen in side elevation. Cross section comprises a view

of the roadway at right angles to the longitudinal section.
Transportation 3: p. 60.

Speed. Federal--GSA. Rate of movement of vehicular traffic, expressed in miles per hour. General 2: p. III-24.

Statewide Multi-Modal Transportation Plan. State--Transportation.
Coordinates plans with ongoing urban-urbanized study areas
to develop comprehensive long-range statewide transportation networks. Transportation 1: p. xv.

Systems Planning. State--Transportation. A procedure for developing an integrated means of providing adequate facilities
for the movement of people and goods. Regional analysis
of transportation needs and the identification of transportation corridors is involved. Transportation 1: p. xv.

Tertiary Facility. Federal--Forest Service. Transportation planning usage. Tertiary facilities are usually intended to
provide access for a specific resource utilization activity, such as a timber sale or recreation site, although
other minor uses may be served. Emphasis is given to resource service rather than travel efficiency. Tertiary
facilities may often be developed and operated for short
term or intermittent service. General 15: p. 217.

Three C Urban-Urbanized Transportation Study. State--Transporta-
tion. A comprehensive, cooperative and continuing study
process established by PPM 50-9 and conducted in major
metropolitan areas, urbanized areas of 50,000 or more and
in selected urban areas with populations approaching 50,000
as required by the 1962 Federal-aid Highway Act. Transportation 1: p. vii.

Traffic. Federal--DOT. Vehicles and/or persons in motion through
an area or along a route or stopped because they are temporarily prevented from moving. Transportation 3: p. 68.

Traffic Burden. Federal--DOT. Traffic volume occurring on a given
roadway or a given region; or the component of total volume
on a roadway which results from a single source. Transportation 3: p. 68.

Traffic Conditions. Federal--GSA. Factors that affect roadway
capacity, including the percentage of trucks and buses in
the traffic stream, parking practices, and the presence of
turning vehicles. General 2: p. III-25.

Traffic Count. Federal--DOT. The number of individuals, vehicles,
or animals passing a given point within a specified time.

The count generally includes the flow in both directions. Transportation 3: p. 68.

Traffic Volumes. Federal--DOT. The number of vehicles that pass over a given section of a lane or roadway during a time period of one hour or more. Volume can be expressed in terms of daily traffic or annual traffic as well as on an hourly basis. Transportation 3: p. 68.

Federal--GSA. Number of vehicles that pass over a given section of a roadway or land during a given time period. General 2: p. III-25.

Transit Service. Federal--GSA. Accessibility (in time or distance) of transit stops to particular origins and destinations, and the frequency of service. General 2: p. III-24.

Transportation. Federal--DOT. The act or means of moving tangible objects (persons or goods) from place to place. Often involves the use of some type of vehicle. Transportation 3: p. 68.

Transportation Plans. Regional--Commission. Each project is to be judged in terms of conflict or conformance with area land use and transportation plans. Because transportation plays such a vital role in community development, this analysis has particular importance. General 10: p. IV-94.

Transportation Terminal. Federal--DOT. A facility where transfer between modes of transportation takes place; also, any facility providing for one or more of the following: the arrival and embarkation of passengers; the receipt, dispatching and temporary storage of goods; the termination point and temporary housing of vehicles. Railroad stations, airports, truck and trailer depots, docks, parking lots and garages are among the varieties of terminal facilities. Transportation 3: p. 69.

Travel Patterns. State--Transportation. Data related to travel patterns collected and analyzed for a study area to determine the overall network of work, shop, social, business and recreational trips generated within the study area. Transportation 1: p. xv.

Trip. Federal--DOT. A one-way journey that proceeds from an origin to a destination by a single type of locomotion. Transportation 3: p. 69.

Turnover. Federal--GSA. Number of vehicles that use a single
         parking space during a day. General 2: p. III-25.

Vehicle Miles Traveled (VMT). The aggregate of total number of
         miles traveled by all vehicles over a given roadway or on
         all roadways within a specified geographic area during a
         given time period; commonly abbreviated as VMT. Transpor-
         tation 3: p. 71.

         State--Board. A term used to describe the number of miles
         traveled by vehicles in a specified area during a given
         time. Air 2: no page number.

Vehicle Mix. Federal--DOT. The relative percentage of specific
         vehicle types contained in a given volume of traffic.
         Transportation 3: p. 71.

Vertical Curvature. Federal--DOT. Curvature in a roadway as seen
         in profile (side view); i.e., the gradient characteristics
         of a roadway. Transportation 3: p. 71.

Volume/Capacity Ratio. Federal--DOT. The ratio of traffic volume
         to roadway capacity; commonly abbreviated v/c. Transporta-
         tion 3: p. 71.

         Federal--GSA. Comparison of traffic volumes with roadway
         capacity; defines a level of service. General 2: p. III-
         25.

10.   ENERGY/UTILITIES

Index

Composite Wastewater Sample
Condenser Cooling Water
Conservation
Construction Waste
Consumer Waste
Contact Filter
Conventional Hydroelectric
    Plant
Conversion
Cooling Tower
Cooling Water Consumption
Cooling Water Requirement
Core
Corridor
Cover Material
Critical Period
Curie
Curtailment
Cycling Load
Cyclone Separator
Dechlorination
Decomposition Of Wastewater
Degree Of Treatment
Demand
Demolition Waste
Demonstration Grant
Dependable Capacity
Desulfurization
Detention Tank
Detention Time
Dewatering
Diatomaceous Earth
Diffused Air
Diffuser
Digested Sludge
Digester
Digestion Chamber
Dilution
Dilution Factor
Disinfected Wastewater
Disinfection
Disposal
Disposal By Dilution
Distribution Line
Distribution Ratio
Dry-Weather Flow
Dump
Economically Recoverable Min-
    eral Reserves
Economic Mineral Reserves

Effluent
Effluent Seepage
Electric Utility Industry Or
    Electric Utilities
Electrodialysis
Electro-Process Industry
Electrostatic Precipitator
Elutriation
Energy
Energy Content Curve
Energy Recovery
Energy Resource Planning
Escherichia Coli (E. coli)
Evaporation Ponds
Exfiltration
Extended Aeration
Ferrofluid
Ferrous
Filter
Filtered Wastewater
Filter Plant
Filtration
Firm Energy
Firm Load
Firm Power
Firm Water Supply
Five-Day BOD
Five-Year Storm
Floc
Flocculation
Flocculation Agent
Flowmeter
Force Main
Fossil Fuels
Front-End System
Fuel
Fuel Oil
Gamma Rays
Garbage
Gasification
Gas Turbine Generating Station
Generating Station
Generation
Generator
Geothermal Generating Station
Grab Sample
Gravity Separation
Grease Trap
Hazardous Waste
Headworks

Heating Season
Heterogeneous Waste
Holding Pond
Homogeneous Waste
Hydro
Hydroelectric Generating
       Station
Hydropower
Imhoff Tank
Incineration
Incinerator
Incinerator Capacity
Indicated Mineral Reserves
Industrial Firm Power
Industrial Waste
Inertial Separator
Inferred Mineral Reserves
Influent
Inorganic Refuse
In-Plant Waste
Institutional Waste
Intercepting Sewer
Interceptor Sewers
Interlocking
Internal Combustion Gener-
       ating Station
Interruptible Power
Junk
Lateral Sewers
Lift
Lift Station
Liquefied Natural Gas (LNG)
Liquid Waste
Litter
Load
Load Curve
Load Diversity
Load Factor
Load Pattern
Load Shape
Magnetic Separator
Main Sewer
Material Balance
Materials Recovery
Maximum Plant Capability--Hydro
Measured Mineral Reserves
Megawatt (MW)
Methane
Mineral Fuels
Mineral Reserves

Mixed Liquor
Most Probable Number (MPN)
Municipal Solid Waste
Natural Gas
Nitrogenous Waste
Non-Renewable Resource
Nuclear Energy
Nuclear Fuel
Nuclear Generating Station
Nuclear Power
Nuclear Power Plant
Nuclear Reactor
Ocean Dumping
Odor Control
Odor Threshold
Open Burning
Organic Refuse
Outfall
Oxidation Pond
Oxidation Rate
Oxygen Demand
Oxygen Depletion
Peak Demand
Peaking Capability
Peaking Plant
Peak Load
Penstock
Petroleum
Phenols
Phenol Wastes
Pig
Pile
Pollution Load
Potable Water
Power
Power Pool
Power Supply Area
Pretreatment
Primary Sedimentation
Primary Settling Tank
Primary Treatment
Primary Waste Treatment
Prime Energy
Process Water
Public Utility
Public Utility District
Pumping Station
Purification
Putrefaction
Quench Tank

Radiation
Radiation Standard
Raw Sewage
Receiving Body Of Water
Receiving Stream
Receiving Waters
Reclamation
Recoverable Resources
Recycling
Refuse
Refuse Reclamation
Reserve Capacity
Residential Waste
Resource Recovery
Resource Recovery Facility
Reverse Osmosis
Rubbish
Run-Of-Canal Plant
Run-Of-River Plant
Sanitary Landfill
Sanitary Landfilling
Sanitary Sewer
Sanitary Wastewater
Sanitation
Secondary Energy
Secondary Power
Secondary Treatment
Secondary Waste Treatment
Secondary Wastewater Treatment
Sedimentation
Sedimentation Basin
Sedimentation Tanks
Separate Sewer System
Septic Sludge
Septic Tank
Service Area
Settleable Solids
Settling Chamber
Sewage
Sewage Sludge
Sewer
Sewer Outfall
Sewer System
Screening
Single Purpose Corridor
Skimming
Sludge
Sludge Bed
Sludge Deposits
Sludge Digestion

Sludge Treatment
Sodium Alkyl Benzene Sulfonate
     (ABS)
Soil Wastes
Solar Energy
Solid Waste
Solid Waste Disposal
Solid Waste Management
Source Separation
Stabilization Pond
Steam Generating Station
Storm Drain
Storm Runoff
Storm Sewer
Strip Mining
Substation
Sump
Surface Dump
Surface Runoff
Suspended Solids (SS)
Tertiary Treatment
Thermal Plant
Thermal Processing
Total Solids
Transfer Station
Transmission Grid
Transmission Line
Trash
Trickling Filter
Ultimate Plant Capability--Hydro
Urban Waste
Usable Energy
Utilities
Waste
Waste Materials
Waste Processing
Waste Stream
Waste Treatment
Wastewater
Wastewater Influent
Wastewater Reclamation
Water Supply System
Wet-Weather Treatment Facility
Winter Peak
Yard Waste

Terms

Activated Carbon. Federal--EPA. A highly adsorbent form of car-
     bon used to remove odors and toxic substances from gaseous
     emissions. In advanced waste treatment, it is used to re-
     move dissolved organic matter from waste water. General
     1: p. 1.

          State--Water Resources. Carbon particles usually obtained
          by carbonization of cellulosic material in the absence of
          air and possessing a high adsorptive capacity. Energy/
          Utility 2: Appendix E.

Activated Sludge. Federal--EPA. Sludge that has been aerated and
     subjected to bacterial action; used to speed breakdown of
     organic matter in raw sewage during secondary waste treat-
     ment. General 1: p. 1.

          Federal--Interior. Material containing a very large amount
          of active microbial population produced in one method of
          sewage disposal by aeration of sewage. General 3: p. 2.

          State--Water Resources. Sludge floc produced in raw or
          settled wastewater by the growth of zooleal bacteria and
          other organisms in the presence of dissolved oxygen and
          accumulated in sufficient concentration by returning floc
          previously formed. Energy/Utility 2: Appendix E.

Activated Sludge Process. State--Water Resources. A biological
     wastewater treatment process in which a mixture of waste-
     water and activated sludge is agitated and aerated. The
     activated sludge is subsequently separated from the treated
     wastewater (mixed liquor) by sedimentation and wasted or
     returned to the process as needed. Energy/Utility 2:
     Appendix E.

Activated Sludge Secondary Treatment. City--Planning Department.
     A biological wastewater treatment process that follows pri-
     mary treatment (sedimentation). A mixture of wastewater,
     bacteria, and other microorganisms is mixed and aerated.
     Suspended matter is later separated from the treated waste-
     water. Energy/Utility 9: p. 1.

Aerated Contact Bed. State--Water Resources. A biological unit
     consisting of stone, cement-asbestos, or other surfaces
     supported in an aeration tank, in which air is diffused up
     and around the surfaces and settled wastewater flows
     through the tank. Also called contact aerator. Energy/
     Utility 2: Appendix E.

Aerated Pond. <u>State--Water Resources</u>. A natural or artificial
    wastewater treatment pond in which mechanical or diffused
    air aeration is used to supplement the oxygen supply. En-
    ergy/Utility 2: Appendix E.

Aeration. <u>Federal--EPA</u>. To circulate oxygen through a substance,
    as in waste water treatment where it aids in purification.
    General 1: p. 1.

    <u>State--Water Resources</u>. The bringing about of intimate
    contact between air and a liquid by one or more of the fol-
    lowing methods: (1) spraying the liquid in the air, (2)
    bubbling air through the liquid, (3) agitating the liquid
    to promote surface adsorption of air. Energy/Utility 2:
    Appendix E. (Revised).

Aeration Tank. <u>State--Water Resources</u>. A tank in which sludge,
    wastewater, or other liquid is aerated. Energy/Utility
    2: Appendix E.

Aerobic Digestion. <u>National--Resources</u>. The utilization of or-
    ganic waste as a substrate for the growth of bacteria
    which function in the presence of oxygen to reduce the
    volume of waste. The products of this decomposition are
    carbon dioxide, water and a remainder consisting of inor-
    ganic compounds and any undigested organic material. En-
    ergy/Utility 5: p. 2.

Aerochlorination. <u>State--Water Resources</u>. The application of
    compressed air and chlorine gas to wastewater for the re-
    moval of grease. Energy/Utility 2: Appendix E.

Agglomeration. <u>State--Water Resources</u>. The coalescence of dis-
    persed suspended matter into larger flocs or particles
    which settle rapidly. Energy/Utility 2: Appendix E.

Agricultural Waste. <u>National--Resources</u>. Waste materials pro-
    duced from the raising of plants and animals for food.
    These materials include such things as animal manure,
    plant stalks, hulls and leaves. Energy/Utility 5: p. 2.

Alpha Particle. <u>Federal--EPA</u>. The least penetrating type of
    radiation, usually not harmful to life. General 1: p. 1.

    <u>National--Soil Conservation</u>. A positively charged particle
    emitted by certain radioactive materials; the least pen-
    etrating of the three common types of radiation (alpha,
    beta, and gamma) and usually not dangerous to plants, ani-
    mals, or man. Earth 2: p. 6g.

Alternating Device. State--Water Resources. A device in a waste-
    water treatment plant whereby wastewater may be delivered
    automatically or manually into different parallel treat-
    ment units in a cycle in accordance with a predetermined
    sequence. Energy/Utility 2: Appendix E.

Anaerobic Digestion. National--Resources. The utilization of or-
    ganic waste as a substrate for the growth of bacteria which
    function in the absence of oxygen to reduce the volume of
    waste. Energy/Utility 5: p. 2. (Revised).

Annual Maximum Demand. National--Electric Institute. The greatest
    of all demands of the load under consideration which oc-
    curred during a prescribed demand interval in a calendar
    year. Energy/Utility 3: p. 24.

Artificial Sludge. State--Water Resources. Substances, such as
    iron, aluminum and manganese hydroxides, and silica gels,
    used experimentally as substitutes for return sludge in
    the activated sludge process. Energy/Utility 2: Appendix
    E.

Atomic Pile. Federal--EPA. A nuclear reactor. General 1: p. 2.

Available Dilution. State--Water Resources. The ratio of the
    quantity of untreated wastewater or partly or completely
    treated effluent to the average quantity of diluting water
    available effective at the point of disposal or at any
    point under consideration; usually expressed in percentage.
    Also called dilution factor. Energy/Utility 2: Appendix
    E.

Average Annual Energy. Regional--River Basin. Average annual
    energy generated by a hydroelectric project or system over
    a specified period. Energy/Utility 6: p. 194.

Average Demand. Federal--GSA. Quantity of flow (i.e., water, sew-
    age, electrical, steam, or gas) required for normal, domes-
    tic, institutional, commercial, or industrial operations.
    General 2: p. VII-19.

        National--Electric Institute. The demand on, or the power
        output of, an electric system or any of its parts over any
        interval of time, as determined by dividing the total num-
        ber of kilowatthours by the number of units of time in the
        interval. Energy/Utility 3: p. 24.

Back-End System. National--Resources. Jargon for any of several
    processes for recovering resources from the organic por-
    tion of the waste stream. Front-end processes separate

and recover the inorganic portion from the incoming refuse. Back-end system operations include refuse-derived fuel recovery, conversion to oil or gas, fiber reclaim, composting, conversion to animal feed, etc. Energy/Utility 5: p. 2.

Baling. <u>Federal--EPA</u>. Compacting solid waste into blocks to reduce volume. General 1: p. 2.

Ballistic Separator. <u>Federal--GSA</u>. Device that drops mixed materials having different physical characteristics onto a high-speed rotary impeller; the materials are hurled at different velocities into separate collecting bins. General 2: p. VI-16.

Bar Screen. <u>Federal--EPA</u>. In waste water treatment, a device that removes large solids. General 1: p. 2.

Base Load. <u>Federal--GSA</u>. Normal conditions. General 2: p. VII-20.

   <u>Federal--Interior</u>. The minimum load in a power system over a given period of time. General 4: Glossary, p. 1.

   <u>National--Electric Institute</u>. The minimum load over a given period of time. Energy/Utility 3: p. 6.

   <u>Regional--River Basin</u>. The minimum load in a stated period of time. Energy/Utility 6: p. 196.

Base Load Station. <u>National--Electric Institute</u>. A generating station which is normally operated to take all or part of the base load of a system and which, consequently, operates essentially at a constant output. Energy/Utility 3: p. 7.

Benefication. <u>Federal--Interior</u>. A general term for converting coal to some other form of energy such as coke, petroleum, or natural gas. General 3: p. 12.

Beta Particle. <u>Federal--EPA</u>. An elementary particle emitted by radioactive decay that may cause skin burns. It is halted by a thin sheet of metal. General 1: p. 2.

Bioassy. <u>State--Water Resources</u>. A method of determining toxic effects of industrial wastes and other wastewaters by using viable organisms or live fish as test organisms. Energy/Utility 2: Appendix E.

Biochemical Oxygen Demand (BOD). State--Water Resources. Abbre-
        viation for biochemical oxygen demand. The quantity of
        oxygen used in the biochemical oxidation of organic matter
        in a specified time, at a specified temperature, and under
        specified conditions. A standard test used in assessing
        wastewater strength. Energy/Utility 2: Appendix E.

Biochemical Oxygen Demand Level. State--Water Resources. The BOD
        content, usually expressed in pounds per unit of time, of
        wastewater passing into a waste treatment system or to a
        body of water. Energy/Utility 2: Appendix E.

Biodegradable Material. National--Resources. Waste material which
        is capable of being broken down by bacteria into basic ele-
        ments. Most organic wastes, such as food remains and
        paper, are biodegradable. Energy/Utility 5: p. 3.

Biodegradation. State--Water Resources. The destruction or min-
        eralization of either natural or synthetic organic mater-
        ials by the microorganisms populating soils, natural bodies
        of water, or wastewater treatment facilities. Energy/
        Utility 2: Appendix E.

Biological Oxidation. Federal--EPA. The way that bacteria and
        microorganisms feed on and decompose complex organic ma-
        terials. Used in self-purification of water bodies and
        activated sludge wastewater treatment. General 1: p. 2.

        State--Water Resources. The process whereby living organ-
        isms in the presence of oxygen convert the organic matter
        contained in wastewater into a more stable or mineral form.
        Energy/Utility 2: Appendix E.

Biological Purification. State--Water Resources. The process
        whereby living organisms convert the organic matter con-
        tained in wastewater into a more stable or a mineral form.
        Energy/Utility 2: Appendix E.

Biological Wastewater Treatment. State--Water Resources. Forms
        of wastewater treatment in which bacterial or biochemical
        action is intensified to stabilize, oxidize, and nitrify
        the unstable organic matter present. Intermittent sand
        filters, contact beds, trickling filters, and activated
        sludge processes are examples. Energy/Utility 2: Appendix
        E.

Biomonitoring. Federal--EPA. The use of living organisms to test
        water quality at a discharge site or downstream. General
        1: p. 3.

Blowdown. Regional--River Basin. Water drawn from boiler systems
     and cold water basins of cooling towers to prevent buildup
     of solids concentrations. Usually contains chemicals used
     for pH adjustment and slime control. Energy/Utility 6: p.
     191.

          State--Water Resources. The water discharged from a boiler
          or cooling tower to dispose of accumulated salts. Energy/
          Utility 2: Appendix E.

Breeder. Federal--EPA. A nuclear reactor that produces more fuel
     than it consumes. General 1: p. 3.

Bulky Waste. National--Resources. Large items of waste material,
     such as appliances, furniture, large auto parts, trees,
     branches, stumps, etc. Energy/Utility 5: p. 3.

Burial Ground. Federal--EPA. A disposal site for unwanted radio-
     active materials that uses earth or water for a shield.
     General 1: p. 3.

Capability. National--Electric Institute. The maximum load which
     a generating unit, generating station, or other electrical
     apparatus can carry under specified conditions for a given
     period of time, without exceeding approved limits of temp-
     erature and stress. Energy/Utility 3: p. 10.

Capacity. Regional--River Basin. The load for which a generator,
     transmission circuit, power plant, or system is rated.
     Capacity is also used synonymously with capability. En-
     ergy/Utility 6: p. 192.

Capacity Factor. Regional--River Basin. The ratio of the average
     load on the generating plant for the period of time con-
     sidered to the capacity rating of the plant. Unless other-
     wise identified, capacity factor is computed on an annual
     basis. Energy/Utility 6: p. 193.

Chemical Coagulation. City--Planning Department. The combining
     of small particles of solid matter suspended in a liquid
     to form larger particles which settle more readily. The
     addition of a chemical provides forces which attract the
     small particles to each other. Energy/Utility 9: p. 1.

Chemical Oxygen Demand (COD). City--Planning Department. The
     quantity of oxygen used in biological and non-biological
     oxidation of materials in water; a measure of water qual-
     ity. Energy/Utility 9: p. 1.

Chemical Primary Treatment. <u>City--Planning Department</u>. The addition of a chemical to assist in the settling of matter
which is suspended in wastewater. Energy/Utility 9: p. 1.

Chlorination. <u>Federal--EPA</u>. The application of chlorine to drinking water, sewage, or industrial waste to disinfect or to
oxidize undesirable compounds. General 1: p. 3.

    <u>Federal--GSA</u>. Addition of chlorine to prevent the spread
of water-borne diseases; also used in wastewater treatment
to oxidize reducing agents (e.g., hydrogen sulfide). General 2: p. IX-17.

    <u>Regional--River Basin</u>. The application of chlorine to
water, sewage, or industrial wastes generally for the purpose of disinfection, but frequently for accomplishing
other biological or chemical results. Energy/Utility 7:
p. 255.

Chlorinator. <u>Federal--EPA</u>. A device that adds chlorine to water
in gas or liquid form. General 1: p. 3.

Chlorine-Contact Chamber. <u>National--Soil Conservation</u>. In a waste
treatment plant, a chamber in which effluent is disinfected
by chlorine before it is discharged to the receiving
waters. Earth 2: p. 12g.

Chlorine Demand. <u>State--Water Resources</u>. The difference between
the amount of chlorine added to water or wastewater and
the amount of residual chlorine remaining at the end of a
specified contact period. The demand for any given water
varies with the amount of chlorine applied, time of contact, and temperature. Energy/Utility 2: Appendix E.

Circulating Water. <u>Regional--River Basin</u>. In a closed-cycle cooling system, this refers to the heated water from the condenser which is cooled, usually by evaporative means, and
recycled through the condenser. Energy/Utility 6: p. 193.

Clarification. <u>Federal--EPA</u>. Clearing action that occurs during
waste water treatment when solids settle out, often aided
by centrifugal action and chemically induced coagulation.
General 1: p. 4.

Clarified Wastewater. <u>State--Water Resources</u>. Wastewater from
which most of the settleable solids have been removed by
sedimentation. Also called settled wastewater. Energy/
Utility 2: Appendix E.

Clarifier. State--Water Resources. A unit of which the primary
          purpose is to secure clarification. Usually applied to
          sedimentation tanks or basins. Energy/Utility 2: Appen-
          dix E.

Coagulant. State--Water Resources. A compound responsible for
          coagulation; a floc-forming agent. Energy/Utility 2:
          Appendix E.

Coagulation. Federal--EPA. A clumping of particles in waste water
          to settle out impurities, often induced by chemicals such
          as lime or alum. General 1: p. 4.

Coal Fired. Federal--Interior. A power generating facility using
          coal as a source of energy. General 3: p. 19.

Coal Gasification. Federal--Interior. The process by which coal
          is converted to a gas, usually methane. General 3: p. 19.

Colloidal Matter. State--Water Resources. Finely divided solids
          which will not settle but may be removed by coagulation or
          biochemical action or membrane filtration. Energy/Utility
          2: Appendix E.

Combined Sewers. Federal--EPA. A system that carries both sewage
          and storm water runoff. In dry weather all flow goes to
          the waste treatment plant. During a storm only part of
          the flow is intercepted due to overloading. The remaining
          mixture of sewage and storm water overflows untreated into
          the receiving stream. General 1: p. 4.

          Federal--GSA. Sewers carrying both sanitary and storm
          sewage. General 2: p. IX-20.

          State--Water Resources. A sewer intended to receive both
          wastewater and storm or surface water. Energy/Utility 2:
          Appendix E.

Combined Sewer System. Federal--GSA. Wastewaters from households
          and industries discharged through separate sanitary sewers
          storm water runoff empties into separate storm sewers.
          General 2: p. VII-19.

Combined Wastewater. State--Water Resources. A mixture of sur-
          face runoff and other wastewater such as domestic or in-
          dustrial wastewater. Energy/Utility 2: Appendix E.

Commercial Waste. National--Resources. Waste material which orig
          inates in wholesale, retail or service establishments, suc
          as office buildings, stores, markets, theaters, hotels and

warehouses. Energy/Utility 5: p. 3.

Composite Wastewater Sample. State--Water Resources. A combination of individual samples of water or wastewater taken at
selected intervals, generally hourly for some specified
period, to minimize the effect of the variability of the
individual sample. Individual samples may have equal
volume or may be roughly proportional to the flow at time
of sampling. Energy/Utility 2: Appendix E.

Condenser Cooling Water. Regional--River Basin. Water required
to condense the steam after its discharge from a steam
turbine. Energy/Utility 6: p. 193.

Conservation. Federal--Interior. Improving the efficiency of
energy use; using less energy to produce the same product.
General 4: Glossary, p. 2.

Construction Waste. National--Resources. Waste materials produced in the construction of homes, office buildings, dams,
industrial plants, schools, etc. The materials usually
include used lumber, miscellaneous metal parts, packaging
materials, cans, boxes, wire, excess sheet metal, etc.
Energy/Utility 5: p. 4.

Consumer Waste. National--Resources. Materials which have been
discarded by the buyer, or consumer, as opposed to "in-
plant waste", or waste created in the manufacturing process. Energy/Utility 5: p. 4.

Contact Filter. State--Water Resources. A filter used in a water
treatment plant for the partial removal of turbidity before
final filtration. Energy/Utility 2: Appendix E.

Conventional Hydroelectric Plant. Regional--River Basin. A hydroelectric power plant which utilizes streamflow only once
as it passes downstream, as opposed to a pumped-storage
plant which recirculates all or a portion of the streamflow in the production of power. Energy/Utility 6: p.
193.

Conversion. National--Resources. A resource recovery method using
chemical or biological processes to change waste materials
into other useful forms. Examples are incineration to produce heat; pyrolysis to produce gas, oil and char; and composting to produce a humus-like soil conditioner. Energy/
Utility 5: p. 4.

Cooling Tower. Federal--EPA. A device that aids in heat removal
from water used as a coolant in electric power generating

plants. General 1: p. 4.

Cooling Water Consumption. Regional--River Basin. The cooling
    water withdrawn from the source supplying a generating
    plant which is lost to the atmosphere. Caused primarily
    by evaporative cooling of the heated water coming from the
    condenser. The amount of consumption (loss) is dependent
    on the type of cooling employed--direct (once-through)
    cooling pond, or cooling tower. When not returned to the
    source of supply, blowdown is also included as a consump-
    tive loss. Energy/Utility 6: p. 193.

Cooling Water Requirement. Regional--River Basin. The amount of
    water needed to pass through the condensing unit in order
    to condense the steam to water. This amount is dependent
    on the type of cooling employed and water temperature.
    Energy/Utility 6: p. 193.

Core. Federal--EPA. The uranium-containing heat of a nuclear
    reactor, where energy is released. General 1: p. 4.

Corridor. Federal--Interior. A narrow strip of land reserved for
    location of transmission lines, pipelines, and service
    roads. General 3: p. 21.

Cover Material. Federal--EPA. Soil used to cover compacted solid
    waste in a sanitary landfill. General 1: p. 4.

Critical Period. Regional--River Basin. Period when the limita-
    tions of hydroelectric power supply due to water conditions
    are most critical with respect to system energy require-
    ments. Energy/Utility 6: pp. 193-194. (Revised).

Curie. State--Water Resources. The basic unit of intensity of
    radioactivity in a sample of material. One curie equals
    37 billion disintegrations per second, or approximately
    the radioactivity of 1 g of radium. Energy/Utility 2:
    Appendix E.

Curtailment. Federal--Interior. Temporary, mandatory load reduc-
    tions reflecting emergency conditions, following after all
    possible conservation action and load management tech-
    niques, and prompted by problems of meeting baseload,
    rather than peaking deficiencies. General 4: p. 3.

Cycling Load. Federal--GSA. Daily peak hours. General 2: p.
    VII-20.

Cyclone Separator. National--Resources. A mechanical separator
    which uses a swirling air flow to sort small particles

according to their size and density. Energy/Utility 5:
p. 4.

Dechlorination. <u>State--Water Resources</u>. The partial or complete
reduction of residual chlorine in a liquid by any chemical
or physical process. Energy/Utility 2: Appendix E.

Decomposition Of Wastewater. <u>State--Water Resources</u>. The breakdown of organic matter in wastewater by bacterial action,
either aerobic or anaerobic. Energy/Utility 2: Appendix
E. (Revised).

Degree Of Treatment. <u>State--Water Resources</u>. A measure of the removal effected by treatment processes with reference to
solids, organic matter, BOD, bacteria, or any other specified matter. Energy/Utility 2: Appendix E.

Demand. <u>National--Electric Institute</u>. The rate at which electric
energy is delivered to or by a system, part of a system,
or a piece of equipment expressed in kilowatts, kilovolt-
amperes or other suitable unit at a given instant or averaged over any designated period of time. Energy/Utility
3: p. 24.

<u>Regional--River Basin</u>. The rate at which electric energy
is delivered to or by a system at a given instant or averaged over any designated period of time, expressed in kilowatts or other suitable units. Energy/Utility 6: p. 194.

Demolition Waste. <u>National--Resources</u>. Waste materials produced
from the destruction of buildings, roads, sidewalks, etc.
The materials usually include large broken pieces of concrete, plaster, bricks and glass. Energy/Utility 5: p.
4.

Demonstration Grant. <u>National--Resources</u>. Authorized by the Solid
Waste Disposal Act of 1965 and the Resources Recovery Act
of 1970, this is a program through which the U.S. Environmental Protection Agency provides funds to municipalities
to: (1) conduct projects to determine the feasibility of
new techniques of solid waste management and serve as
models for other communities to adopt; (2) demonstrate resource recovery systems; and (3) construct new improved
solid waste disposal facilities representing an advance-
ment in the state-of-the-art. Energy/Utility 5: p. 4.

Dependable Capacity. <u>Regional--River Basin</u>. The load-carrying
ability of a station or system under adverse conditions
for the time interval and period specified when related
to the characteristics of the load to be supplied. For

hydro projects the term refers to the capability in the
most adverse month in the critical period--January 1932
in the case of the 1928-32 critical period. Energy/Util-
ity 6: p. 192.

Desulfurization. Federal--EPA. Removal of sulfur from fossil
fuels to cut pollution. General 1: p. 5.

Detention Tank. State--Water Resources. A tank used in water or
wastewater treatment to provide adequate time for chemical
or physical reactions to take place in the body of liquid
being treated. Energy/Utility 2: Appendix E.

Detention Time. State--Water Resources. The theoretical time required to displace the contents of a tank or unit at a
given rate of discharge (volume divided by rate of discharge). Energy/Utility 2: Appendix E.

Dewatering. National--Resources. The removal of water by filtration, centrifugation, pressing, coagulation or other
methods. Dewatering makes sewage sludge suitable for disposal by burning or landfilling. The term is also applied
to removal of water from pulp. Energy/Utility 5: p. 4.

Diatomaceous Earth. Federal--EPA. A chalk-like material used to
filter out solid wastes in waste water treatment plants,
also found in powdered pesticides. General 1: p. 5.

Diffused Air. Federal--EPA. A type of aeration that forces oxygen
into sewage by pumping air through perforated pipes inside
a holding tank. General 1: p. 5.

Diffuser. City--Planning Department. A section of pipe with
openings through which effluent is discharged. Energy/
Utility 9: p. 1.

Digested Sludge. State--Water Resources. Sludge digested under
either aerobic or anaerobic conditions until the volatile
content has been reduced to the point at which the solids
are relatively nonputrescible and inoffensive. Energy/
Utility 2: Appendix E.

Digester. Federal--EPA. In wastewater treatment a closed tank,
sometimes heated to 95 degrees F, where sludge is subjected
to intensified bacterial action. General 1: p. 5.

Digestion Chamber. State--Water Resources. A sludge-digestion
tank. Frequently refers specifically to the lower or
sludge-digestion compartment of an Imhoff tank. Energy/
Utility 2: Appendix E.

Dilution. <u>State--Water Resources</u>. Disposal of wastewater or
        treated effluent by discharging it into a stream or body
        of water. Energy/Utility 2: Appendix E.

Dilution Factor. <u>State--Water Resources</u>. The ratio of the quan-
        tity of untreated wastewater or partly or completely treat-
        ed effluent to the average quantity of diluting water avail-
        able at the point of disposal or at any point under consid-
        eration; usually expressed in percentage. Also called
        available dilution. Energy/Utility 2: Appendix E.

Disinfected Wastewater. <u>State--Water Resources</u>. Wastewater to
        which chlorine or other disinfecting agents has been added,
        during or after treatment, to destroy pathogenic organisms.
        Energy/Utility 2: Appendix E.

Disinfection. <u>Federal--EPA</u>. A chemical or physical process that
        kills organisms that cause infectious disease. Chlorine
        is often used to disinfect sewage treatment effluent. Gen-
        eral 1: p. 5.

Disposal. <u>National--Resources</u>. The final disposition of waste,
        usually through burning or burying. Energy/Utility 5:
        p. 5.

Disposal By Dilution. <u>State--Water Resources</u>. A method of dis-
        posing of wastewater or treated effluent by discharging it
        into a stream or body of water. Energy/Utility 2: Appen-
        dix E.

Distribution Line. <u>Federal--Interior</u>. Low voltage electric power-
        line, usually 69 kilovolts or less. General 3: p. 26.

Distribution Ratio. <u>State--Water Resources</u>. The ratio of the
        daily average rate of application of wastewater to a high-
        rate filter that would exist if wastewater were applied to
        the entire surface of the filter at the mean maximum momen-
        tary rate of application continuously for 24 hours to the
        actual average daily application. Energy/Utility 2: Appen-
        dix E.

Dry-Weather Flow. <u>City--Planning Department</u>. The flow of waste-
        water in a combined sewer during dry weather. Energy/Util-
        ity 9: p. 1.

Dump. <u>Federal--EPA</u>. A site used to dispose of solid wastes with-
        out environmental controls. General 1: p. 5.

        <u>Federal--GSA</u>. Land site where solid waste is disposed of
        in a manner that does not protect the environment. General

2: p. VI-14.

National--Resources. An open land site where waste materials are burned, left to decompose, rust or simply remain.
Because of the problems which they create, such as air and
water pollution, unsanitary conditions, and general un-
sightliness, dumps have been declared illegal (with varying moratorium dates) in all states. Energy/Utility 5:
p. 5.

Economically Recoverable Mineral Reserves. Regional--River Basin.
Extractable over a reasonable period of time at a cost
which allows for a return on investment, plus a reasonable
profit. Earth 1: p. 376.

Economic Mineral Reserves. Regional--River Basin. Profitable
to mine under the technical and economic conditions existing at time of production. Earth 1: p. 376.

Effluent. Federal--Commerce. A discharge of pollutants into the
environment; untreated or partially or completely treated.
General 6: p. 311.

Federal--EPA. Waste material discharged into the environment, it can be treated or untreated. Generally refers to
water pollution. General 1: p. 6.

National--Resources. Solid, liquid or gas wastes which
enter the environment as a by-product of chemical or biological processes, usually from man-oriented processes.
Energy/Utility 5: p. 5.

State--Board. The mixture of substances, gases and liquids and suspended matter, discharged into the atmosphere,
ground, river or ocean as the result of given process.
Air 2: no page number.

State--Water Resources. Wastewater or other liquid,
partially or completely treated, or in its natural state,
flowing out of a reservior, basin, treatment plant, or industrial treatment plant, or part thereof. Energy/Util-
ity 2: Appendix E.

City--Planning Department. Partially or completely treated
water flowing out of a treatment plant process. Energy/
Utility 9: p. 1.

Effluent Seepage. National--Soil Conservation. Diffuse discharge
onto the ground of liquids that have percolated through
solid waste or another medium; they contain dissolved or

suspended materials.   Earth 2: p. 21g.

Electric Utility Industry Or Electric Utilities.   National--Elec-
    tric Institute.   All enterprises engaged in the production
    and/or distribution of electricity for use by the public,
    including investor-owned electric utility companies; coop-
    eratively-owned electric utilities; government-owned elec-
    tric utilities (municipal systems, Federal agencies, state
    projects, and public power districts); and, where the data
    are not separable, those industrial plants contributing to
    the public supply.   Energy/Utility 3: pp. 31-32.

Electrodialysis.   Federal--EPA.   A process that uses electrical
    current applied to permeable membranes to remove minerals
    from water.   Often used to desalinize salt or brackish
    water.   General 1: p. 6.

    National--Resources.   A system for removing unwanted part-
    icles from a solution by using an electric charge across
    a membrane.   It is used as a spearation device in some
    waste treatment processes.   Energy/Utility 5: p. 5.

Electro-Process Industry.   Regional--River Basin.   An industry
    which requires very large amounts of electricity in man-
    ufacturing for heat or chemical processes (as distinguished
    from wheel-turning or mechanical applications).   Examples
    are electric furnace steel, aluminum, and chlorine.   En-
    ergy/Utility 6: p. 194.

Electrostatic Precipitator.   National--Resources.   A system for
    removing unwanted colloidal particles from a solution by
    passing the particles through an electrostatic field and
    then collecting the charged particles on collecting plate
    or pipe.   Sometimes used in incinerators. furnaces and
    treatment plants to collect or separate dust particles.
    Energy/Utility 5: p. 5.

Elutriation.   National--Resources.   The separation of finer, light-
    er particles from coarser, heavier particles in a mixture
    by means of a usually slow upward stream of fluid so that
    the lighter particles are carried upward.   Energy/Utility
    5: p. 5.

    State--Water Resources.   A process of sludge conditioning
    whereby the sludge is washed with either fresh water or
    plant effluent to reduce the demand for conditioning chem-
    icals and to improve settling or filtering characteristics
    of the solids.   Excessive alkalinity is removed in this
    process.   Energy/Utility 2: Appendix E.

Energy. <u>Regional--River Basin</u>. That which does or is capable of
doing work. It is measured in terms of the work it is capable of doing; electric energy is commonly measured in
kilowatt-hours or average megawatts. Energy/Utility 6:
p. 194.

Energy Content Curve. <u>Regional--River Basin</u>. A seasonal guide
to the use of reservoir storage for at-site and downstream
power generation. Energy/Utility 6: p. 195. (Revised).

Energy Recovery. <u>National--Resources</u>. A form of resource recovery
in which the organic fraction of waste is converted to some
form of usable energy. Recovery may be achieved through
the combustion of processed or raw refuse to produce steam,
through the pyrolysis of refuse to produce oil or gas; and
through the anaerobic digestion of organic wastes to produce methane gas. Energy/Utility 5: p. 5. (Revised).

Energy Resource Planning. <u>Federal--Forest Service</u>. Functional
planning for the development and use of energy resources.
General 15: p. 70.

Escherichia Coli (E. coli). <u>State--Water Resources</u>. One of the
species of bacteria in the coliform group. Its presence
is considered indicative of fresh fecal contamination.
Energy/Utility 2: Appendix E.

Evaporation Ponds. <u>Federal--EPA</u>. Areas where sewage sludge is
dumped and allowed to dry out. General 1: p. 6.

Exfiltration. <u>State--Water Resources</u>. The quantity of wastewater
which leaks to the surrounding ground through unintentional
openings in a sewer. Also, the process whereby this leak-
ing occurs. Energy/Utility 2: Appendix E.

Extended Aeration. <u>State--Water Resources</u>. A modification of the
activated sludge process which provides for aerobic sludge
digestion within the aeration system. The concept envis-
ages the stabilization of organic matter under aerobic conditions and disposal of the end products into the air as
gases and with the plant effluent as finely divided suspended matter and soluble matter. Energy/Utility 2: Appendix E. p. 12.

Ferrofluid. <u>National--Resources</u>. A stabilized colloidal suspension of ferromagnetic particles in an aquaeous or non-
aqueous liquid, and subjected to a magnetic field, the
strength of which changes the apparent density of the suspension. Thus density separations of non-magnetic substances can be achieved. Energy/Utility 5: p. 6.

Ferrous. National--Resources. Metals which are predominantly
        composed of iron. Most common ferrous metals are magnetic.
        In the waste materials stream, these usually include steel
        or "tin" cans, automobiles, old refrigerators, stoves, etc.
        Energy/Utility 5: p. 6.

Filter. State--Water Resources. A device or structure for re-
        moving solid or colloidal material, usually of a type that
        cannot be removed by sedimentation, from water, wastewater,
        or other liquid. The liquid is passed through a filtering
        medium, usually a granular material but sometimes finely
        woven cloth, unglazed porcelain, or specially prepared
        paper. There are many types of filters used in water or
        wastewater treatment. Energy/Utility 2: Appendix E.

Filtered Wastewater. State--Water Resources. Wastewater that has
        passed through a mechanical filtering process but not
        through a trickling filter bed. Energy/Utility 2: Appen-
        dix E.

Filter Plant. State--Water Resources. In wastewater treatment
        units, the devices and structures required to provide
        trickling filtration. Energy/Utility 2: Appendix E.

Filtration. Federal--EPA. Removing particles of solid materials
        from water, usually by passing it through sand. General
        1: p. 7.

Firm Energy. Regional--River Basin. Electric energy which is
        considered to have assured availability to the customer
        to meet all or any agreed upon portion of his load re-
        quirements. Energy/Utility 6: p. 194. (Revised).

Firm Load. Regional--River Basin. That part of the system load
        which must be met with firm power. Energy/Utility 6: p.
        196.

Firm Power. Regional--River Basin. Power which is considered to
        have assured availability to the customer to meet all or
        any agreed upon portion of his load requirements. It is
        firm energy supported by sufficient capacity to fit the
        load pattern. The availability of firm power is based
        on the same probability considerations as is firm energy.
        Energy/Utility 6: p. 198.

Firm Water Supply. Federal--Interior. An assured minimum supply
        of water under the most adverse water year supply condi-
        tions. General 3: p. 35.

Five-Day BOD. State--Water Resources. That part of oxygen demand associated with biochemical oxidation of carbonaceous, as distinct from nitrogeneous, material. It is determined by allowing biochemical oxidation to proceed, under conditions specified in Standard Methods, for five days. Energy/ Utility 2: Appendix E.

Five-Year Storm. City--Planning Department. A storm which occurs on the average of once in five years. Most of San Francisco's sewers (but not treatment plants) are sized to handle this level of flow. Energy/Utility 9: p. 2.

Floc. Federal--EPA. A clump of solids formed in sewage by biological or chemical action. General 1: p. 7.

State--Water Resources. Small gelatinous masses formed in a liquid by the reaction of a coagulant added thereto, through biochemical processes, or by agglomeration. Energy/Utility 2: Appendix E.

Flocculation. Federal--EPA. Separation of suspended solids during waste water treatment by chemical creation of clumps of flocs. General 1: p. 7.

State--Water Resources. In water and wastewater treatment, the agglomeration of colloidal and finely divided suspended matter after coagulation by gentle stirring by either mechanical or hydraulic means. In biological wastewater treatment where coagulation is not used, agglomeration may be accomplished biologically. Energy/Utility 2: Appendix E.

Flocculation Agent. State--Water Resources. A coagulating substance which, when added to water, forms a flocculent precipitate which will entrain suspended matter and expedite sedimentation. Examples are alum, ferrous sulfate, and lime. Energy/Utility 2: Appendix E.

Flowmeter. Federal--EPA. A gauge that shows the speed of waste water moving through a treatment plant. General 1: p. 7.

Force Main. City--Planning Department. Flows must be "forced" through a force-main by pumping. Sewage flows downhill under the influence of gravity in ordinary sewers. Force-mains are usually smaller in cross-section than ordinary sewers. Energy/Utility 9: p. 2.

Fossil Fuels. Federal--EPA. Combustibles derived from the remains of ancient plants and animals, like coal, oil and natural gas. General 1: p. 7.

Regional--Commission. Coal, oil, natural gas, and other fuels originating from fossilized geologic deposits and depending on oxidation for release of energy. General 13: p. 588.

Front-End System. National--Resources. Jargon referring to processing of municipal solid waste for recovery of material--metals, glass and paper. A front-end system also prepares the organic portion in a form readily usable in energy recovery, or back-end systems. Energy/Utility 5: p. 6.

Fuel. City--Commission. Solid, liquid, or gaseous material such as, but not limited to natural or manufactured gas, gasoline, oil, coal, or wood, used to produce heat or power by burning. Air 1: no page number.

Fuel Oil. City--Commission. A liquid petroleum product derived directly or indirectly from crude oil. Air 1: no page number.

Gamma Rays. Federal--EPA. The most penetrating waves of radiant nuclear energy. They can be stopped by dense materials like lead. General 1: p. 7.

Garbage. National--Resources. Waste materials which are likely to decompose or putrefy. Usually contain food wastes from a kitchen, restaurant, grocery store, slaughter house or food processing plant. Energy/Utility 5: p. 7.

Gasification. Federal--EPA. Conversion of a solid material, such as coal, into a gas for use as fuel. General 1: p. 7.

Gas Turbine Generating Station. National--Electric Institute. An electric generating station in which the prime mover is a gas turbine engine. Energy/Utility 3: p. 41.

Generating Station. National--Electric Institute. A station at which are located prime movers, electric generators, and auxiliary equipment for converting mechanical, chemical, and/or nuclear energy into electric energy. Energy/Utility 3: p. 40.

Generation. Regional--River Basin. The act or process of producing electric energy from other forms of energy; also the amount of electric energy so produced. Energy/Utility 6: p. 196.

Generator. Federal--EPA. A device that converts mechanical energy into electrical energy. General 1: p. 7.

Geothermal Generating Station. National--Electric Institute.
    An electric generating station in which the prime mover
    is a steam turbine. The steam is generated in the earth
    by heat from the earth's magma. Energy/Utility 3: p. 41.

Grab Sample. State--Water Resources. A single sample of waste-
    water taken at neither set time nor flow. Energy/Utility
    2: Appendix E.

Gravity Separation. National--Resources. The collection of sub-
    stances immersed in a liquid by taking advantage of dif-
    ferences in density. Energy/Utility 5: p. 7.

Grease Trap. State--Water Resources. A device for separation of
    grease from wastewater by floatation so that it can be re-
    moved from the surface. Energy/Utility 2: Appendix E.

Hazardous Waste. National--Resources. Waste materials which by
    their nature are dangerous to handle or dispose of. These
    materials include old explosives, radioactive materials,
    some chemical and some biological wastes, usually produced
    in industrial operations or in institutions. Not meant to
    imply that other wastes are non-hazardous. Energy/Utility
    5: p. 7.

Headworks. City--Planning Department. The pumps and transport
    facilities associated with intake pumping within the treat-
    ment plant. Energy/Utility 9: p. 2.

Heating Season. Federal--EPA. The coldest months of the year,
    when pollution increases in some areas because people burn
    fossil fuels to keep warm. General 1: p. 8.

Heterogeneous Waste. National--Resources. A body of waste mate-
    rial made up of dissimilar components (e.g., municipal
    refuse, which contains metals, glass, paper, plastics,
    food wastes, yard wastes, etc.). Energy/Utility 5: p. 7.

Holding Pond. Federal--EPA. A pond or reservoir usually made of
    earth built to store polluted runoff. General 1: p. 8.

Homogeneous Waste. National--Resources. A body of waste material
    made up of similar components (e.g., newspapers or aluminum
    cans). Energy/Utility 5: p. 8.

Hydro. National--Electric Institute. A term used to identify a
    type of generating station or power or energy output in
    which the prime mover is driven by water power. Energy/
    Utility 3: p. 45.

Hydroelectric Generating Station. National--Electric Institute.
     An electric generating station in which the prime mover is
     a water wheel. The water wheel is driven by falling water.
     Energy/Utility 3: p. 41.

Hydropower. Federal--Interior. A term used to identify a type of
     generating station, or power, or energy output in which
     the prime mover is driven by water power. General 4: p. 6.

Imhoff Tank. State--Water Resources. A deep, two-storied waste-
     water tank originally patened by Karl Imhoff. It consists
     of an upper continuous-flow sedimentation chamber and a
     lower sludge-digestion chamber. The floor of the upper
     chamber slopes steeply to trapped slots through which
     solids may slide into the lower chamber. The lower chamber
     receives no fresh wastewater directly, but is provided with
     gas vents and with means for drawing digested sludge from
     near the bottom. Energy/Utility 2: Appendix E.

Incineration. Federal--GSA. Controlled process by which solid,
     liquid, or gaseous combustible wastes are burned and
     changed into gases; residue produced contains little or
     no combustible material. General 2: p. VI-14.

Incinerator. National--Resources. A plant designed to reduce
     waste volume by combustion. Incinerators consist of refuse
     handling and storage facilities, furnaces, subsidence cham-
     bers, residue handling and removal facilities, chimneys and
     other air pollution control equipment. Energy/Utility 5:
     p. 8.

Incinerator Capacity. Federal--GSA. Number of tons of solid
     waste that a designer anticipates the facility will be
     able to process in a 24-hour period if specified criteria
     are met. General 2: p. VI-13.

Indicated Mineral Reserves. Regional--River Basin. Tonnage and
     grade computed from measurements, samples, or production
     data, and from projection on geological evidence. Earth
     1: p. 376.

Industrial Firm Power. Federal--Interior. Power intended to have
     assured availability to the industrial customer on a con-
     tract demand basis. General 4: p. 6.

Industrial Waste. National--Resources. Those waste materials
     generally discarded from industrial operations or derived
     from manufacturing processes. Energy/Utility 5: p. 8.

State--Water Resources. The liquid wastes from industrial processes, as distinct from domestic or sanitary wastes. Energy/Utility 2: Appendix E.

Inertial Separator. Federal--GSA. Material separation device that relies on ballistic or gravity separation of materials having different physical characteristics. General 2: p. VI-16.

Inferred Mineral Reserves. Regional--River Basin. Quantitative estimates based upon broad knowlege of the geology of the deposit for which there are few samples or actual measurements. Earth 1: p. 376.

Influent. City--Planning Department. Water or raw sewage entering a treatment plant. Energy/Utility 9: p. 2.

Inorganic Refuse. National--Resources. Waste material made from substances composed of matter other than plant, animal, or certain chemical compounds of carbon. Examples are metals and glass. Energy/Utility 5: p. 8.

In-Plant Waste. National--Resources. Waste generated in manufacturing processes. Such might be recovered through internal recycling or through a salvage dealer. Energy/Utility 5: p. 8.

Institutional Waste. National--Resources. Waste materials originating in schools, hospitals, research institutions and public buildings. The materials include packaging materials, certain hazardous wastes, food wastes, disposable products, etc. Energy/Utility 5: p. 8.

Intercepting Sewer. State--Water Resources. A sewer that receives dry-weather flow from a number of transverse sewers or outlets and frequently additional predetermined quantities of storm water, if from a combined system, and conducts such waters to a point for treatment or disposal. Energy/Utility 2: Appendix E.

Interceptor Sewers. Federal--EPA. The collection system that connects main and trunk sewers with the wastewater treatment plant. In a combined sewer system interceptor sewers allow some untreated wastes to flow directly into the receiving streams so the plant won't be overloaded. General 1: p. 9.

Interlocking. Federal--GSA. Means by which electrical utilities are often connected to form a regional grid; power can then be "borrowed" to supply service area needs during peak conditions. General 2: p. VII-20.

Internal Combustion Generating Station. National--Electric Institute. An electric generating station in which the prime
mover is an internal combustion engine. Energy/Utility
3: p. 41.

Interruptible Power. Regional--River Basin. Nonfirm power; power
made available under agreements which permit curtailment
or cessation of delivery by the supplier. Energy/Utility
6: p. 198. (Revised).

Junk. National--Resources. Waste materials, such as rags, paper,
metals, broken furniture, toys, equipment, etc. The term
usually implies that the materials can be recovered or
converted for reuse. Energy/Utility 5: p. 8.

Lateral Sewers. Federal--EPA. Pipes running underneath city
streets that collect sewage. General 1: p. 9.

Lift. Federal--EPA. In a sanitary landfill, a compacted layer
of solid waste and the top layer of cover material. General 1: p. 9.

Lift Station. State--Water Resources. A small wastewater pumping
station that lifts the wastewater to a higher elevation
when the continuance of the sewer at reasonable slopes
would involve excessive depths of trench, or that raise
wastewater from areas too low to drain into available
sewers. These stations may be equipped with pnematic
ejectors or centrifugal pumps. Energy/Utility 2: Appendix E.

Liquefied Natural Gas (LNG). Federal--Interior. A clean flamable
liquid existing under very cold conditions; that is, almost
pure methane. General 4: Glossary, p. 7.

Liquid Waste. National--Soil Conservation. A general term de-
noting pollutants such as soap, chemicals, or other substances in liquid form. Earth 2: p. 32g.

Litter. National--Resources. Solid waste discarded outside the
established collection-disposal system. Solid waste prop-
erly placed in containers is often referred to as trash
and garbage; uncontainerized, it is referred to as litter.
Litter accounts for about two percent of municipal solid
waste. Energy/Utility 5: p. 8.

Load. National--Electric Institute. The amount of electric power
delivered or required at any specified point or points on

a system. Load originates primarily at the power consuming
equipment of the customer. Energy/Utility 3: p. 49.

Regional--River Basin. The amount of power delivered to
a given point. Energy/Utility 6: p. 196.

Load Curve. National--Electric Institute. A curve on a chart
showing power, kilowatts, supplied, plotted against time
of occurrence, and illustrating the varying magnitude of
the load during the period covered. Energy/Utility 3:
p. 50.

Load Diversity. Regional--River Basin. Literally refers to the
difference between (1) the sum of the separate peak loads
of two or more load areas and (2) the actual coincident
peak load of the combined areas. Energy/Utility 6: p. 197.
(Revised).

Load Factor. Regional--River Basin. The ratio of the average
load over a designated period to the peak load occurring
in that period. Energy/Utility 6: p. 197. (Revised).

Load Pattern. Regional--River Basin. The characteristic variation
in the magnitude of the power load with respect to time.
This can be for a daily, weekly, or annual period. Energy/
Utility 6: p. 197.

Load Shape. Regional--River Basin. The characteristics variation
in the magnitude of the power load with respect to time.
This can be for a daily, weekly, or annual periods. En-
ergy/Utility 6: p. 197.

Magnetic Separator. Federal--GSA. Any device that removes ferrous
metals by means of magnets. General 2: p. VI-16.

Main Sewer. State--Water Resources. In larger sewer systems, the
principal sewer to which branch sewers and submains are
tributary, also called trunk sewers. In small systems, a
sewer to which one or more branch sewers are tributary.
In plumbing, the public sewer to which the house or building is connected. Energy/Utility 2: Appendix E.

Material Balance. Federal--GSA. Accounting of the weights of materials entering and leaving a processing unit (e.g., incinerator), usually on an hourly basis. General 2: p. VI-
14.

Materials Recovery. National--Resources. The initial phase,
front-end, of a resource recovery system where recyclable
and reuseable materials are extracted from waste for sale.

Energy/Utility 5: p. 9.

Maximum Plant Capability--Hydro. <u>Regional--River Basin</u>. The maximum load which a hydroelectric plant can supply under optimum head and flow conditions without exceeding approved limits of temperature and stress. This may be less than the overload rating of the generators due to encroachment of tailwater on head at high discharge. Energy/Utility 6: p. 191.

Measured Mineral Reserves. <u>Regional--River Basin</u>. Tonnage computed from dimensions revealed in outcrops, trenches, workings, and drill holes with grade derived from detailed sampling. Earth 1: p. 376.

Megawatt (MW). <u>National--Electric Institute</u>. 1,000 kilowatts. Energy/Utility 3: p. 52.

Methane. <u>Federal--EPA</u>. A colorless, nonpoisonous, flammable gas emitted by marshes and dumps undergoing anaerobic decomposition. General 1: p. 9.

> <u>National--Resources</u>. An odorless, colorless, flammable gas which can be formed by the anaerobic decomposition of organic waste matter or by chemical synthesis. It is the principal constituent of natural gas. Energy/Utility 5: p. 9.

Mineral Fuels. <u>Regional--River Basin</u>. Naturally occurring carbonaceous minerals that include petroleum, coal, and natural gas. Earth 1: p. 376.

Mineral Reserves. <u>Regional--River Basin</u>. Discovered ore, coal, petroleum, or natural gas of established extent and grade producible but not yet produced. Earth 1: p. 376.

Mixed Liquor. <u>Federal--EPA</u>. Activated sludge and water containing organic matter being treated in an aeration tank. General 1: p. 9.

Most Probable Number (MPN). <u>Federal--Interior</u>. A statistical estimate of bacterial numbers in polluted water based on probability formulas. General 3: p. 53.

> <u>State--Water Resources</u>. That number of organisms per unit volume that, in accordance with statistical theory, would be more likely than any other number to yield the observed test result or that would yield the observed test result with the greatest frequency. Expressed as density of organisms per 100 ml. Results are computed from the number

of positive findings of coliform-group organisms resulting
from multiple-portion decimal-dilution plantings. Energy/
Utility 2: Appendix E.

Municipal Solid Waste. National--Resources. The combined residential and commercial waste materials generated in a given
municipal area. The collection and disposal of these
wastes are usually the responsibility of local government.
Energy/Utility 5: p. 9.

Natural Gas. Federal--EPA. A natural fuel containing methane and
hydrocarbons that occurs in certain geologic formations.
General 1: p. 10.

Nitrogenous Waste. Federal--EPA. Animal or plant residues that
contain large amounts of nitrogen. General 1: p. 10.

Non-Renewable Resource. Federal--Army. A natural, normally nonliving, resource such as a mineral which is present in
finite supply and is not renewed by natural systems. General 9: p. 31.

Nuclear Energy. National--Electric Institute. Energy produced in
the form of heat during the fission process in a nuclear
reactor. When released in sufficient and controlled quantity, this heat energy may be used to produce steam to
drive a turbine-generator and thus be converted to electric
energy. Energy/Utility 3: p. 56.

Nuclear Fuel. National--Electric Institute. Material containing
fissionable materials of such composition and enrichment
that when placed in a nuclear reactor will support a self-
sustaining fission chain reaction and produce heat in a
controlled manner for process use. Energy/Utility 3: p.
56.

Nuclear Generating Station. National--Institute. An electric
generating station in which the prime mover is a steam
turbine. The steam is generated in a reactor by heat from
the fissioning of nuclear fuel. Energy/Utility 3: p. 41.

Nuclear Power. National--Electric Institute. Power released in
exothermic (a reaction which gives off heat) nuclear reactions which can be converted to electric power by means
of heat transformation equipment and a turbine-generator
unit. Energy/Utility 3: p. 57.

Nuclear Power Plant. Federal--EPA. A device that converts atomic
energy into usable power; heat produced by a reactor makes
steam to drive electricity-generating turbines. General 1

p. 10.

Nuclear Reactor. <u>Federal--Interior</u>. A device in which a fission
chain reaction can be initiated, maintained, and controlled. Its essential component is a core with fissionable
fuel. General 4: Glossary, p. 8.

> <u>National--Electric Institute</u>. An apparatus in which the
> nuclear fission chain may be initiated, maintained and
> controlled, so that the accompanying energy is released at
> a specific rate. It includes fissionable material (con-
> tained in fuel) such as uranium or plutonium, fertile ma-
> terial, moderating material (unless it is a fast reactor),
> a heavy-walled pressure vessel, shielding to protect per-
> sonnel, provision for heat removal, control elements and
> instrumentation. Energy/Utility 3: p. 58. (Revised).

Ocean Dumping. <u>National--Resources</u>. The dumping of raw or treated
wastes into the sea beyond the continental shelf. Energy/
Utility 5: p. 9.

Odor Control. <u>State--Water Resources</u>. In wastewater treatment,
the prevention or reduction of objectionable odors by
chlorination, aeration, or other processes or by masking
with chemical aerosols. Energy/Utility 2: Appendix E.

Odor Threshold. <u>State--Water Resources</u>. The point at which, after
successive dilutions with odorless water, the odor of a
water sample can just be detected. The threshold odor is
expressed quantitatively by the number of times the sample
is diluted with odorless water. Energy/Utility 2: Appendix E.

Open Burning. <u>National--Resources</u>. The burning of waste materials
in the open or in a dump--sometimes intentionally to reduce
volume, sometimes accidentally. Open burning produces
smoke, odor and other objectionable air pollutants. En-
ergy/Utility 5: p. 9.

Organic Refuse. <u>National--Resources</u>. Waste material made from
substances composed of chemical compounds of carbon and
generally manufactured in the life processes of plants
and animals. These materials include paper, wood, food
wastes, plastic, and yard wastes. Energy/Utility 5: p. 9.

Outfall. <u>Federal--EPA</u>. The place where an effluent is discharged
into receiving waters. General 1: p. 10.

> <u>State--Water Resources</u>. The point, location, or structure
> where wastewater or drainage discharges from a sewer,

drain, or other conduit. The conduit leading to the ultimate disposal area. Energy/Utility 2: Appendix E.

Oxidation Pond. Federal--EPA. A holding area where organic wastes are broken down by aerobic bacteria. General 1: p. 11.

Oxidation Rate. State--Water Resources. The rate at which the organic matter in wastewater is stabilized. Energy/ Utility 2: Appendix E.

Oxygen Demand. State--Water Resources. The quantity of oxygen utilized in the biochemical oxidation of organic matter in a specified time, at a specified temperature, and under specified conditions. Energy/Utility 2: Appendix E.

Oxygen Depletion. State--Water Resources. Loss of dissolved oxygen from water or wastewater resulting from biochemical or chemical action. Energy/Utility 2: Appendix E.

Peak Demand. Federal--GSA. Quantity of flow (i.e., water, sewage, electrical, steam, or gas) required for domestic, institutional, commercial, or industrial operations during extreme conditions (e.g., seasonal periods, fire-fighting procedures). General VII-19.

Peaking Capability. Regional--River Basin. The maximum peak load that can be supplied by a generating unit, station, or system in a stated period of time. Energy/Utility 6: p. 191. (Revised).

Peaking Plant. Regional--River Basin. A power plant which is normally operated to provide all or most of its generation during maximum load periods. Energy/Utility 6: p. 197.

Peak Load. Federal--GSA. Extreme conditions (usually peak hours + seasonal demand). General 2: p. VII-20.

Regional--River Basin. Literally, the maximum load in a stated period of time. Sometimes the term is used in a general sense to describe that portion of the load above the base load. Energy/Utility 6: p. 196.

Penstock. Regional--River Basin. A conduit to carry water to the turbines of a hydroelectric plant (usually refers only to conduits which are under pressure). Energy/Utility 6: p. 197.

Petroleum. Federal--DOT. An oily, flammable bituminous liquid that may vary from almost colorless to black. It occurs in many places in the upper strata of the earth and is a complex mixture of hydrocarbons with small amounts of other substances, and is prepared for use as gasoline, naphtha or other products by varying refining processes. Transportation 3: pp. 51-52.

Phenols. Federal--EPA. Organic compounds that are byproducts of petroleum refining, tanning, textile, dye, and resin manufacture. Low concentrations can cause taste and odor problems in water, higher concentrations can kill aquatic life. General 1: p. 11.

Phenol Wastes. State--Water Resources. Industrial wastes containing phenols, derived chiefly from coking processes and oil refineries. Energy/Utility 2: Appendix E.

Pig. Federal--EPA. A container, usually lead, used to ship or store radioactive materials. General 1: p. 11.

Pile. Federal--EPA. A nuclear reactor. General 1: p. 11.

Pollution Load. State--Water Resources. The quantity of material in a waste stream that requires treatment or exerts an adverse effect on the receiving system. Energy/Utility 2: Appendix E. (Revised).

Potable Water. Federal--EPA. Appetizing water that is safe for drinking and use in cooking. General 1: p. 12.

Power. Regional--River Basin. The time rate of transferring energy. Note, the term is frequently used in a broad sense, as a commodity of capacity and energy, having only general association with classic or scientific meaning. Energy/Utility 6: p. 198.

Power Pool. National--Electric Institute. A power pool is two or more interconnected electric systems planned and operated to supply power in the most reliable and economical manner for their combined load requirements and maintenance program. Energy/Utility 3: p. 64.

Power Supply Area. Regional--River Basin. Geographic grouping of electric power supplies as established by the Federal Power Commission in accordance with utility service areas. Energy/Utility 6: p. 198.

Pretreatment. Federal--EPA. Processes used to reduce the amount of pollution in water before it enters the sewers or the treatment plant. General 1: p. 12.

Primary Sedimentation. <u>City--Planning Department</u>. Removal of
    suspended matter by settling. Energy/Utility 9: p. 2.

Primary Settling Tank. <u>State--Water Resources</u>. The first set-
    tling tank for the removal of settleable solids through
    which wastewater is passed in a treatment works. Energy/
    Utility 2: Appendix E.

Primary Treatment. <u>Federal--EPA</u>. The first stage of waste water
    treatment; removal of floating debris and solids by screen-
    ing and sedimentation. General 1: p. 12.

    <u>Federal--GSA</u>. Treatment of sewage by detention in settling
    basins; removes approximately one-third of the BOD and
    two-thirds of the suspended solids. General 2: p. IX-19.

    <u>State--Water Resources</u>. The first major, sometimes only,
    treatment in a wastewater treatment works, usually sed-
    imentation. The removal of a substantial amount of sus-
    pended matter but little or no colloidal and dissolved
    matter. Energy/Utility 2: Appendix E.

Primary Waste Treatment. <u>Regional--Commission</u>. The removal of
    settleable, suspended, and floatable solids from waste
    water by the application of mechanical and/or gravitational
    forces. General 13: p. 590.

Prime Energy. <u>Regional--River Basin</u>. Hydroelectric energy which
    is assumed to be available 100 percent of the time; specif-
    ically, the average energy generated during the critical
    period. Energy/Utility 6: p. 195.

Process Water. <u>State--Water Resources</u>. Water that comes in con-
    tact with an end product or with materials incorporated in
    an end product. Energy/Utility 2: Appendix E.

Public Utility. <u>Federal--DOT</u>. A private enterprise so essential
    to the public interest as to justify granting an exclusive
    franchise in return for submitting to regulation and the
    obligation to serve its customers without discrimination.
    The most common examples are gas, electric, railroad and
    telephone companies. Rates and service standards are usu-
    ally controlled by a public-service commission or other
    agency. Transportation 3: p. 56.

Public Utility District. <u>National--Electric Institute</u>. A polit-
    ical subdivision (quasi-public corporation of a state),
    with territorial boundaries embracing an area wider than
    a single municipality (incorporated as well as unincorp-
    orated) and frequently covering more than one county for

the purpose of generating, transmitting, and distributing electric energy. Energy/Utility 3: p. 66.

Pumping Station. Federal--EPA. A machine installed on sewers to pull the sewage uphill. In most sewer systems waste water flows by gravity to the treatment plant. General 1: p. 12.

Purification. State--Water Resources. The removal of objectionable matter from water by natural or artificial methods. Energy/Utility 2: Appendix E.

Putrefaction. State--Water Resources. Biological decomposition of organic matter with the production of ill-smelling products associated with anaerobic conditions. Energy/Utility 2: Appendix E.

Quench Tank. Federal--EPA. A water-filled tank used to cool incinerator residues, or hot materials during industrial processes. General 1: p. 12.

Radiation. Federal--EPA. The emission of particles or rays by the nucleus of an atom. General 1: p. 12.

Radiation Standard. Federal--EPA. Regulations that govern exposure to permissible concentrations of and transportation of radioacitve materials. General 1: p. 12.

Raw Sewage. Federal--EPA. Untreated waste water. General 1: p. 12.

Receiving Body Of Water. State--Water Resources. A natural watercourse, lake, or ocean into which treated or untreated wastewater is discharged. Energy/Utility 2: Appendix E.

Receiving Stream. National--Soil Conservation. The body of water into which effluent is discharged. Earth 2: p. 41g.

Receiving Waters. Federal--EPA. Any body of water where untreated wastes are dumped. General 1: p. 12.

Reclamation. National--Resources. The restoration to usefulness or productivity of materials found in the waste stream. The reclaimed materials may be used for purposes which are different from their original usage. Energy/Utility 5: p. 11.

Recoverable Resources. Federal--GSA. Materials that still have useful physical or chemical properties after serving a specific purpose and that can, therefore, be reused or

> recycled for the same or other purposes. General 2: p.
> VI-15.

Recycling. <u>Federal--GSA</u>. Process by which waste materials are
transformed into new products in such a manner that the
original products may lose their identity. General 2:
p. VI-15.

> <u>National--Resources</u>. A resource recovery method involving
> the collection and treatment of a waste product for use as
> raw material in the manufacture of the same or a similar
> product, e.g., ground glass used in the manufacture of new
> glass. Energy/Utility 5: p. 11.

Refuse. <u>National--Resources</u>. A generally used term for solid
waste materials. Energy/Utility 5: p. 11.

Refuse Reclamation. <u>Federal--EPA</u>. Conversion of solid waste into
useful products, e.g., composting organic wastes to make
a soil conditioner. General 1: p. 12.

Reserve Capacity. <u>Regional--River Basin</u>. Extra generating capacity available to meet unanticipated demands for power
or to generate power in the event of loss of generation
resulting from scheduled or unscheduled outages of regularly used generating capacity. Energy/Utility 6: p.
192.

Residential Waste. <u>National--Resources</u>. Waste materials generated in houses and apartments. The materials include
paper, cardboard, beverage and food cans, plastics, food
wastes, glass containers, old clothes, garden wastes, etc.
Energy/Utility 5: p. 11.

Resource Recovery. <u>National--Resources</u>. A term describing the
extraction and utilization of materials and values from
the waste stream. Materials recovered, for example, would
include metals and glass which can be used as "raw materials" in the manufacture of new products. Energy/Util-
ity 5: p. 12. (Revised).

Resource Recovery Facility. <u>Federal--GSA</u>. Any physical plant
that: (1) processes residential, commercial, or institutional solid wastes biologically, chemically, or physically, and recovers useful products (e.g., shredded fuel,
combustible oil or gas, steam, metal, glass) for recycling
and (2) creates a relatively inert process residue for disposal. General 2: p. VI-15.

Reverse Osmosis. Federal--EPA. An advanced method of waste treat-
        ment that uses a semi-permeable membrane to separate water
        from pollution. General 1: p. 13.

Rubbish. Federal--EPA. Solid waste, excluding food wastes and
        ashes, from homes, institutions, and workplaces. General
        1: p. 13.

        National--Resources. A general term for solid waste that
        does not contain food waste. Energy/Utility 5: p. 12.

Run-Of-Canal Plant. Regional--River Basin. A hydroelectric plant
        similar to a run-of-river plant but located on an irriga-
        tion canal or waterway instead of a stream. Energy/Util-
        ity 6: p. 199.

Run-Of-River Plant. Regional--River Basin. A hydroelectric plant
        which depends chiefly on the flow of a stream as it occurs
        for generation, as opposed to a storage project, which has
        sufficient storage capacity to carry water from one season
        to another. Some run-of-river projects have a limited
        storage capacity (poundage) which permits them to regulate
        streamflow on a daily or weekly basis. Energy/Utility 6:
        p. 199.

Sanitary Landfill. Federal--EPA. Protecting the environment when
        disposing of solid waste. Waste is spread in thin layers,
        compacted by heavy machinery and covered with soil daily.
        General 1: p. 13.

        National--Resources. A method of disposing of refuse on
        land without creating nuisances or hazards to public health
        or safety. Careful preparation of the fill area and con-
        trol of water drainage are required to assure proper land-
        filling. To confine the refuse to the smallest practical
        area and reduce it to the smallest practical volume, heavy
        tractor-like equipment is used to spread, compact, and usu-
        ally cover the waste daily with at least six inches of com-
        pacted dirt. After the area has been completely filled and
        covered with a final two to three foot layer of dirt, and
        has been allowed to settle an appropriate period of time,
        the reclaimed land may be turned into a recreational area
        such as a park or golf course. Under certain highly con-
        trolled conditions the land may be used as a plot on
        which some types of buildings can be constructed. Energy/
        Utility 5: p. 12.

Sanitary Landfilling. Federal--GSA. Engineering method of dis-
        posing solid waste on land in a manner that protects the
        environment. Waste is spread in thin layers, compacted

to the smallest practical volume, and covered with soil
by the end of each working day. General 2: p. VI-16.

Sanitary Sewer. Federal--EPA. Underground pipes that carry only
domestic or commercial waste, not stormwater. General 1:
p. 13.

Federal--GSA. Sewers carrying only household and industrial wastes. General 2: p. IX-20.

State--Water Resources. A sewer that carries liquid and
water-carried wastes from residences, commercial buildings,
industrial plants, and institutions, together with minor
quantities of ground, storm, and surface waters that are
not admitted intentionally. Energy/Utility 2: Appendix E.

Sanitary Wastewater. State--Water Resources. The water supply
of a community after it has been used and discharged into
a sewer. Energy/Utility 2: Appendix E. (Revised).

Sanitation. Federal--EPA. Control of the physical factors in the
human environment that can harm development, health, or
survival. General 1: p. 13.

Federal--GSA. Control of all factors in the physical environment that can exercise a deleterious effect on human
physical development, health, and survival. General 2:
p. VI-16.

Secondary Energy. Federal--Interior. Electric energy surplus to
the needs of a supplier, the delivery of which may be interrupted for any reason by the supplier. General 4: p.
9.

Regional--River Basin. All hydroelectric energy other
than prime energy; specifically, the difference between
average annual energy and prime energy. Energy/Utility
6: p. 195.

Secondary Power. Federal--Interior. Power not having the assured
availability of firm power; power that is available from
a system intermittently and that is used to serve markets
that can accomodate such power. General 4: p. 9.

Secondary Treatment. Federal--EPA. Biochemical treatment of
wastewater after the primary stage, using bacteria to consume the organic wastes. Use of trickling filters or the
activated sludge process, removes floating and settleable
solids and about 90 percent of oxygen demanding substances
and suspended solids. Disinfection with chlorine is the

final stage of secondary treatment. General 1: p. 13.

Federal--GSA. Biological treatment of primary-treated
sewage, generally using activated sludge or trickling
filter processes in which micro-organisms degrade the
organic compounds present in the sewage. BOD and suspended solids removals of 85-90 percent are achieved.
General 2: p. IX-19.

Secondary Waste Treatment. Regional--Commission. The removal
of dissolved and colloidal materials that in their unaltered state, as found in waste water, are not amenable
to separation through the application of primary treatment. General 13: p. 591.

Secondary Wastewater Treatment. State--Water Resources. The
treatment of wastewater by biological methods after primary treatment by sedimentation. Energy/Utility 2: Appendix E.

Sedimentation. Federal--EPA. Letting solids settle out of waste
water by gravity during waste water treatment. General 1:
p. 13.

Sedimentation Basin. State--Water Resources. A basin or tank in
which water or wastewater containing settleable solids is
retained to remove by gravity a part of the suspended
matter. Also called sedimentation tank, settling basin,
settling tank. Energy/Utility 2: Appendix E.

Sedimentation Tanks. Federal--EPA. Holding areas for waste water
where floating wastes are skimmed off and settled solids
are pumped out for disposal. General 1: p. 13.

Separate Sewer System. Federal--GSA. Wastewaters from households
and industries discharged through separate sanitary sewers;
storm water runoff empties into separate storm sewers.
General 2: p. VII-19.

Septic Sludge. State--Water Resources. Sludge from a septic tank
or partially digested sludge from an Imhoff tank or sludge-
digestion tank. Energy/Utility 2: Appendix E.

Septic Tank. Federal--EPA. An enclosure that stores and (processes) wastes where no sewer system exists, as in rural
areas or on boats. Bacteria decompose the organic matter
into sludge, which is pumped off periodically. General 1:
p. 13.

State--Water Resources. A settling tank in which settled sludge is in immediate contact with the wastewater flowing through the tank and the organic solids are decomposed by anaerobic bacterial action. Energy/Utility 2: Appendix E.

Service Area. National--Electric Institute. Territory in which a utility system is required or has the right to supply electric service to ultimate customers. Energy/Utility 3: p. 72.

Settleable Solids. Federal--EPA. Materials heavy enough to sink to the bottom of waste water. General 1: p. 13.

State--Water Resources. That matter in wastewater which will not stay in suspension during a preselected settling period, such as one hour, but either settles to the bottom or floats to the top. Energy/Utility 2: Appendix E. (Revised).

Settling Chamber. Federal--EPA. A series of screens placed in the way of flue gases to slow the stream of air, thus helping gravity to pull particles out of the emission into a collection area. General 1: p. 13.

Sewage. Federal--EPA. The organic waste and waste water produced by residential and commercial establishments. General 1: p. 14.

Federal--GSA. Wastewater. General 2: p. IX-20.

State--Water Resources. The spent water of a community. Term now being replaced in technical usage by preferable term wastewater. Energy/Utility 2: Appendix E.

Sewage Sludge. National--Soil Conservation. Settled sewage solids combined with varying amounts of water and dissolved materials that is removed from sewage by screening, sedimentation, chemical precipitation, or bacterial digestion. Earth 2: p. 46g.

Sewer. Federal--EPA. A channel that carries waste water and stormwater runoff from the source to a treatment plant or receiving stream. Sanitary sewers carry household and commercial waste. Storm sewers carry runoff from rain or snow. Combined sewers are used for both purposes. General 1: p. 14.

Sewer Outfall. State--Water Resources. The outlet or structure through which wastewater is finally discharged. Energy/Utility 2: Appendix E.

Sewer System. <u>State--Water Resources</u>. Collectively, all of the property involved in the operation of a sewer utility. It includes land, wastewater lines and appurtenances, pumping stations, treatment works, and general property. Occasionally referred to as a sewerage system. Energy/Utility 2: Appendix E.

Screening. <u>Federal--EPA</u>. Use of racks of screens to remove coarse floating and suspended solids fr.m sewage. General 1: p. 13.

Single Purpose Corridor. <u>Federal--Interior</u>. A corridor developed for one primary use such as pipelines or transmission lines. Single purpose corridors are sited to accommodate one utility and boundaries are set accordingly. They offer few opportunities for joint access. General 4: p. 3.

Skimming. <u>Federal--EPA</u>. Using a machine to remove oil or scum from the surface of the water. General 1: p. 14.

Sludge. <u>Federal--EPA</u>. The concentration of solids removed from sewage during waste water treatment. General 1: p. 14.

   <u>Federal--GSA</u>. Solids removed by wastewater treatment plants. Since sludge contains most of the offensive characteristics found in sewage, its treatment and disposal are an integral part of waste treatment. Sludge is generally concentrated, broken down by digestion, conditioned, dried, and disposed. General 2: p. IX-20.

   <u>National--Resources</u>. Waste materials in the form of a concentrated suspension of waste solids in water. One type of sludge is produced from the treatment of sewage. Energy/Utility 5: p. 13.

   <u>State--Water Resources</u>. The accumulated solids separated from liquids, such as water or wastewater, during processing, or deposits on bottoms of streams or other bodies of water. Energy/Utility 2: Appendix E. (Revised).

Sludge Bed. <u>State--Water Resources</u>. An area comprising natural or artificial layers of porous material on which digested wastewater sludge is dried by drainage and evaporation. A sludge bed may be open to the atmosphere or covered, usually with a greenhouse-type superstructure. Also called sludge drying bed. Energy/Utility 2: Appendix E.

Sludge Deposits. <u>National--Soil Conservation</u>. Accumulations of settled, usually rapidly decomposing organic material in the aquatic system. Earth 2: p. 47g.

Sludge Digestion. <u>State--Water Resources</u>. The process by which
    organic or volatile matter in sludge is gasified, liq-
    uified, mineralized, or converted into more stable organic
    matter through the activities of either anaerobic or aero-
    bic organisms. Energy/Utility 2: Appendix E.

Sludge Treatment. <u>State--Water Resources</u>. The processing of
    wastewater sludges to render them innoccuous. This may
    be done by aerobic or anaerobic digestion followed by dry-
    ing on sand beds, filtering and incineration, filtering
    and drying, or wet air oxidation. Energy/Utility 2:
    Appendix E.

Sodium Alkyl Benzene Sulfonate (ABS). <u>State--Water Resources</u>.
    Customary abbreviation of sodium alkyl benzene sulfonate
    (hard detergent). Energy/Utility 2: Appendix E.

Soil Wastes. <u>Federal--Interior</u>. Garbage, refuse, sludges, and
    other discarded solid materials resulting from industrial,
    commercial, and agricultural operations and from community
    activities. Does not include solids or dissolved material
    in domestic sewage or other significant pollutants in water
    resources, such as silt, dissolved or suspended solids in
    industrial wastewater effluents, dissolved materials in
    irrigation flows, or other common water pollutants. Gen-
    eral 3: p. 72.

Solar Energy. <u>Federal--EPA</u>. Power collected from sunlight, used
    most often for heating purposes but occasionally to gen-
    erate electricity. General 1: p. 14.

Solid Waste. <u>Federal--EPA</u>. Useless, unwanted, or discarded ma-
    terial with insufficient liquid to be free-flowing. Gen-
    eral 1: p. 14.

    <u>National--Resources</u>. Discarded solid materials. Includes
    agricultural waste, mining waste, industrial waste and
    municipal waste. Energy/Utility 5: p. 13. (Revised).

Solid Waste Disposal. <u>Federal--EPA</u>. The final placement of re-
    fuse that cannot be salvaged or recycled. General 1: p.
    14.

Solid Waste Management. <u>National--Resources</u>. Conduct and reg-
    ulation of the entire process of generation, storage,
    collection, transportation, processing, reclamation and
    disposal of refuse. Energy/Utility 5: p. 13.

Source Separation. <u>Federal--GSA</u>. Setting aside of recyclable ma-
    terials at the point of generation by the generator.

General 2: p. VI-16.

Stabilization Pond. State--Water Resources. A type of oxidation pond in which biological oxidation of organic matter is effected by natural or artificially accelerated transfer of oxygen to the water from air. Energy/Utility 2: Appendix E.

Steam Generating Station. National--Electric Institute. An electric generating station in which the prime mover is a steam turbine. The steam is generated in a boiler by heat from burning fossil fuels. Energy/Utility 3: p. 41.

Storm Drain. State--Water Resources. A drain used for conveying rainwater, groundwater, subsurface water, condensate, cooling water, or other similar discharge to a storm sewer or combined sewer. Energy/Utility 2: Appendix E.

Storm Runoff. State--Water Resources. That portion of the total runoff that reaches the point of measurement within a relatively short period of time after the occurrence of precipitation. Also called direct runoff. Energy/Utility 2: Appendix E.

Storm Sewer. Federal--EPA. A system that collects and carries rain and snow runoff to a point where it can soak back into the groundwater or flow into surface waters. General 1: p. 14.

Federal--GSA. Sewers carrying storm water from roof and street surfaces. General 2: p. IX-20.

State--Water Resources. A sewer that carries storm water and surface water, street wash and other wash waters, or drainage, but excludes domestic wastewater and industrial wastes. Also called storm drain. Energy/Utility 2: Appendix E.

Strip Mining. Federal--EPA. A process that uses machines to scrape soil or rock away from mineral deposits just under the Earth's surface. General 1: p. 15.

Federal--Forest Service. Any operation in connection with prospecting for, excavating, or mining minerals which results in a large-scale surface or stream bottom disturbance from stripping, trenching, dredging, rim cutting or open-pit digging. General 15: p. 130.

Substation. Federal--Interior. An electric power station which serves as a control-and-transfer point on an electrical

transmission system. Substations route and control electrical power flow, transform a voltage to a higher or lower
voltage, and serve as delivery points to individual customers. General 4: p. 10.

Sump. Federal--EPA. A depression or tank that catches liquid
runoff for drainage or disposal, like a cesspool. General 1: p. 15.

Surface Dump. National--Soil Conservation. A land site where
solid waste is disposed of in a manner that does not protect the environment. Earth 2: p. 55g.

Surface Runoff. State--Water Resources. That portion of the runoff of a drainage basin that has not passed beneath the
surface after deposition. Energy/Utility 2: Appendix E.
(Revised).

Suspended Solids (SS). Federal--EPA. Tiny pieces of pollutants
floating in sewage that cloud the water and require special treatment to remove. General 1: p. 15.

   State--Water Resources. Solids that either float on the
   surface of, or are in suspension in, water, wastewater, or
   other liquids, and which are largely removable by laboratory filtering. Energy/Utility 2: Appendix E.

Tertiary Treatment. Federal--EPA. Advanced cleaning of waste
water that goes beyond the secondary or biological stage.
It removes nutrients such as phosphorus and nitrogen and
most suspended solids. General 1: p. 15.

   Federal--GSA. Advanced waste treatment to remove chemical
   constituents (e.g., nitrates, phosphates) that are not removed in primary or secondary treatments. General 2: p.
   IX-20.

   Regional--River Basin. Selective application of biological, physical, and chemical separation processes to effect
   removal or organic and inorganic substances that resist
   conventional treatment practices. Energy/Utility 8: p.
   530.

   City--Planning Department. Treatment of secondary sewage
   effluent to remove specific pollutants. The process used
   varies with the nature of the pollutants remaining after
   secondary treatment. Energy/Utility 9: p. 2.

Thermal Plant. Regional--River Basin. A power generating plant
which uses heat to produce energy. Such plants may burn

fossil fuels or use nuclear energy to produce the necessary
thermal energy. Energy/Utility 6: p. 199.

Thermal Processing. Federal--GSA. Processing of waste material
by means of heat. General 2: p. VI-16.

Total Solids. State--Water Resources. The sum of dissolved and
undissolved constituents in water or wastewater, usually
stated in milligrams per liter. Energy/Utility 2: Appendix E.

Transfer Station. Federal--GSA. Site at which solid waste is
concentrated and then taken to a processing facility or
sanitary landfill. General 2: p. VI-16.

Transmission Grid. Regional--River Basin. An interconnected system of electric transmission lines and associated equipment for the movement or transfer of electric energy in
bulk between points of supply and points of demand. En-
ergy/Utility 6: p. 200.

Transmission Line. Federal--Interior. A facility for transmitting electrical energy at high voltage from one point to
another point. Transmission line voltages are normally
115-kV or larger. General 3: p. 78.

Trash. National--Resources. Waste materials which usually do
not include garbage but may include other organic materials, such as plant trimmings. Energy/Utility 5: p. 14.

Trickling Filter. Federal--EPA. A biological treatment device:
wastewater is trickled over a bed of stones covered with
bacterial growth, the bacteria break down the organic
wastes in the sewage and produce cleaner water. General
1: p. 15.

Ultimate Plant Capability--Hydro. Regional--River Basin. The
maximum plant capability of a hydroelectric plant when all
contemplated generating units have been installed. En-
ergy/Utility 6: p. 191.

Urban Waste. National--Resources. A general term used to catego-
rize the entire waste stream from an urban area. It is
sometimes used in contrast to "rural wastes". Energy/
Utility 5: p. 14.

Usable Energy. Regional--River Basin. All hydroelectric energy
which can be used in meeting system firm and secondary
loads. Energy/Utility 6: p. 195. (Revised).

Utilities. Federal--DOT. The basic service systems required by
     a developed area water supply, sanitary and storm sewers,
     electricity, gas and telephone service. Sometimes public
     transportation and garbage collection are included. Trans-
     portation 3: pp. 70-71.

Waste. Federal--Army. Refuse from places of human or animal hab-
     itation; a solid, liquid, or gaseous by-product derived
     from human activities. General 9: p. 49.

     Federal--EPA. Unwanted materials left over from manufac-
     turing processes, refuse from places of human or animal
     habitation. General 1: p. 16.

Waste Materials. National--Resources. A wide variety of solid
     materials that may even include liquids in containers,
     which are discarded or rejected as being spent, useless,
     worthless, or in excess. Does not usually include waste
     solids found in sewage systems, water resources or those
     emitted from smoke stacks. Energy/Utility 5: p. 15.

Waste Processing. National--Resources. An operation in which the
     physical or chemical properties of waste are changed to
     reduce size and/or volume to facilitate handling. Ex-
     amples of waste processing are shredding, separation, com-
     paction and incineration. Energy/Utility 5: p. 15.

Waste Stream. National--Resources. A general term used to denote
     the waste material output of an area, location or facility.
     Energy/Utility 5: p. 15.

Waste Treatment. State--Water Resources. Any process to which
     wastewater or industrial waste is subjected to make it
     suitable for subsequent use. Energy/Utility 2: Appendix
     E.

Wastewater. Federal--Army. Water derived from a municipal or in-
     dustrial waste treatment plant. General 9: p. 49.

     Federal--EPA. Water carrying dissolved or suspended solids
     from homes, farms, businesses, and industries. General 1:
     p. 16.

     State--Water Resources. The spent water of a community.
     From the standpoint of source, it may be a combination of
     the liquid and water-carried wastes from residences, com-
     mercial buildings, industrial plants, and institutions, to-
     gether with any groundwater, surface water, and storm water
     that may be present. In recent years, the word wastewater
     has taken precedence over the word sewage. Energy/Utility

2: Appendix E.

Wastewater Influent. State--Water Resources. Wastewater as it enters a wastewater treatment plant or pumping station. Energy/Utility 2: Appendix E.

Wastewater Reclamation. State--Water Resources. Processing of wastewater for reuse. Energy/Utility 2: Appendix E.

Water Supply System. Federal--EPA. The collection, treatment, storage and distribution of potable water from source to consumer. General 1: p. 16.

Wet-Weather Treatment Facility. City--Planning Department. Facilities that treat a mixture of wastewater and storm-water, the volume of which exceeds the capacity of dry-weather treatment facilities. Energy/Utility 9: p. 2.

Winter Peak. National--Electric Institute. The greatest load on an electric system during any prescribed demand interval in the winter or heating season, usually between December 1 of a calendar year and March 31 of the next calendar year. Energy/Utility 3: p. 81.

Yard Waste. National--Soil Conservation. Grass clippings, prunings, and other discarded material from yards and gardens. Earth 2: p. 61g.

11. HEALTH

Index

Terms

Acute Toxicity. Federal--EPA. Any poisonous effect produced by
     a single short-term exposure, that results in severe bio-
     logical harm or death. General 1: p. 1.

Aerotoxicology. Federal--HEW. The study of health effects re-
     sulting from inhalation of foreign matter in the air.
     Health 1: p. 34.

Airway Resistance. State--Committee. The narrowing of the air
     passages of the respiratory system in response to the pre-
     sence of irritating substances. Air 3: no page number.

Alveoli. State--Committee. The tiny air spaces at the end of the
     terminal bronchioles of the lungs, where the exchange of
     oxygen and carbon dioxide takes place. Air 3: no page
     number.

Asbestos. Federal--EPA. A mineral fiber that can pollute air or
     water and cause cancer if inhaled or ingested. General 1:
     p. 2.

Asbestosis. Federal--HEW. A chronic lung condition characterized
     by fibrosis or a fine scarring of the lung; fibrosis phys-
     ically obstructs the lung's air passages and leads to dif-
     ficulty in breathing, poor oxygen absorption, and poor car-
     bon dioxide removal. Health 1: p. 34.

Background Incidence. Federal--HEW. The level or incidence of a
     disease occuring from causes other than the one in ques-
     tion. Health 1: p. 34.

Beryllium. Federal--EPA. A metal that can be hazardous to human
     health when inhaled. It is discharged by machine shops,
     ceramic and propellant plants, and foundries. General 1:
     p. 2.

Bioassy. Federal--Interior. A method of determining the quantity
     of a substance necessary to affect test organisms under
     specified laboratory conditions. General 3: p. 12.

Biomedical Research. Federal--HEW. Investigation or experimenta-
     tion aimed at the discovery and interpretation of facts to
     be used for application in the human population. Health 1:
     p. 34.

Cancer. State--Committee. An abnormal, potentially unlimited,
     disorderly new cell growth. Air 3: no page number.

Carbon Monoxide (CO). <u>Federal--HEW</u>. A colorless, odorless, highly toxic gas that is a normal by-product of incomplete
combustion of carbon-containing substances such as coal,
gasoline, oil, and natural gas; one of the major air pollutants; can be harmful in small amounts if breathed over
a long period of time. Health 1: p. 34.

Carbonyl Compounds. <u>Federal--HEW</u>. Oxidation products (products
formed after combination with oxygen) of normal blood
constituents; normally not present in the blood and so
when they are present they are easy to detect by simple
chemical techniques. Health 1: p. 34.

Carcinogen. <u>Federal--Interior</u>. Any substance capable of causing
cancer. General 3: p. 16.

Carcinogenic. <u>State--Committee</u>. Cancer producing. Air 3: no
page number.

Carcinogenicity. <u>Federal--HEW</u>. The ability to produce or incite
cancer. Health 1: p. 34.

Causative Agents. <u>Federal--HEW</u>. Something that affects a result;
as a particular agent causes cancer. Health 1: p. 34.

Chloroflurocarbons. <u>Federal--HEW</u>. Gases formed of chlorine,
fluorine, and carbon whose molecules normally do not react with other substances; they are therefore useful, when
compressed, as spray can propellants because they do not
alter the material being sprayed. Health 1: p. 35.

Chronic. <u>State--Committee</u>. Marked by long duration or frequent
recurrence, as a chronic disease. Air 3: no page number.

Chronic Disease Epidemiology. <u>Federal--HEW</u>. The study of population patterns to determine possible causes of diseases
which may exhibit a prolonged latency or dormant period
before they are clinically identifiable. Health 1: p. 35.

Chronic Toxicity. <u>Federal--GSA</u>. Toxicity marked by a long duration that may produce death, although the usual effects
are sublethal (e.g., slows growth, inhibits reproduction).
General 2: p. IX-18.

Cilia. <u>State--Committee</u>. Hairlike cells that line the airways
and by their seeping movement propel the dirt and germ-
filled mucus out of the respiratory tract. Air 3: no page
number.

Clinical Evaluation. <u>Federal--HEW</u>. Examination and judgment based on direct observation of patients. Health 1: p. 35.

Cochlea. <u>Federal--HUD</u>. The essential organ of hearing; a spirally wound tube, resembling a small shell, which forms part of the inner ear. Health 1: p. 35.

Dermal Toxicity. <u>Federal--EPA</u>. The ability of a pesticide or toxic chemical to poison people or animals by touching the skin. General 1: p. 5.

Detoxication. <u>Federal--HEW</u>. The process whereby a toxic agent is either removed or metabolically converted to a less toxic compound. Health 1: p. 35.

Dioxin. <u>Federal--HEW</u>. Common name for tetrachlorodibenzo-p-dioxin or TCDD; can occur as a highly toxic contaminant in the production of some chlorinated phenols, or chemicals derived from them, such as the herbicide 2,4,5-T which is widely used in agriculture. Health 1: p. 35.

Direct Toxicity. <u>National--Soil Conservation</u>. Toxicity that has an effect on organisms themselves instead of having an effect by actual alteration of their habitat or interference with their food supply. Earth 4: p. 19g.

Disease. <u>Federal--Interior</u>. A particular destructive process in an organism with a specific cause and characteristic symptoms. General 3: p. 26.

Dose. <u>Federal--EPA</u>. In radiology, the quantity of energy or radiation absorbed. General 1: p. 5.

Dose-Response Relationship. <u>Federal--HEW</u>. The correlation of relationship between the amount of an agent (drug, chemical, etc.) to which one is exposed and the resulting effect on the body. Health 1: p. 35.

Dosimeter. <u>Federal--EPA</u>. An instrument that measures exposure to radiation. General 1: p. 5.

Emphysema. <u>Federal--HEW</u>. A condition marked by rupture of the air sacs of the lung and consequently by impairment of heart action. Health 1: p. 35.

   <u>State--Committee</u>. A disease which causes a change in the lungs, characterized by a breakdown of the walls of the alveoli which can become enlarged, lose their resilience and disintegrate. Air 3: no page number.

Environmental Agents. Federal--HEW. External influences and con-
      ditions--particularly physical, biological, and man-made
      and natural chemical substances found in the environment.
      Health 1: p. 36.

Environmental Health Sciences. Federal--HEW. A broad range of
      fields of study including chemistry, toxicology, pharm-
      acology, biometry, biophysics, biology, etc., aimed at
      detecting and understanding the action in man and animals
      of agents in the environment which have an adverse effect
      on man. Health 1: p. 36.

Environmentally-Related Diseases. Federal--HEW. Illnesses brought
      about or aggravated by environmental agents. Health 1: p.
      36.

Gene. Federal--HEW. An element in the chromosome that transmits
      hereditary characters or that bears the hereditary "blue
      print". Health 1: p. 36.

Genetic Disorders. Federal--HEW. Diseases involving alterations
      in the hereditary material; may include mental retardation,
      physical diseases (hemophilia), mental diseases. Health 1:
      p. 36.

Genetic Risks. Federal--HEW. Changes in the hereditary material
      in the germ cells (cells of reproduction) that will be
      passed on to the offspring and, therefore, increase the
      risk of occurrence of genetic disorders. Health 1: p.
      36.

Heavy Metals. Federal--HEW. Metallic elements with high molec-
      ular weights; generally toxic in low concentrations to
      plant and animal life; examples include mercury, chromium,
      cadmium, arsnic, and lead. Health 1: p. 36.

Indirect Toxicity. National--Soil Conservation. Toxicity that
      affects organisms by interfering with their food supply
      or modifying their habitat instead of directly acting on
      the organisms themselves. Earth 2: p. 28g.

LC50. Federal--EPA. Median lethal concentration, a standard
      measure of toxicity. It tells how much of a substance is
      needed to kill half of a group of experimental organisms.
      General 1: p. 9.

Lead. Federal--EPA. A heavy metal that may be hazardous to health
      if breathed or swallowed. General 1: p. 9.

Mechanism Of Action. Federal--HEW. How a substance acts or pro-
        duces its effect. Health 1: p. 37.

Mercury. Federal--EPA. A heavy metal, highly toxic if breathed
        or swallowed. It can accumulate in the environment. Gen-
        eral 1: p. 9.

Mesothelioma. Federal--HEW. A rare form of cancer with malignant
        tumors occurring on the surface of the lung or abdominal
        cavities. Health 1: p. 37.

Morbidity. Federal--Army. In medical ecology, the incidence
        (measured frequency) of disease in a population; the ill-
        ness rate. General 9: p. 29.

Mortality. Federal--Army. Death in a population; the death rate.
        General 9: p. 29.

Mutagenic. City--Planning. Causing permanent, inheritable genetic
        effects. General 8: p. 3.

Oncogenic. Federal--EPA. A substance that causes tumors, whether
        benign or malignant. General 1: p. 10.

Ossicular Chain. Federal--HEW. The small bones of the middle ear
        (called the malleus, incus, and stapes) which transmits
        the vibrations from the tympanic membrane to the oval win-
        dow. Health 1: p. 38.

Pathogen. Federal--GSA. Organism or virus that causes a disease.
        General 2: p. IX-19.

Pathogenic. Federal--Commerce. Capable of causing disease. Gen-
        eral 6: p. 313.

Pathology. Federal--HEW. The study of the essential nature of
        diseases and especially of the structural and functional
        changes produced by them. Health 1: p. 38.

Physical Environmental Agents. Federal--HEW. Influences and con-
        ditions characterized by the forces and operations of
        physics; as noise, microwave radiation, vibration. Health
        1: p. 38.

Polycyclic Aromatic Hydrocarbons. Federal--HEW. Organic compounds
        that contain carbon and hydrogen in particular combination
        and that often occur in petroleum, natural gas, coal, and
        bitumens; some have been shown to cause cancer. Health 1:
        p. 38.

Rem. Federal--EPA. A measurement of radiation by biological
      effect on human tissue. General 1: p. 12.

Rep. Federal--EPA. A measurement of radiation by energy devel-
      opment in human tissue. General 1: p. 13.

Safety Testing. Federal--HEW. Experiments on environmental
      agents to determine whether they are safe or whether they
      could have adverse effects on human health. Health 1:
      p. 39.

Sulfuric Acid Mist. Federal--HEW. An aerosol of sulfuric acid
      that either results from interaction of sulfur dioxide
      with other compounds or is emitted as an effluent from
      automobiles on which catalytic converters have been in-
      stalled; may cause health problems particularly for those
      afflicted with respiratory diseases. Health 1: p. 39.

Teratogenic. Federal--EPA. Substances that are suspected of
      causing malformations or serious deviations from the nor-
      mal type, which can't be inherited, in or on animal embryos
      or fetuses. General 1: p. 15.

Teratogenic Effects. Federal--HEW. Birth defects produced in
      the developing embryo (the offspring before birth).
      Health 1: p. 39.

Threshold Dose. Federal--EPA. The minimum application of a given
      substance required to produce a measurable effect. General
      1: p. 15.

Toxic Agents. Federal--HEW. Substances that kill or injure an
      organism through their chemical or physical action or
      through altering its environment. Health 1: p. 39.

Toxicant. National--Soil Conservation. A substance that through
      its chemical or physical action, kills, injures, or im-
      pairs an organism; any environmental factor which, when
      altered, produces a harmful biological effect. Earth 2:
      p. 57g.

Toxicity. Federal--EPA. The degree of danger posed by a substance
      to animal or plant life. General 1: p. 15.

      Federal--HEW. The degree or quality of being poisonous
      or harmful to plant or animal life. Health 1: p. 39.

Toxicology. Federal--HEW. A science that deals with poisons,
      their actions, their detection, and the treatment of the
      conditions they produce. Health 1: p. 40.

Toxic Substance. Federal--EPA. A chemical or mixture that may
     present an unreasonable risk or injury to health or the
     environment. General 1: p. 15.

Toxin. Federal--Interior. A poisonous substance generally of
     plant or animal origin. General 3: p. 78.

Vector. Federal--EPA. An organism, often an insect, that carries
     disease. General 1: p. 16.

12. RECREATION

Index

Lease Back
Limited Access
Limiting Factor
Long Term Camping
Master Plan
Maximum Capacity
Mini Park
Multiple-Use Concept
Multiple-Use Recreation Area
National Battlefields
National Cultural Park
National Forest
National Lakeshore
National Memorials
National Outdoor Recreation
    Plan
National Park
National Parkway
National Recreation Areas
National Riverway
National Seashore
National Trails System
National Wild And Scenic
    River System
Nature Center
Neighborhood Park
Non-Power Boating
Non-Resource Based
Observation Unit
Optimum Recreation Carrying
    Capacity
Other General Activities
Outdoor Recreation
Outdoor Recreation Activity
Outdoor Recreation Area
Outdoor Recreation Facility
Outdoor Recreation Resources
Outdoor Recreation Site
Outdoor Recreation Unit
Outing
Overlook
Overnight Use
Park Furniture
Park Values
Participation
Participation Rate
Passive-Appreciative
Peak Load Day
Peak Recreation Season
Persons At One Time

Phase Development
Physical Capacity
Physical Recreation Resource
    Carrying Capacity
Playground
Population Density
Portage
Power Boating
Primitive Area
Public Access
Public Benefits
Recreational Development
Recreational Experience
Recreation Area
Recreation Benefits
Recreation Capacity
Recreation Complex
Recreation Day
Recreation Demand
Recreation Demand Allocation
Recreation Design Load
Recreation Development
Recreation Easement
Recreation Fee Program
Recreation Land Classification
    System
Recreation Market Area
Recreation Modeling
Recreation Need
Recreation Opportunity
Recreation Resources
Recreation Site
Recreation Standard
Recreation Supply
Recreation Vehicle Camping
Regional Park
Resource Based
Resource-Expressive
Resource Interpretive Program
Restricted Zones
RNR Coordination
Sanitary Facilities
Scenic Easement
Separable Recreation Lands
Service Area
Shoreline Mooring
Sightseeing
Site Amenities
Social Recreational Carrying
    Capacity

Spacial Qualities
Special Interest Area
Specialized Recreation Day
State Park
Statewide Comprehensive Outdoor
    Recreation Plan
Study Rivers
Support Facilities
Tot Lot
Transfer Of Demand
Travel Cost Method
Turnover Rate
Undeveloped Recreation Area
Unit Day Value

Vacation-Use Zone
Visit
Visitation
Visitor
Visitor Center
Visitor Day
Visitor Hours
Vista Clearing
Visual Corridor
Walk-In Access
Water Based Recreation
Water Related Recreation Activity
Weekend-Use Zone

Terms

Action Program. Federal--Interior. Actions which accomplish def-
        inite objectives during a specific period of time. Rec-
        reation 3: p. 3.

Active-Appreciative. Federal--Army. Those recreation activities
        requiring some physical effort which are pursued mainly
        for the appreciation of environmental qualities of an area,
        (i.e., mountain climbing, cycling, hiking, ect.). Recre-
        ation 2: p. A-29.

Activities. Federal--Interior. The kinds of recreation in which
        individuals participate. Recreation 3: p. 3.

Activity Day. Federal--Army. A measure of recreation use by one
        person on one facility or area for one day or any part of
        a day. Recreation 2: p. A-29.

        Federal--Interior. Twelve activity hours, which may be
        aggregated continuously, intermittently or simultaneously
        by one or more persons. Recreation 3: p. 3.

Activity Hour. Federal--Interior. An accumulation of 60 minutes
        by one or more persons attributable to a specific recre-
        ational activity. Recreation 3: p. 3.

Activity Occasion. Federal--Interior. Participation by one per-
        son in an activity without relation to the duration of such
        participation. Recreation 3: p. 3.

        Regional--River Basin. Participation by an individual in
        any one recreation activity during any part of a 24-hour
        period. Recreation 1: p. 319.

Adverse Use. Federal--Interior. Use of recreation areas or their
        resources which conflicts with the purposes for which the
        area was established. Recreation 3: p. 3.

Annual Capacity. Federal--Army. The total capacity for use within
        a given year, considering the daily, weekly and seasonal
        variations in density of use. Recreation 2: p. A-29.

Back Country. Federal--Interior. A general reference to remote
        areas. Recreation 3: p. 4.

Bathhouse. Federal--Army. A recreation structure for changing and
        storing clothes, showering and other purposes in relation
        to use of a swimming beach. Recreation 2: p. A-29.

Beach. Federal--Army. Natural or constructed facility consisting
      of a sand or concrete blanket along a lake, river or ocean
      shoreline which extends both above and below the waterline
      on a slight grade for the purpose of swimming or sunbath-
      ing. Recreation 2: p. A-30.

Beautification. Federal--Army. Improvement of project lands for
      the primary purpose of aesthetic quality usually involving
      landscaping, restoring construction scars, etc. Recreation
      2: p. A-30.

Boat Access. Federal--Army. Denotes that a particular recreation
      area or site is accessible by boat only. Recreation 2: p.
      A-30.

Boat Basin. Federal--Army. A protected anchorage for watercraft
      with facilities for launching and retrieval. The basin
      may be excavated from the shoreline or created by break-
      waters. Recreation 2: p. A-30.

Boat Dock. Federal--Army. A structural facility located in or
      adjacent to the water which is designed and constructed to
      provide for the mooring and/or storage of watercraft. Rec-
      reation 2: p. A-30.

Boat House. Federal--Army. A recreation structure built on the
      shoreline of a lake for the storage of boats. May be built
      on pilings or may be floating. Recreation 2: p. A-30.

Boat Slip. Federal--Army. The water spaces between docks where
      watercraft are moored. Recreation 2: p. A-31.

Boundary Plans. Federal--Army. Division/District wide maps clear-
      ly delineating the limits of each regional recreation mar-
      ket area (area from which the majority of use originates)
      for one or more Civil Works water resource projects. Rec-
      reation 2: p. A-31.

Braille Trail. Federal--Interior. A nature trail designed spe-
      cifically for use and enjoyment of the blind. Recreation
      3: p. 5.

Camp. Federal--Interior. A place of temporary accommodation for
      persons remaining outside of permanent habitations for
      periods of one night or longer. Recreation 3: p. 5.

Campground. Federal--Army. A portion of land within a recreation
      area or water resources project which is designated and at
      least partially developed for camping use. May include
      other facilities such as swimming beach, playgrounds, and

boat launching ramps. Recreation 2: p. A-31.

Federal--Interior. An area of land designated and developed for camping. Recreation 3: p. 5.

Camping Unit. Federal--Army. The sum total of those facilities
which are normally provided at a campsite such as a trailer space, tent pad, stove, fire ring, picnic table, trash
receptacle and landscaping. Other facilities such as
electric outlets, water spigots or sanitary drains may be
included on intensively developed sites. Recreation 2:
p. A-32.

Campsite. Federal--Army. A portion of a campground or camping
area which is developed to accomodate a singular camper
or family with those elements which comprise a camping
unit. Recreation 2: p. A-32.

Carrying Capcity. Federal--Army. The maximum use of a portion of
land or ecosystem in an area beyond which no significant
increase can occur without damage occurring to that land
or ecosystem. Recreation 2: p. A-32.

Federal--EPA. In recreation, the amount of use a recreation area can sustain without deterioration of its quality. General 1: p. 3.

Federal--Interior. The capacity of a given recreation area
can be defined in four ways: (1) ability of existing natural resources to withstand use, (2) engineering capacity
of installed facilities, (3) desired quality of recreation
experience, and (4) public health and safety. Any one of
the four capacities may be the limiting factor on a specific area, at a specific time, and while three of the ways
are generally recognized as being quantifiable, there is
less agreement on the determination of a desired quality
of experience. Recreation 3: p. 6.

Circulation Road. Federal--Army. A public road used for vehicular
traffic within a recreation area. Recreation 2: p. A-32.

Commercial Recreation. Federal--Interior. Recreation conducted
by a business enterprise for profit and open to the public
on a fee or charge basis. Recreation 3: p. 6.

Community Park. Federal--Interior. An area whose purpose is to
provide recreation opportunities for two or more neighbor-
hoods. Recreation 3: p. 6.

Community Recreation. __Federal--Interior__. Recreation services and
     facilities provided for residents within a common geograph-
     ical area encompassing two or more neighborhoods. Recre-
     ation 3: p. 6.

Compartmentalize. __Federal--Army__. Schematic division of land and
     water areas according to similar natural resources char-
     acteristics for the purpose of analysis. Recreation 2:
     p. A-32.

Composites. __Federal--Interior__. Specifically designated recreation
     areas within authorized national forest boundaries which
     justify concentrated priority purchases using Land and
     Water Conservation Fund monies; approved jointly by the
     Forest Service and the Bureau of Outdoor Recreation. Rec-
     reation 3: p. 7.

Controlled Access. __Federal--Army__. An access road on which traffic
     is regulated or monitored by an entrance station prior to
     entering a recreation area. Recreation 2: p. A-32.

Cooperative Management. __Federal--Interior__. The administration of
     one public agency's property by another for public recre-
     ation purposes. Recreation 3: p. 8.

Daily Capacity. __Federal--Army__. The number of recreationists who
     can use an area in a single day considering the facilities
     available and the turnover rate. Recreation 2: p. A-33.

Daily Design Capacity. __National--Soil Conservation__. The number
     of people that a recreation area or facility is designed
     to accomodate, including turnover, on an average Sunday
     during the normal heavy use season. Earth 4: p. 17g.

Day-Use. __Federal--Army__. A type of recreation use lasting for
     one day or less such as picnicking, water sports, sight-
     seeing. Recreation 2: p. A-33.

Day-Use Area. __Federal--Army__. A recreation area which is construc-
     ted and designated for day-use type activities. Recreation
     2: p. A-33.

     __Federal--Interior__. An area developed for recreation in
     which overnight use is not allowed. Recreation 3: p. 8.

Day Use Zone. __Federal--Interior__. That area generally within a
     50-mile or a 1-hour travel radius of a designated place.
     Recreation 3: p. 8.

Depreciative Behavior. Federal--Interior. Violation of manage-
          ment restrictions, accepted social norms or both, by rec-
          reation visitors. Recreation 3: p. 8.

Design Analysis. Federal--Army. Stage of design when the func-
          tional relationship diagram is developed from the program
          of facilities and all uses and circulation routes are
          evaluated in relation to each other. Recreation 2: p.
          A-33.

Design Concept Stage. Federal--Army. Stage of design when the
          functional diagram is transposed upon the topographic map
          and is modified to fit the terrain. Recreation 2: p. A-33.

Design Load. Federal--Army. The theoretical number of people
          which a recreation area can accommodate at a given moment.
          Recreation 2: p. A-33.

Design Recreation Load. Regional--River Basin. The maximum number
          of recreationists expected to use an area at any one time
          on an average weekend day during the peak month of annual
          visitation for which facilities and land or water would be
          provided. Recreation 1: p. 319.

Developed Recreation Area. Regional--River Basin. Land that is
          developed relatively intensively with any type of recre-
          ation facilities, recreation roads, or other visitor im-
          provements. Also included is land adjacent to facilities
          that receive intensive human use. All or most land in
          recreation "sites" is considered to fall in this category.
          Earth 1: p. 378.

Development Plan. Federal--Army. A plan which locates and orients
          to scale all features and facilities included in a project.
          The refinement of the design concept stage. Recreation 2:
          p. A-33.

          Federal--Interior. A detailed design for providing facil-
          ities and services for a recreation area. Recreation 3:
          p. 8.

Economic Recreation Demand. Federal--Interior. A measure of the
          quantity of outdoor recreation opportunities people are
          willing and able to pay for at one point in time with a
          given level of opportunity conditions. Recreation 3: p.
          21.

Entrance Station. Federal--Army. A structural facility located
          on or adjacent to an access road to a recreation area at
          which visitors are required to register or pay fees prior

to entering the area. Recreation 2: p. A-34.

Environmental Corridor. Federal--Interior. Any linear natural
and scenic area which links other recreation areas. Recreation 3: p. 9.

Exhibit R. Federal--Interior. That portion of a project proposal
from an applicant for a Federal Power Commission license
which addresses plans for recreation development at the
project. Recreation 3: p. 10.

Extensive Recreation. Federal--Interior. Activities that are
usually dispersed over a large area and require few or no
facilities. Recreation 3: p. 10.

Extractive-Challenge. Federal--Army. Recreational activities requiring varying degrees of physical exertion in which the
primary attraction is the opportunity to extract "trophies"
from the environment, i.e., fishing, hunting, rock collecting. Recreation 2: p. A-35.

Facilities. Federal--Interior. Developments provided for use,
accommodation and convenience of visitors to a recreation
area, or for its administration. Recreation 3: p. 10.

Facility-Expressive. Federal--Army. Recreation activities which
are largely focused on the instruments in which they take
place, i.e., court games, field games, weapon ranges.
Recreation 2: p. A-35.

Fixed Courtesy Pier. Federal--Army. A structural facility located
on the shoreline which may or may not be attached to a
boat launching ramp for the convenience of those launching
or retrieving watercraft. Recreation 2: p. A-33.

Floating Courtesy Pier. Federal--Army. A structural floating
facility anchored in the water adjacent to a boat launching ramp provided for loading and unloading watercraft.
Recreation 2: p. A-33.

Free-Flowing Streams. Federal--Interior. Streams or portions of
streams unmodified by works of man or, if modified, still
retain natural scenic qualities and recreation opportunities. Recreation 3: p. 11.

General Recreation Day. Federal--Army. A recreation day which involves primarily those activities attractive to the majority of outdoor recreationists and which generally require
the development of access and facilities. Recreation 2:
p. A-37.

Group Camp. Federal--Army. A recreation area developed for over-
      night use by organized groups on a first come, first serve
      basis. Recreation 2: p. A-37.

Hiking. Federal--Army. The action of walking substantial dis-
      tances on improved trails for the primary purposes of ap-
      preciating environmental qualities, physical fitness,
      education and social interaction. Usually accomplished
      in conjunction with other pursuits such as flora and fauna
      observation camping, etc. Recreation 2: p. A-38.

Impact Control. Federal--Army. The lessening of unavoidable ad-
      verse impacts to significant resource functions and values
      through management measures to offset losses during the
      development of water resources projects. Incorporates
      traditional notions of mitigation, replacement, and main-
      tenance. Recreation 2: p. A-38.

Instant Capacity. Federal--Interior. The number of people that
      a recreation facility or area can reasonably accomodate
      at one time; some researchers refer to this as PAOT (per-
      sons at one time) or OTU (one time use). Recreation 3:
      p. 12.

Intensive Recreation. Federal--Interior. Activities that are or
      can be enjoyed in a limited amount of space; a relatively
      high concentration of participants or spectators often are
      present. Recreation 3: p. 12.

Interpretation. Federal--Interior. The presentation of informa-
      tional, esthetic and philosophical values of an area's
      resources to visitors. Recreation 3: p. 12.

Interpretive Ecological Area. Federal--Army. A particular area
      located on a water resources project in which the natural
      resources, events or forces which comprise the ecosystem
      are analyzed and presented to the public through a variety
      of media for educational or aesthetic purposes. Recreation
      2: p. A-39.

Inventory. Federal--Interior. A catalog of areas identifying,
      tabulating and listing land, water, recreation and park
      facilities and equipment. Recreation 3: p. 12.

Land Oriented Recreation Activity. Regional--River Basin. A rec-
      reation activity that is essentially dependent upon a land
      area for fulfillment. Recreation 1: p. 319.

Latent Demand. Regional--River Basin. That recreational demand
      which is inherent in the population but not reflected in

the use of existing facilities--preferred participation
which is yet unrevealed. Recreation 1: p. 319.

Lease Back. Federal--Interior. A procedure whereby land is purchased by a recreation agency and leased to the former
owner, usually with certain restrictions as to its use.
Recreation 3: p. 13.

Limited Access. Federal--Army. A recreation area which through
its location or design limits ingress and egress for its
use by the public. Recreation 2: p. A-40.

Limiting Factor. Federal--Army. A particular element of the
natural or man-made resources which limits the development
or use of a recreation or natural resource. Recreation
2: p. A-40.

Long Term Camping. Federal--Army. Overnight recreation use which
involves sleeping in the out-of-doors for more than one
night in either a tent or a recreation vehicle. Recreation
2: p. A-31.

Master Plan. Federal--Interior. A long-term guide for the future
preservation, selection and development of recreation facilities and services at a given area. Recreation 3: p.
13.

Maximum Capacity. Federal--Army. The highest possible level of
use of a land or water area or recreation area limited
only by physical size or facility number. Recreation 2:
p. A-40.

Mini Park. Federal--Interior. A relatively small recreation area
in a densely populated or highly developed section of a
community. Recreation 3: p. 13.

Multiple-Use Concept. Federal--Army. Term describing the management of water resource projects by which several land and
water uses, as authorized for the project by Congress, are
planned, developed and administered to meet the public
need while conserving the natural resource. Recreation
2: p. A-41.

Multiple-Use Recreation Area. Regional--River Basin. Land which
is, or can be developed and managed for recreation in combination with other uses. Earth 1: p. 378.

National Battlefields. Federal--Interior. Battlefields of national significance preserved in part, or in entirely, for the
inspiration and benefit of the people. Recreation 3: p. 13.

National Cultural Park. Federal--Interior. An area and facil-
    ities providing a wide range of cultural activities, such
    as folk art and the performing arts. Recreation 3: p. 14.

National Forest. Federal--Interior. A Federal forest, under a
    program of multiple use and sustained yield for timber,
    range, watershed, wildlife and outdoor recreation purposes.
    Recreation 3: p. 14.

National Lakeshore. Federal--Interior. Natural shoreline areas
    of national significance bordering large bodies of fresh
    water, such as the Great Lakes. Recreation 3: p. 14.

National Memorials. Federal--Interior. Structures or areas desig-
    nated to commemorate ideas, events or persons of national
    significance. Recreation 3: p. 14.

National Outdoor Recreation Plan. Federal--Interior. A planning
    document which identifies critical outdoor recreation
    issues and recommends solutions and desirable actions to
    be taken at each level of government and by private in-
    terests. Recreation 3: p. 16.

National Park. Federal--Interior. Areas of special scenic, his-
    torical or scientific importance set aside and maintained
    by the Federal Government especially for recreation or
    study; conserved in such a manner that they will be un-
    impaired for use of future generations. Recreation 3: p.
    14.

National Parkway. Federal--Interior. An elongated national park
    embracing a road designated for pleasure travel. Rec-
    reation 3: p. 15.

National Recreation Areas. Federal--Interior. Nationally signif-
    icant areas developed and managed to provide a variety of
    public recreation opportunities. Recreation 3: p. 15.

National Riverway. Federal--Interior. A federally administered
    river and associated land established to protect and pre-
    serve an outstanding free-flowing river area and its re-
    lated environment. Recreation 3: p. 15.

National Seashore. Federal--Interior. Coastal areas set aside
    for preservation and public recreation use of nationally
    significant scenic, scientific, historic, or recreation
    values, or a combination of such values. Recreation 3:
    p. 15.

National Trails System. Federal--Interior. A network of nation-
     ally significant scenic and recreation trails: (1) Scenic;
     extended trails which provide outdoor recreation opportun-
     ities and conserve nationally significant scenic, historic,
     natural or cultural qualities of areas through which they
     pass, (2) Recreation; trails which provide a variety of
     outdoor recreation uses in or reasonably accessible to
     urban areas. Recreation 3: p. 15.

National Wild And Scenic Rivers System. Federal--Interior. Rivers
     and their immediate environments which possess outstanding
     scenic, recreational, geologic, fish and wildlife, histor-
     ic, cultural and other similar values, and are preserved
     in a free flowing condition: (1) Recreation; rivers or
     sections of rivers readily accessible by road or railroad,
     that may have some development along their shoreline and
     that may have undergone some impoundment or diversion in
     the past, (2) Scenic; rivers or sections of rivers free
     of impoundments, with shorelines or watersheds still large-
     ly undeveloped, but accessible in places by roads, (3)
     Wild; rivers or sections of rivers free of impoundments
     and generally inaccessible except by trails, with water-
     sheds or shorelines essentially primitive and waters un-
     polluted. Recreation 3: p. 16.

Nature Center. Federal--Army. A structural visitor facility which
     is planned, designed and constructed for the purpose of
     displaying, interpreting and disseminating information
     pertaining to the natural resources of a project. Recre-
     ation 2: p. A-41.

Neighborhood Park. Federal--Interior. An area whose purpose is
     to provide outdoor recreation opportunities within walking
     distance of residents in the service area. Recreation 3:
     p. 17.

Non-Power Boating. Federal--Army. Watercraft which are propelled
     by wind or manual means such as sailboats, canoes, kayaks,
     row boats and rafts. Recreation 2: p. A-30.

Non-Resource Based. Federal--Army. Recreation activities which
     can be provided irrespective of available natural resources
     at a project. Infers structural facilities which would
     function in any environment regardless of the natural re-
     sources. Recreation 2: p. A-42.

Observation Unit. Federal--Army. An area of origin of recreation
     use associated with a project area destination. Usually
     within the market area. Recreation 2: p. A-42.

Optimum Recreation Carrying Capacity. Federal--Army. The amount
        of recreation use of a recreation resource which reflects
        the level of use appropriate for both the protection of
        the resource and the satisfaction of the participants.
        Recreation 2: p. A-42.

Other General Activities. Regional--River Basin. Recreation ac-
        tivities not primarily dependent on bodies of water in-
        cluding driving for pleasure, mountain climbing, walking
        for pleasure, playing outdoor games and sports, bicycling,
        attending outdoor sports and drama, and horseback riding.
        Recreation 1: p. 319.

Outdoor Recreation. Federal--Army. Outdoor activities beyond
        those required for personal or family maintenance or for
        material gain which are engaged in for refreshment of the
        spirit, enjoyment of nature, and/or restoring a sense of
        well being. Recreation 2: p. A-42.

        Regional--River Basin. Leisure time activities which
        utilize an outdoor setting. Recreation 1: p. 319.

Outdoor Recreation Activity. Regional--River Basin. A specific
        leisure time pursuit of an outdoor recreation opportunity.
        Recreation 1: p. 319.

Outdoor Recreation Area. Regional--River Basin. A land and/or
        water area where outdoor recreation is recognized as the
        dominant or primary resource management purpose. Earth
        1: p. 377.

Outdoor Recreation Facility. Regional--River Basin. Recreation
        structures or conveniences supplied in a designated area
        for outdoor recreation activities. Recreation 1: p. 319.

Outdoor Recreation Resources. Federal--Interior. Land and water
        areas and associated developments which provide opportun-
        ities for outdoor recreation. Recreation 3: p. 17.

        Regional--River Basin. Land and water resources capable
        of providing outdoor recreation opportunity. Recreation
        1: p. 319.

Outdoor Recreation Site. Regional--River Basin. A tract of land
        developed for specific recreation activities. Recreation
        1: p. 319.

Outdoor Recreation Unit. Regional--River Basin. A facility or
        group of complementary facilities normally in a camp, site
        or park, designed to accommodate a family or other small

group.   Recreation 1: p. 319.

Outing.   Federal--Army.   A visit by one individual to a public
        use area or project for recreation purposes during all or
        any part of a 24-hour day--the same day. Recreation 2:
        p. A-42.

Overlook.   Federal--Army. A visitor structure or area designed
        and constructed for the purpose of providing visitors a
        place from which to view, observe, photograph and/or in-
        terpret project facilities and/or significant landscapes.
        Recreation 2: p. A-42.

Overnight Use.   Federal--Army. Recreational activity which in-
        cludes an overnight stay at a project and normally involves
        the use of cabins or camping facilities. Recreation 2: p.
        A-43.

Park Furniture.   Federal--Army. Recreational structures such as
        picnic benches, fireplaces, benches, trash receptacles,
        etc., which are provided for visitor use, other than en-
        closed buildings. Recreation 2: p. A-43.

Park Values.   Federal--Interior. Scenic, inspirational, esthetic,
        educational and recreational qualities which contribute to
        visitor enjoyment. Recreation 3: p. 17.

Participation.   Federal--Interior. Present or projected use of a
        recreation area or facility expressed in terms of visits,
        visitor hours, visitor days, activity hours or activity
        days. Recreation 3: p. 18.

Participation Rate.   Federal--Army. The average number of times
        an individual engages in outdoor recreation activity during
        the period of a year. Recreation 2: p. A-43.

        Federal--Interior. A numerical expression of the number of
        times a person takes part in a given recreation activity
        over a specific period of time. Recreation 3: p. 18.

Passive-Appreciative.   Federal--Army. Recreation activities re-
        quiring little physical effort which are pursued primarily
        for the enjoyment of social interaction, nature study, or
        combination thereof. Recreation 2: p. A-43.

Peak Load Day.   Federal--Army. The day of maximum use of a recre-
        ation area, project or site. Usually the 4th of July.
        Recreation 2: p. A-43.

Peak Recreation Season. <u>Federal--Army</u>. The period of days in a
    year in which the highest amount of visitor use is real-
    ized. Usually Memorial Day through Labor Day. Recreation
    2: p. A-43.

Persons At One Time. <u>Federal--Forest Service</u>. Public recreational
    usage measurement term. The number of people in an area
    or using a facility at the same time. Generally used as
    "maximum PAOT" to indicate capacity of an area or facility
    to support peak usage loads. General 15: p. 151.

Phase Development. <u>Federal--Army</u>. The construction of recreation
    facilities in stages based on recreation demand, funds,
    etc. Recreation 2: p. A-43.

Physical Capacity. <u>Federal--Army</u>. The material capability of a
    natural or man-made resource to accommodate the level of
    use intended. Only limited by size, number, durability,
    stability or sensitivity. Recreation 2: p. A-43.

Physical Recreation Resource Carrying Capacity. <u>Federal--Army</u>.
    The level of use which a recreation resource can support
    without causing an unacceptable change in the environment
    beyond which the resource cannot restore itself by natural
    means. Recreation 2: p. A-44.

Playground. <u>Federal--Army</u>. A particular recreation site which
    consists of a conglomerate of children's play apparatus
    including swings, climbers, slides, earth mounds, sand
    area, or general day-use area. Recreation 2: p. A-44.

Population Density. <u>Federal--Army</u>. A numerical function which
    expresses the total number of people per square mile of
    land surface. Indicates the intensity of residential de-
    velopment of an area and the potential number of recreation
    users concentrated around a project. Recreation 2: p.
    A-44.

Portage. <u>Federal--Army</u>. An established trail over which water-
    craft are carried or otherwise transported between two
    lakes or rivers or around obstructions in a river or a
    stream. Recreation 2: p. A-45.

Power Boating. <u>Federal--Army</u>. Watercraft which are propelled by
    gasoline or electric motors for recreation purposes. Rec-
    reation 2: p. A-30.

Primitive Area. <u>Federal--Army</u>. Any recreation area at which the
    level of facility development is held to the minimum stan-
    dards of health and safety and the natural resources of

the area are least disturbed. Recreation 2: p. A-45.

Public Access. Federal--Army. An approach, trail or road, over
    land to a lake, stream or river which is publicly owned
    or in which the public holds a vested interest for the
    purpose of recreation use of that water resource. May
    include improvements or conveniences on the shoreline.
    Recreation 2: p. A-46.

Public Benefits. Federal--Army. The tangible and intangible gains
    to society directly attributable to a water resource pro-
    ject that satisfies the expressed or observed needs of
    society. Recreation 2: p. A-46.

Recreational Development. Federal--Army. Any type of facility or
    improvement which are planned, designed, developed and
    managed for recreational purposes. Recreation 2: p. A-46.

Recreational Experience. Federal--Army. The physical and psycho-
    logical benefits or liabilities which are derived from the
    pursuit or recreational activities. Recreation 2: p. A-46.

Recreation Area. Federal--Army. A tract of land and water area
    of substantial size which may contain one or several rec-
    reational activities on a project. Usually reached by a
    single access road for control purposes. Recreation 2:
    p. A-46.

Recreation Benefits. Federal--Army. The tangible and intangible
    gains to the public directly attributable to recreation
    activities at a water resources project. Recreation 2:
    p. A-30.

    Federal--Interior. An expression of monetary value that
    accrues to people from use of resources and facilities for
    recreation. Recreation 3: p. 20.

Recreation Capacity. Regional--River Basin. Annual number of rec-
    reation days that can be accommodated without causing se-
    vere damage to the resource. Recreation 1: p. 320.

Recreation Complex. Federal--Interior. An area containing a va-
    riety of resources and facilities providing different types
    of recreation. Recreation 3: p. 20.

Recreation Day. Federal--Army. A measure of recreation use con-
    sisting of a visit by one individual to a recreation site,
    area or project for recreation purposes during all or any
    portion of a 24-hour day. May consist of several activity
    days. Recreation 2: p. A-46.

Recreation Day. Regional--River Basin. An individual's partici-
     pation in recreation activities for a reasonable portion
     or all of a 24-hour period. Averages 2.5 activities per
     day. Recreation 1: p. 320.

Recreation Demand. Federal--Interior. The quantity of outdoor
     recreation opportunity which an individual desires or
     needs and could be realized if all impediments were re-
     moved, i.e., mobility, social or emotional. The amount of
     additional outdoor recreation opportunity which would be
     utilized if impediments were removed is referred to as
     latent demand. Recreation 3: p. 21.

     Regional--River Basin. The total participation in outdoor
     recreation activities which would occur if opportunities
     to participate were available. Recreation 1: p. 320.

Recreation Demand Allocation. Federal--Army. The apportioning of
     recreation demand among several water resources projects
     according to a users inclination to use the projects, the
     projects ability to serve the user, and the competition be-
     tween several projects for users participation. Recreation
     2: p. A-46.

Recreation Design Load. Federal--Interior. The desired maximum
     number of people during a specific time period for which
     a recreation area is planned and developed. Recreation 3:
     p. 21.

Recreation Development. Federal--Interior. Any kind of facility
     or improved area used for recreation. Recreation 3: p. 21.

Recreation Easement. Federal--Interior. A leagally enforceable
     interest in land, whereby the holder or the beneficiaries
     of the easement are assured certain uses or restrictions
     covering a specific property for recreation purposes. Rec-
     reation 3: p. 21.

Recreation Fee Program. Federal--Interior. The collection of fees
     and charges for entrance to or use of certain Federal rec-
     reation areas or facilities. Recreation 3: p. 21.

Recreation Land Classification System. Federal--Interior. The
     systematic array of recreation lands by factors such as
     natural characteristics, intensity of development or types
     of use. Recreation 3: p. 21.

Recreation Market Area. Federal--Army. That geographic area, mea-
     sured in road miles, or driving time which encompasses a
     project or projects and which contributes 90% of the day

use and 50% of the overnight use to the projects. Recreation 2: p. A-40.

> Regional--River Basin. The zone of program or project influence from which 80 percent or more of the people are drawn on one day outings and/or overnight trips. Recreation 1: p. 320.

Recreation Modeling. Federal--Army. The development of equations which reproduce as nearly as possible the survey estimates of recreation use. Recreation 2: p. A-46.

Recreation Need. Federal--Interior. The difference between current outdoor recreation demand and available supply of recreation opportunities expressed in terms of resources or facilities. Recreation 3: p. 22.

> Regional--River Basin. The difference between demand and supply. Expressed in terms of recreation days or acres of land and water surface. Recreation 1: p. 320.

Recreation Opportunity. Regional--River Basin. The combination of resources favorable for recreation use. Recreation 1: p. 320.

Recreation Resources. Federal--Interior. Land and water areas and associated facilities, people, organizations and financial support that provide opportunities for outdoor recreation. Recreation 3: p. 22.

Recreation Site. Federal--Interior. A specific tract of land or water within a recreation area used for particular recreation activities. Recreation 3: p. 22.

Recreation Standard. Federal--Interior. The measure of quantity and quality considered as the desirable goal in the provision of outdoor recreation areas and facilities. Recreation 3: p. 22.

Recreation Supply. Federal--Army. The quantity of outdoor recreation available to the public at a given time based on resources and facility capability. Recreation 2: p. A-46.

> Regional--River Basin. The resources and facilities capable of providing outdoor recreation opportunities. Developed supply refers to sites and areas. Undeveloped supply refers to potential supply. Recreation 1: p. 320.

Recreation Vehicle Camping. Federal--Army. Overnight recreation use consisting of sleeping within a trailer, motorhome, or vehicle designed for or modified for that purpose within a designated campground. Recreation 2: p. A-31.

Regional Park. Federal--Interior. A park operated by one or more cities, counties or States. Recreation 3: p. 23.

Resource Based. Federal--Army. Recreational activities which are dependent upon the existence of certain defined natural resources at a project, i.e., nature study, mountain climbing, hunting, etc. Recreation 2: p. A-47.

Resource-Expressive. Federal--Army. Recreation activities, often times competitive, in which a natural resource provides the vehicle and/or challenge of the event, but man-made facilities are essential to its success. Recreation 2: p. A-47.

Resource Interpretive Program. Federal--Army. A planned course of action which sets the objectives of interpretation, lays out basic concepts and themes, estimates costs, and establishes a phasing framework for implementation. Recreation 2: p. A-47.

Restricted Zones. Federal--Army. Those areas on a water resources project into which public access is prohibited for reasons of fire prevention, public health and safety. May be water and/or land areas. Recreation 2: p. A-48.

RNR Coordination. Federal--Army. Active process by which recreation and natural resources information is solicited from Federal and non-Federal interests during planning and understandings reached on plans to insure that necessary construction, funding and management agreements are fulfilled as reflected in written response to reports. Recreation 2: p. A-48.

Sanitary Facilities. Federal--Army. Those facility developments or services which are necessary to protect public health at a project, i.e., restroom, sewers, sewage treatment plant, solid wastes disposal, etc. Recreation 2: p. A-48.

Scenic Easement. Federal--Army. Lands in which a limited interest is acquired for the primary purpose of preserving the natural characteristics of the environment by restricting the types of development which can occur. Recreation 2: p. A-34.

Federal--Interior. The right to control use of land, including airspace above it for the purpose of protecting natural qualities; normally such controls do not affect

any regular, non-destructive use exercised prior to acquisition of the easement. Recreation 3: p. 23.

Separable Recreation Lands. <u>Federal--Army</u>. Those lands recommend-
ed/authorized for acquisition on a water resources project
for the specific purpose of recreation use and facility
development that would not otherwise be requested for the
project. Recreation 2: p. A-48.

Service Area. <u>Federal--Army</u>. A specific area allocated for the
facilities used in the routine operation and maintenance
of recreation and operational facilities. Recreation 2:
p. A-48.

Shoreline Mooring. <u>Federal--Army</u>. An area at water's edge designated for boat tie-up which may require grading and mooring posts. Recreation 2: p. A-48.

Sightseeing. <u>Federal--Army</u>. A recreational activity involving
the observation of natural and man-made features while
driving or walking from point to point. Sightseer may
utilize such facilities as restrooms, visitor center,
nature center, overlooks, and roadside rest areas. Recreation 2: p. A-49.

Site Amenities. <u>Federal--Army</u>. The natural or man-made features
of a site which provide for convenient or attractive use
or observation of the site such as restrooms, roads, vegetation, rock formations, running water, etc. Recreation
2: p. A-49.

Social Recreational Carrying Capacity. <u>Federal--Army</u>. The level
of visitation to a recreation area, project or site beyond
which the recreational experiences of the visitors is di-
minished. Recreation 2: p. A-49.

Spacial Qualities. <u>Federal--Army</u>. The characteristics, both physical and psychological, which a defined area conveys to its
users and which effects their use and enjoyment of the area
by stimulating the senses and emotions. Recreation 2: p.
A-49.

Special Interest Area. <u>Federal--Forest Service</u>. U.S. Forest Service usage. Areas managed to make recreation opportunities
available for the understanding of the earth and its geological, historical, archaeological, botanical, and memo-
rial features. General 15: p. 203.

Specialized Recreation Day. <u>Federal--Army</u>. A recreation day involving primarily those activities for which opportunities

are limited in the region of the project and are generally resource based. Recreation 2: p. A-49.

State Park. <u>Federal--Interior</u>. A State area designated for a variety of recreation uses and generally characterized by natural or historic features. Recreation 3: p. 24.

Statewide Comprehensive Outdoor Recreation Plan. <u>Federal--Interior</u>. A guide to outdoor recreation actions focusing on needs, trends, problems and policies and encompassing all outdoor recreation activities, resources and programs that are significant in providing outdoor recreation opportunities within a state. Recreation 3: p. 25.

Study Rivers. <u>Federal--Interior</u>. Those rivers named by the Congress for determination of their suitability for inclusion in the National Wild and Scenic Rivers System. Recreation 3: p. 25.

Support Facilities. <u>Federal--Interior</u>. Those facilities that are not themselves used for recreation but are, nevertheless, required for public recreation use or management of an area, such as service buildings, access roads, parking areas, water systems and toilet facilities. Recreation 3: p. 25.

Tot Lot. <u>Federal--Interior</u>. A relatively small area in a highly populated section developed primarily for use of young children. Recreation 3: p. 26.

Transfer Of Demand. <u>Federal--Interior</u>. Shifts in expected recreation use between areas or regions. Recreation 3: p. 26.

Travel Cost Method. <u>Federal--Army</u>. An economic method of determining both visitation and estimates of value of recreation by determining the vehicle expenses incurred by people who will utilize the recreation facilities proposed. Recreation 2: p. A-50.

Turnover Rate. <u>Federal--Army</u>. The average number of times a given recreation facility or site is used during a given period of time. Recreation 2: p. A-50.

Undeveloped Recreation Area. <u>Regional--River Basin</u>. Land adjoining a developed recreation area that provides recreation activities such as hunting, hiking, and nature walks. Recreation will be the primary use of such land. Earth 1: p. 378.

Unit Day Value. <u>Federal--Army</u>. A monetary value per visitor day
of use which is assigned to a recreation activity or activities to determine the recreation benefits generated by an
overall plan of those activities. May be applied to recreation opportunities foregone or proposed for comparison.
Recreation 2: p. A-50.

Vacation-Use Zone. <u>Regional--River Basin</u>. Area beyond the weekend use zone which normally requires more than 3 hours of
travel. Recreation 1: p. 320.

Visit. <u>Federal--Interior</u>. The entry of one person into a recreation area or site to carry on one or more recreation
activities. Recreation 3: p. 26.

Visitation. <u>Federal--Interior</u>. The total number of persons entering and using a recreation area over a specified period
of time. Recreation 3: p. 26.

Visitor. <u>Federal--Interior</u>. One who enters a recreation area for
enjoyment of the opportunities provided. Recreation 3:
p. 27.

Visitor Center. <u>Federal--Army</u>. A structure designed and constructed specifically at a central location within a project for
the purpose of providing visitors with interpretive facilities and programs which explain the project features such
as natural, man-made and cultural resources. Recreation
2: p. A-51.

Visitor Day. <u>Federal--Army</u>. A measure of recreation use by one
person for one day or part of a day. Recreation 2: p. A-
51.

Federal--Interior. Twelve visitor hours, which may be aggregated continuously, intermittently or simultaneously by
one or more persons. Recreation 3: p. 27.

Visitor Hours. <u>Federal--Interior</u>. The presence of one or more
persons on lands or waters, generally recognized as providing outdoor recreation, for continuous, intermittent
or simultaneous periods of time aggregating 60 minutes.
Recreation 3: p. 27.

Vista Clearing. <u>Federal--Army</u>. The removal of tree and shrubs
species from a vantage point in order to open up views of
significant natural or man-made features. Recreation 2:
p. A-51.

Visual Corridor. Federal--Interior. Land adjacent to or within
    eyesight of a riverway, trail, highway or other route of
    travel. Recreation 3: p. 27.

Walk-In Access. Federal--Army. A recreation area or site to which
    the only access is by foot traffic. Usually applied to
    primitive or wilderness camping areas. Recreation 2: p.
    A-51.

Water Based Recreation. Federal--Army. Recreation which occurs
    on, at, in, under or because of the presence of standing
    or running water such as boating, scuba diving, fishing,
    water skiing, sailing, etc. Recreation 2: p. A-51.

Water Related Recreation Activity. Regional--River Basin. A rec-
    reation activity dependent on, or enhanced by water includ-
    ing swimming, all boating, water skiing, fishing, picnick-
    ing, camping, sightseeing, hiking, and nature walks. Rec-
    reation 1: p. 320.

Weekend-Use Zone. Regional--River Basin. That area between 40
    and 125 miles from an SMSA, which requires 1 to 3 hours
    of travel time. Recreation 1: p. 320.

13. ARCHAEOLOGY/HISTORY

Index

Terms

Archaeological Area. Federal--Interior. Any location containing
    significant relics and artifacts of past cultures. Rec-
    reation 3: p. 4.

Archaeological Resources. Federal--DOT. Objects or areas made or
    modified by man, and data associated with these artifacts
    and features. Transportation 3: p. 13.

    Federal--GSA. Objects or areas made or modified by man
    (including their settings, if they provide a context in
    which to interpret the resource). These include occupa-
    tion sites, work areas, evidence of farming or hunting and
    gathering, burials and other funerary remains, artifacts,
    and structures of all types--usually dating from prehis-
    toric or aboriginal periods, or from historic periods and
    non-aboriginal activities for which only vestiges remain.
    General 2: p. XIII-41.

Architectural Resources. Federal--GSA. Structures, landscaping,
    or other human constructions that possess artistic merit,
    and particularly representative of their class or period,
    or represent achievements in architecture, engineering,
    technology, design, or scientific research and development;
    such resources may be important for their archaeological
    or historical value as well. General 2: p. XIII-41.

Artifact. Federal--Interior. In the broadest sense, any product
    or by-product of human activity. Recreation 3: p. 4.

    Federal--Interior. Something made by man, especially prim-
    itive man. General 3: p. 8.

Cultural History Resource. Federal--Forest Service. U.S. Forest
    Service usage. Potential knowlege about human cultural
    systems, in the form of historic and prehistoric products
    and by-products of man. General 15: p. 56.

Cultural Resource. Federal--Army. Any building, site, district,
    structure, object, data or other material significant in
    history, architecture, science, archaeology, or culture.
    Recreation 2: p. A-33.

    Federal--Interior. The works of man of educational, in-
    spirational or esthetic value. Recreation 3: p. 8.

Districts. Federal--GSA. Geographically definable areas, urban
    or rural, possessing a significant concentration or linkage

of sites, structures, or objects, unified by past events
or aesthetically by plan, physical developments, or similarity of occupation. General 2: p. XIII-41.

Historical Area. Federal--Forest Service. U.S. Forest Service
usage. Sites and areas which have been designated by the
Forest Service as containing important evidence and remains of the life and activities of early settlers and
others who used or visited the area or the sites where important events took place. Examples are battlegrounds,
remnants of mining camps, old cemeteries, important pioneer roads and trails, and early trading sites. General
15: p. 93.

Historic Preservation. Federal--DOT. The protection, rehabilitation, restoration, or reconstruction of districts, sites,
buildings, structures and objects of historic significance.
Transportation 3: p. 35.

Historic Resources. Federal--GSA. Evidence of human activities
that represent facets of the history of the Nation, State
or locality; places where significant historical or unusual
events occurred even though no evidence of the event remains; or places associated with a personality important
in history. Cultural resources can also include districts,
sites, structures, and objects important to an indigenous
culture, a subculture, or a community for traditional spir-
itual, religious, or magical reasons, as well as places important for the artistic, recreational, or other community
activities that take place there. General 2: p. XIII-41.

Historic Sites Act (49 Stat. 666). Federal--Forest Service. The
Historic Sites Act of 1935 authorizes the programs that
are known as the Historic American Buildings Survey, the
Historic American Engineering Record, and the National Survey of Historic Sites and Buildings; authorizes the establishment of National Historic Sites and otherwise authorizes the preservation of properties "of national historical
or archaeological significance"; authorizes the designation
of National Historic Landmarks; establishes criminal sanc-
tions for violation of regulations pursuant to the act;
authorizes interagency, intergovernmental, and interdisciplinary efforts for the preservation of cultural resources;
and other provisions. The program is administered by the
U.S. National Park Service. General 15: p. 94.

National Historic Preservation Act. Federal--Historic Preservation. Means Public Law 89-665, approved October 15, 1966,
an "Act to establish a program for the preservation of
additional historic properties throughout the Nation and

for other purposes", 80 Stat. 204 (1970) and 87 Stat. 139 (1973) hereinafter referred to as "the Act". Legal Jargon 34: p. 3367.

National Historic Sites. Federal--Interior. Historic sites, buildings or objects designated in recognition of thier national significance. Recreation 3: p. 14.

National Monuments. Federal--Interior. Nationally significant landmarks, structures, objects or areas of scientific or prehistoric interest designated by the Federal Government for preservation and public use. Recreation 3: p. 14.

National Register. Federal--Historic Preservation. The National Register of Historic Places, which is a register of districts, sites, buildings, structures, and objects, significant in American history, architecture, archaeology, and culture, maintained by the Secretary of the Interior under authority of section 2(b) of the Historic Sites Act of 1935 and of the National Historic Preservation Act. The National Register is published in its entirety in the Federal Register each year in February. Addenda are published on the first Tuesday of each month. Legal Jargon 34: p. 3367.

National Register Of Historic Places. Federal--DOT. A listing maintained by the National Park Service of architectural, historical, archaeological and cultural sites of local, state, or national significance. Transportation 3: p. 47.

National Register Property. Federal--Historic Preservation. Means a district, site building, structure, or object included in the National Register. Legal Jargon 34: p. 3367.

Natural History Resource. Federal--Forest Service. U.S. Forest Service usage. Natural phenomena which reference the development of the earth's surface and the evolution of life. Two interrelated categories of natural features are recognized. One, the geological category, results from forces and processes acting on the earth's surface to produce land forms and other nonliving entities. The other, the ecological category, involves living entities and processes between biological forms and their environment. General 15: p. 139.

Objects Or Artifacts. Federal--GSA. Material things of functional, aesthetic, cultural, symbolic or scientific value, usually by nature of design. General 2: p. XIII-42.

Paleontological Area. <u>Federal--Forest Service</u>. U.S. Forest Service usage. Areas which have been designated by the Forest Service as containing significant remains (usually
fossilized) of flora and fauna (nonhuman) of geologic time
periods before the appearance of man. General 15: p. 147.

Property Eligible For Inclusion In The National Register. <u>Federal
Historic Preservation</u>. Means any district, site, building, structure, or object which the Secretary of the Interior determines is likely to meet the National Register
Criteria. As these determinations are made, a listing is
published in the Federal Register on the first Tuesday of
each month, as a supplement to the National Register.
Legal Jargon 34: p. 3367.

Sites. <u>Federal--GSA</u>. Distinguishable pieces of ground, or areas
of historic, prehistoric, or symbolic importance, upon
which important historic or prehistoric events occurred;
or which are importantly associated with historic or prehistoric events, persons or cultures; or which were subject to sustained historic or prehistoric activity of man,
sometimes featuring changes in topography produced by human activity. Examples are battlefields, historic campgrounds, ancient trails, gathering places, middens (refuse
heaps, dumps), historic farms, fire pits, storage pits,
burial pits, hard-packed house floors, pot holes, mounds
and caves. General 2: p. XIII-42.

State Historic Preservation Officer. <u>Federal--Historic Preserva-
tion</u>. Means the official within each State, authorized by
the State at the request of the Secretary of the Interior,
to act as liaison for purposes implementing the Act, or
his designated representative. Legal Jargon 34: p. 3367.

Structures. <u>Federal--GSA</u>. Works of man, either prehistoric or
historic, created to serve human activity, usually immovable by nature of design. Examples are buildings of various kinds, dams, canals, bridges, fences, military earth-
works, Indian mounds, gardens, historic roads, and mills.
General 2: p. XIII-42.

14. GEOGRAPHIC AREA
15. POPULATION
16. HOUSING

INTRODUCTION

Themes 14-16 are a reprint of the U.S. Department of Commerce, Bureau of the Census, <u>1970 Census Users' Guide, Part 1</u>, "Census Users' Dictionary". Much of EIS data concerning geographic areas, population and housing themes are developed from census data and it seems appropriate to include the "Census Users' Dictionary" in its original form. At the publication time of this book, <u>EIS Glossary</u>, the 1980 census material was unavailable, and it appears that the majority of terms will remain intact. Unlike themes 1-13, the index to geographic, population and housing terms are located at the end of the housing theme.

CENSUS USERS' DICTIONARY

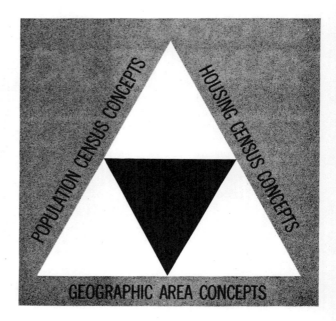

POPULATION CENSUS CONCEPTS

HOUSING CENSUS CONCEPTS

GEOGRAPHIC AREA CONCEPTS

Prepared by:
Data Access and
Use Laboratory

Introduction

The Census Users' Dictionary is a comprehensive dictionary of geographic, population, and housing concepts for which data are collected and presented by the Bureau of the Census. The Dictionary is designed to be a convenient standard reference to facilitate accurate communication among users, between users and the Census Bureau, and within the Bureau itself. The Dictionary may also serve users as a general guide to available census tabulations.

Population and housing concept titles in this Dictionary reflect terminology used in the technical documentation of 1970 census summary tapes. Specific concept title wording which will be used in census printed reports is, in some cases, still being developed at the time of this writing and may differ from that in the technical documentation. Similarly, the wording of concept definitions which will appear in printed reports may reflect modifications resulting from continued review during the coming months.

Census Bureau Statistical Programs

The Bureau is responsible for conducting all censuses (complete enumerations) authorized by Federal law, including the censuses of:

Population. Taken every 10 years in years ending with zero. (First census in 1790.) Definitions of population concepts in the Census Users' Dictionary apply to the 1960 census and the 1970 census.

Housing. Taken every 10 years in years ending with zero. (First census in 1940.) Housing concepts definitions apply to the 1960 and 1970 censuses.

Governments. Taken every 5 years in years ending with 2 and 7. (First census in 1850.)

Agriculture - Taken every 5 years in years ending with 4 and 9. (First census in 1840.)

Economic Censuses

Construction. Taken every 5 years in years ending with 2 and 7. (First census in 1967.)

Business. Formerly taken every 5 years in years ending with 3 and 8. (First census in 1930.) Beginning in 1967 taken every 5 years in years ending with 2 and 7.

These same comments apply to the census of:

Manufactures. (First census in 1870.)

Mineral Industries. (First census in 1840.)

Transportation. (First census in 1963.)

Definitions in the Census Users' Dictionary largely concern those subject concepts employed in the population and housing censuses. Concepts relating to the other censuses may be added at a later date. The Dictionary also includes a section (Part I) which presents definitions of geographic areas recognized in tabulations of all censuses.

Census Procedures: Implications for Concept Definitions

Several aspects of 1970 census collection and processing procedures affect concept definitions and merit a brief mention at this point.

Collection. Certain questions on the census form are not designed strictly for tabulation purposes. Some of these questions, such as the respondent's name, are included only to aid in checking completeness of enumeration. Other questions may be worded so as to increase the probability of reliable responses, for example, asking data of birth as well as age in years. Finally, some concepts tabulated are derived not from a direct question but inferred from one or more items more readily understood by the respondent, i.e., family-type tabulations are a product of questions on an individual's age and relationship to the head of the household.

Processing. Census questionnaires are not simply processed as they stand. Extensive editing procedures (computerized for the most part) are employed to render the data as complete and accurate as possible. Inconsistent answers are reconciled according to fixed editing rules. Missing entries are filled in according to set criteria. Characteristics of the universe are estimated from sample information.

Computers read edited responses onto basic record tapes, so called because they contain information about individual units enumerated (persons, households, and housing units in the case of the census of population and housing). Basic record data are then tabulated or summarized on summary tapes which are used to produce the final printed results. Because of the summarization of data items, data categories

74

carried in the end products of the census may differ from the categories carried in the basic record tapes.

Implications for Census Data Users

The collection and processing procedures involved in an operation of such massive proportions as the 1970 Census of Population and Housing are necessarily complex. The implications for concept definitions, such as those just mentioned, will be of differing importance to data users depending upon their plans for employing the data.

First, all users require concise, basic definitions of census concepts which appear in the tabulations the Bureau makes generally available; i.e., in printed reports, on summary tapes, and microfilm.

Second, there is a subcategory of users which finds knowledge of changes or additions in concept definitions throughout the census process important as a guide to the availability of additional information. The Census Bureau can make available, at user expense, special tabulations from the basic record tapes to produce different breakdowns or combinations of data categories to user specifications. However, no data is released that violates the confidentiality of an individual.

Finally, a small segment of data users requires detailed information regarding operational considerations affecting census concepts in order to apply sophisticated techniques of analysis to the data or to relate their own data to census statistics. This group may need to know, for instance, the percentage of nonresponse to a question and what was done about the missing information.

This Dictionary meets the general need for basic definitions of census concepts and furnishes information of value to persons planning requests for special tabulations from the Bureau. Requirements for more detailed information on census procedures as they affect concept definitions will be met later through papers and reports published by the Bureau and responses to inquiries from data users.

Using the Dictionary

The Census Users' Dictionary is organized to facilitate user understanding of census statistics. There is a part for geographic area definitions and separate parts for concept definitions associated with population and housing data. Each part includes an introductory discussion of collection and processing procedures and other considerations which affect concept meanings and the availability of information. Within each part, concept definitions are organized into broad subjects, such as family structure or occupancy status, generally in the order in which they appear in census publications. Each part is assigned a series of numbers to be used with each definition. Those numbers not used are available for future concept additions. For example, Part I is assigned concepts 1 through 49 but the concepts run only through number 35.

Note that only concepts which appear in connection with tabulated results receive identifying numbers. Additional concepts or categories carried on basic record tapes or census schedules are indicated in the text of appropriate numbered concepts' definitions. Numbers are assigned to indicate the conceptual logic and structure of census categorization. Concepts which logically stand alone and do not constitute subcategories of other concepts, for example, "sex" or "type of foundation in a housing unit," receive whole numbers (59 and 170 respectively). Subcategories of these concepts, such as "male" and "female" for sex, are indented under the main concept and receive suffix numbers (59.1 and 59.2 respectively). It is possible for a concept to be broken down in this manner into many sublevels of categorization. Words and phrases which appear in bold or are underlined are, in most cases, census concepts (vacancy status, family type, urbanized area) or sub-categories of concepts (vacant year-round units, husband-wife families, urban fringe). To aid the user in quickly locating a desired concept or category, an Alphabetical Index is appended to the Dictionary.

The text of concept definitions usually proceeds from the basic to the complex. Users who only require a general idea of a concept, such as "household relationship" or "tenure status of occupied housing units," need not look further than the first sentence or paragraph in most instances. Users who want to know what questionnaire categories the concept is derived from, what additional categories are available, and so on, must look further.

Concept definitions include information derived from instructions to respondents and enumerators which affect concept meanings and, in many cases, information about the progress of a concept from questionnaire categories to processing categories to final tabulations. However, definitions are not completely "operational." Precise details on editing and allocation procedures are not supplied.

Part I. Geographic Areas

(Concepts 1 through 49)

Introduction

This section of the Census Users' Dictionary describes the geographic areas recognized in census tabulations. Definitions specify:

The defining characteristics of the area. Both general and detailed descriptions are presented. Users who only require a general idea of a standard metropolitan statistical area, for example, need not look further than the first sentence or paragraph of the definition.

The agency which defines the area. Many areas are political entities such as States, counties, and municipalities with legally established boundaries. Others are identified by the Census Bureau or other governmental agencies based upon statistical criteria to satisfy particular information needs.

The number of units in each geographic category. For example, 50 States, 1,500 unincorporated places in 1960, 233 SMSA's in 1970. Many areas have increased in number over the past decade because of population growth (more units now meet certain population size minimums), such as urban places, or because they are now identified in larger portions of the country, such as tracts.

The censuses which recognize the area. This includes the population and housing, governments, agriculture, or economic censuses. Population and housing census tabulations recognize more types of areas than other censuses, particularly more types of smaller areas, primarily because the universe enumerated in this census--people and households--is larger, making it possible to present data without violating confidentiality requirements.

The type of geographic codes assigned to the area. Geographic codes, ranging in length from 1-digit to multi-digit characters, have been assigned to various political and statistical areas for control and tabulation purposes. For example, each township or equivalent area (including census county

divisions) is assigned a numeric code in alphabetic sequence within the county.

The area definitions do not include a statement of the subject matter which is tabulated for each type of area. Generally speaking, the larger the area the greater the number and detail of the tabulations produced and published. In many cases, summary tapes contain data for areas smaller than are recognized in the printed reports. In addition, it is possible, in some cases, to obtain information for areas not recognized on the summary tapes or the printed reports on a contract basis. Smaller areas generally mean smaller numbers of reporting units. Hence, to avoid disclosure about individual units, data for these small areas sometimes must be suppressed.

In addition to taking into account the fact of suppression, users need to interpret small-area tabulations with caution, particularly if the information was collected on a sample basis. Smaller numbers of reporting units in tally cells may lessen the reliability of the figures because of sampling fluctuations.

Changes in boundaries from one census to another can be expected. Therefore, users desiring to analyze characteristics of particular areas--large or small--over time should be prepared to cope with the problem of area comparability.

Geographic Areas and Concepts

1. States--The major political units of the United States. The 1970 State codes, appearing on the summary tapes and related geographic products, are two-digit numbers assigned in sequence to States listed alphabetically. A listing of these codes is contained in the Federal Information Processing Standard Publication (FIPSPUB), Nos. 5 and 6.

The 1960 census State codes, also a two-digit numeric, were assigned by geographical divisions. The first digit indicates the geographic division within which the State is located and the second digit the State.

75

431

76

1.1 **Quasi-State or Pseudo-State**-- A portion of a large State which is identified only for data processing purposes at the Census Bureau when data exceeds capacity of a single work unit.   Two-digit numeric codes are assigned to Quasi-State areas when they appear in census summary tapes.

1.2 **United States**--This designation includes the 50 States and the Distirict of Columbia.

1.21 **Conterminous United States**--The 48 contiguous States and the District of Columbia.   Alaska, Hawaii, and outlying areas are excluded.

1.3 **Puerto Rico and other outlying areas**-- Information for the Commonwealth of Puerto Rico, the Virgin Islands of the United States, and Guam is published in the reports of the censuses of agriculture, population, housing, business, manufacturers, and mineral industries.   In addition, some census of population and housing reports show information for the Canal Zone and American Samoa. Population and housing totals are also available for the small outlying areas of Midway, Wake, Canton and Enderbury Islands, Johnston Island and San Island, the Swan Islands, the Corn Islands, and the Trust Territory of the Pacific Islands.

2. **Geographic division**--This is an area composed of contiguous States, with Alaska and Hawaii also included in one of the divisions. There are 9 geographic divisions and these have been used largely unchanged for the presentation of summary statistics since the 1910 census. See Figure 1.

3. **Region**--A unit composed of two or more geographic divisions.   There are 4 regions, although for some purposes the Northeast and North Central Regions have been combined into the North Region. See Figure 1.

4. **State Economic Areas (SEA's)**-- These are single counties or groups of counties within a State which are relatively homogeneous with respect to economic and social characteristics. Boundaries are drawn in such a manner that each economic area has certain significant characteristics which distinguish it from adjoining areas. There are 509 SEA's.

The larger SMSA's are recognized as SEA's. In 1960, all SMSA's of 1,000,000 or more population constituted SEA's except: (1) in New England (SMSA's in New England are groups of cities and towns rather than counties); and (2) in cases where SMSA boundaries cross State lines, thereby necessitating designation of each State's part of the SMSA as a separate SEA.

In 1970, the SEA's of 1960 will be used without change.   SEA's are identified in census tabulations by a two-digit numeric code or a one-digit alphabetic code and are assigned sequentially within the State.

5. **Economic Sub-Regions (ESR's)**-- The 121 ESR's are combinations of the 509 SEA's, each grouping bringing together those SEA's which are most closely related in terms of their economic and social characteristics. In order to achieve such homogeneity, State lines are frequently crossed. A three-digit numeric code is assigned to each Economic Sub-Region.

6. **Counties**--Counties are the primary political administrative divisions of the States, except in Louisiana where such divisions are called parishes, and in Alaska where 29 census divisions have been recently established as county equivalents.   In 1960, census statistics for Alaska were shown for 24 election districts (reduced to 19 in 1961).

A number of cities (e.g., Baltimore, St. Louis, and many Virginia cities) are independent of any county organization and thereby constitute primary divisions of their States.   A three-digit numeric code unique within State is assigned to each county.   The codes used are those defined in the Federal Information Processing Standard Publications (FIPSPUBS) No. 6.

On the summary tapes, each summary record geographic identification carries two FIPSPUBS codes for county - the 1970 county code and the 1970 county of tabulation code. These two codes usually agree. However, users should note that when the 1970 county of tabulation code differs from the 1970 county code, a record had been assigned an incorrect 1970 county code, and the 1970 county of tabulation code represents a correction which should be followed in the aggregation of records.

In 1960, county tabulations included the District of Columbia and the three parts of Yellowstone National Park in Idaho, Montana, and Wyoming. In 1970, only the District of Columbia and the segment of Yellowstone National Park in Montana are included in county tabulations.

There were 3,134 counties and county equivalents in the U.S. in 1960. County maps are available which identify the minor civil division (or census county division), tract, place, and enumeration district boundaries.

Figure 1. CENSUS REGIONS AND GEOGRAPHIC DIVISIONS OF THE UNITED STATES

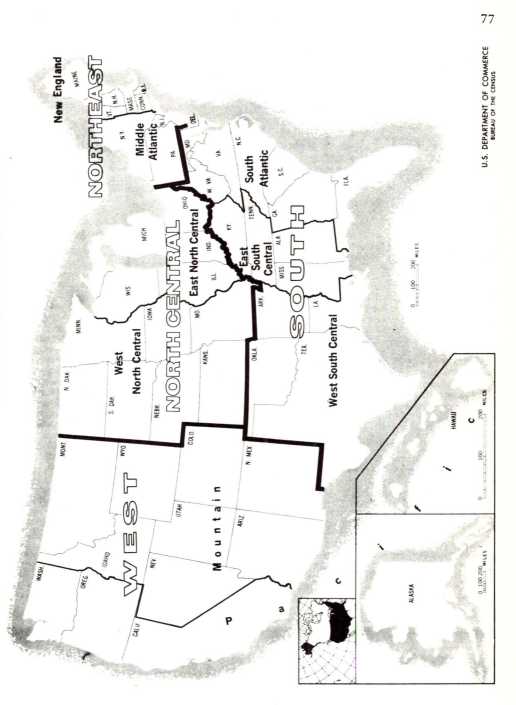

U.S. DEPARTMENT OF COMMERCE
BUREAU OF THE CENSUS

78

On the 1970 summary tapes, a one-digit code, known as the central county code, indicates those counties that contain central cities of SMSA's. The insertion of an "O" in this field indicates all those counties which contain SMSA central cities and a "1" for all counties which do not.

6.1 **Census division**--In Alaska there are no counties; for this State, census statistics were shown in 1960 for 24 election districts. Statistics for 1970 are presented for the 10 boroughs and 29 census divisions which generally conform to the 1960 election districts outside of the boroughs.

7. **Congressional districts**--These areas are defined by State legislatures for the purpose of electing congressmen to the U.S. House of Representatives and may change after each decennial census. Congressional districts are identified by a two-digit numeric code which corresponds to the number assigned in State legislation, except on occasion "01" is used to identify areas in which members of Congress are elected at large, rather than by district.

Population and housing are the only censuses which tabulate statistics for the 435 congressional districts. Published census reports include only population totals for each district. These are found, along with much other census data, in the *Congressional District Data Book* and its supplements. Additional information from other censuses is presented only for districts made up of whole counties or for the smallest combination of split-county congressional districts following county lines. A series of reports from the 1964 Census of Agriculture shows data for congressional districts made up of whole counties.

Several population and housing data items are tabulated for each congressional district on the First Count summary tape.

8. **School districts**--Tabulations of the census of governments recognize independent school districts; "dependent" school systems are regarded as agencies of other governments and are therefore excluded. A drop in number from over 34,500 in 1962 to over 21,500 in 1967 is due primarily to consolidation and reorganization. School districts are not recognized for regular tabulations of the 1970 Census of Population and Housing.

9. **Special purpose districts**

9.1 **Water locations**--Areas established to provide tabulations useful in analyzing the population growth of SMSA's near coasts,

lakes and rivers. Water locations first appeared in the 1960 population census report titled *Standard Metropolitan Statistical Areas,* PC(3)-1D.

9.2 **Production areas**--These are essentially single SMSA's or clusters of SMSA's selected to represent relatively large but geographically compact concentrations of industrial activity. They are utilized in some reports from the census of transportation. There are 25 of these areas.

9.3 **Industrial water-use regions**--Twenty of these units (defined by a Federal interagency committee) are recognized in a subject report from the census of manufactures, *Water Use in Manufacturing.* Each region is a combination of counties grouped to recognize major drainage basins.

9.4 **Fishing regions**--Ten of these (defined by the Department of the Interior) have statistics printed in the report from the 1963 Census of Commercial Fisheries.

9.5 **Petroleum regions**--Statistics for 8 of these regions are presented in a report from the census of business on one of the whole sale trade businesses. They are defined by the Departments of Defense and the Interior and by the Executive Office of the President.

9.6 **Lumber industry regions**--Statistics for the 10 regions are shown in the annual Current Industrial Report, *Lumber Production and Mill Stocks.*

9.7 **Regional marketing areas**--Statistics for the areas for brick and structural clay tile (except surfacing tile) appear in the monthly Current Industrial Report *Clay Construction Products.*

9.8 **Oil and gas districts**--These regions are located in Louisiana, Texas, and New Mexico. In Louisiana, they are composed of parishes, and in New Mexico and Texas they are composed of counties. Statistics for the 17 districts are shown in the reports on petroleum and natural gas industries in the census of mineral industries.

10. **Foreign trade statistical areas**-- Statistics on U.S. imports and exports are published for many different areas. Information is shown for foreign countries, foreign ports, Puerto Rico, U.S. possessions (Virgin Islands, Wake-Island, Guam, and American Samoa), U.S. coastal districts, U.S. customs districts, U.S. ports (including Great Lakes ports), and for combinations of trading areas.

Four classification schedules show the specific areas used in compiling the statistics. Foreign ports grouped into 20 major trading areas with 31 subdivisions are listed in Schedule K; definitions of these areas are in Schedule R. Foreign country designations made by the Census Bureau frequently include adjacent provinces, territories, islands and other areas; these are found in Schedule C. The American ports included in the 25 U.S. customs districts are in Schedule D. The schedules are available from the Census Bureau.

11. **Ward**--Wards are political subdivisions of cities used for voting and representation purposes. These areas are usually reported in the population and housing census tabulations in cities of 3,000 or more which have provided boundary information. It is planned that 1970 census population totals for wards of cities with 10,000 or more will be published in the census reports; unpublished statistics for wards are available at the cost of photocopying the census tabulations. Ward statistics appeared in a Supplementary Report, PC (S1), of the 1960 Census of Population.

The ward code is a two-digit number for each ward within a place of 3,000 or more population which contains wards.

12. **Municipalities and townships**--In the census of governments reports, statistics are shown for types of government rather than for types of places, and the statistics for individual cities and towns are shown for either municipality or township governments. The term "municipality" includes all active governmental units officially designated "cities," "boroughs," "villages," or "towns" (except in New England, New York and Wisconsin). This concept generally corresponds to the incorporated places that are recognized in the population and housing censuses.

The term "township" as used in the census of governments refers to over 17,000 organized governments located in 17 States. The designation includes governments known officially as "towns" in New England, New York, and Wisconsin; some "plantations" in Maine; and "locations" in New Hampshire; as well as all governmental units officially called townships in other areas having this type of government.

13. **Minor Civil Divisions (MCD's)**--These are the primary political and administrative subdivisions of a county; for example, towns, townships, precincts, magisterial districts, and gores. MCD tabulations are made for the census of population and housing. Each township or equivalent area (including census county divisions) is assigned a three-digit numeric code in alphabetic sequence within a county. Codes are not consecutive; gaps of five were allowed for addition of new units. In 1960, over 31,000 MCD's were recognized. Almost two-thirds of these were townships.

For those States in which MCD's are not suitable for presenting statistics, census county divisions (CCD's) are established by the Census Bureau.

In 1960, territories in counties that were not organized into MCD's were reported as a single unit in each county although they may have been split into several discontiguous pieces. In 1970, each separate discontiguous territory will be reported in one or more pieces and given a name. If the piece of unorganized territory in the county is large in area or population, it may be divided into named parts in a manner similar to the delineation of census county divisions. This program is limited to South Dakota, Minnesota, and Maine.

The publication code indicated on the summary tapes is a one-digit number assigned to specific minor civil divisions (MCD's) or census county divisions (CCD's) to indicate whether or not they will be included in printed reports. The codes include: 0 = MCD/CCD records which are to be published; 1 = MCD/CCD records which are independent coextensive incorporated places and are not to be published; 2 = the 19 Connecticut MCD's which are coextensive with dependent incorporated places and are to be published.

MCD-CCD maps are available by State and show township and city boundaries.

13.1 **MCD - place**--This term applies to a unit of tabulation appearing in file B of the First Count summary tape. MCD-places occur in the following situations in most States, the incorporated places form sub-units within minor civil divisions in which they are located; in other States, all or some of the incorporated places are themselves also minor civil divisions; and incorporated places, as well as unincorporated places, may be located in two or more minor civil divisions. An MCD-place is, therefore, any place which is either tabulated in segments, if the place straddles MCD boundaries, or tabulated as a whole, if the place is an MCD itself or a sub-unit of an MCD.

14. **Census County Divisions (CCD)**--In the 21 States for which MCD's are not suitable for presenting statistics, either because the areas have lost their original significance, are too

80

small, have frequent boundary changes, or have indefinite boundaries, the Census Bureau has established relatively permanent statistical areas and designated them as CCD's.

The 18 States with CCD's in 1960 were: Alabama, Arizona, California, Colorado, Florida, Georgia, Hawaii, Idaho, Kentucky, Montana, New Mexico, Oregon, South Carolina, Tennessee, Texas, Utah, Washington, and Wyoming. In 1970, three additional States, Delaware, North Dakota, and Oklahoma, will have CCD's defined.

The population, housing, and agriculture censuses are the only ones for which CCD data have been tabulated. MCD's and CCD's are not recognized in tabulations of the 1969 Census of Agriculture, however. CCD's are defined with boundaries that seldom require change and can be easily located--e.g., roads, highways, power lines, streams, and ridges. The larger incorporated places are recognized as separate CCD's even though their boundaries may change as a result of annexations. Cities with 10,000 or more inhabitants generally are separate CCD's, and some incorporated places with as few as 1,000 population may be separate CCD's.

CCD boundaries were reviewed by county officials and various State agencies and were approved by either the governors or their representatives. Consideration was given to the trade or service areas of the principal settlements and in some instances to major land or physiographic differences.

Unincorporated enclaves within a city are included in the same CCD as the city. In tracted areas, each CCD is normally a single tract or group of tracts, or the combination of two CCD's represents one tract.

For 1970, most CCD counties with small populations that were single CCD's in 1960 have been split into two; also, some CCD's have been consolidated in SMSA counties where central cities have annexed all or major portions of surrounding small CCD's, and other CCD's have been modified or completely changed to agree with newly established census tracts.

MCD-CCD maps are issued by State and include township and city boundaries.

15. **Place (Cities and other incorporated and unincorporated places)**--The term place, as used in the decennial population and housing census, refers to a concentration of population, regardless of the existence of legally prescribed units, powers, or functions. However, most of the places identified in the census are incorporated as cities, towns, villages, or boroughs. In addition, the larger unincorporated places are delineated.

A four-digit numeric code is assigned to each place in alphabetic sequence within State. Place codes are unique within States but place boundaries can cross county, MCD, or CCD lines. These codes are gapped at intervals of five digits to permit insertion of codes for additional places. There are about 20,000 places. In the six New England States, a four-digit New England town code, which is essentially a pseudo-place code, is used to assign New England towns in alphabetic sequence with places within the State. The New England town codes were assigned at intervals of ten (larger than the place code intervals) to provide for insertion of new towns and places.

Since there are no incorporated places in Hawaii, there has always been a problem of recognizing and delimiting places in this State. Only two places, Honolulu and Hilo, have had legal boundaries. Since 1960, a program has been developed under the direction of the State legislature whereby the State Department of Planning and Economic Development has delineated boundaries of places with an estimated population of 300 or more, in cooperation with the Geography Division of the Census Bureau. The Bureau has agreed to treat these places, which are identified as cities, towns, and villages, in the same manner as incorporated places in other States.

A one-digit numeric code, which identifies places by type, appears on the summary tapes and is called the place description code. The codes are: (1) central city of an SMSA only, (2) central city of an urbanized area only, (3) central city of both an SMSA and an urbanized area, (4) other incorporated place, (5) unincorporated place, and (7) not a place. Code 6 is no longer used.

Places are classified on the summary tapes according to a two-digit place size code which identifies the size group (16 groups) into which a place falls on the basis of actual 1970 population. The size codes are: (00) under 200; (01) 200 to 499; (02) 500 to 999; (03) 1,000 to 1,499; (04) 1,500 to 1,999; (05) 2,000 to 2,499; (06) 2,500 to 4,999; (07) 5,000 to 9,999; (08) 10,000 to 19,999; (09) 20,000 to 24,999; (10) 25,000 to 49,999; (11) 50,000 to 99,999; (12) 100,000 to 249,999; (13) 250,000 to 499,999; (14) 500,000 to 999,999; (15) 1,000,000 or more.

The New England town size codes, also identified on the tapes, consist of the same codes and size groupings shown above.

Place maps may be purchased showing streets and containing enumeration district boundaries and also tract boundaries where applicable.

15.1 Incorporated places-- These are political units incorporated as cities, boroughs (excluding Alaska), villages, or towns (excluding New England States, New York, and Wisconsin). Most incorporated places are subdivisions of the minor civil divisions in which they are located; for example, an incorporated village located in an unincorporated township. Some incorporated places, however, constitute MCD's or cross MCD and county lines. Incorporated places never cross State lines since they are chartered by a State. In 1960, they numbered over 18,000.

Statistics for incorporated places of all types and sizes are given in the population and housing census reports; the figures for larger cities are quite detailed. The other censuses provide information for incorporated places of larger than a specified size:

2,500 in the census of governments and the retail trade and selected services segments of the census of business

5,000 in the wholesale trade segment of the census of business

10,000 in the census of manufactures

In the census of business reports, statistics are shown for towns in New England and townships in Pennsylvania and New Jersey (not usually classified as incorporated places) with an urban population of 2,500 or more (5,000 for the wholesale trade segment) or a total population of 10,000 or more.

In the 1970 Census of Population and Housing, boroughs in Alaska are not included as incorporated places because they may include incorporated places within their limits and also they may include large areas with little population. Similarly excluded are towns in New England, New York, and Wisconsin. All townships are excluded.

15.11 Annexed areas-- Areas annexed to incorporated places of 2,500 or more inhabitants since the preceding census are recognized separately in certain decennial census tabulations.

A one-digit annexation code on the summary tapes indicates areas annexed to cities since the previous census; 0 no and 5 • yes. A code "9" is used in some instances to identify areas annexed to smaller cities.

15.2 Unincorporated places-- These are densely settled population centers which are not incorporated. Each has a definite residential nucleus, and boundaries are drawn by the Census Bureau to include insofar as possible, all the closely settled area. In the publications of the census of population and housing, statistics are shown, in all except urbanized areas, for unincorporated places of 1,000 or more population. In 1960, there were over 1,500 of these unincorporated places.

In order to recognize all unincorporated places of 1,000 or more, outside the urbanized area, in unincorporated places are being enumerated separately in 1970 for settlements estimated to have at least 800 inhabitants, as was done in 1960. The Bureau has received varying degrees of cooperation from all the State highway departments in identifying and delineating these places and in providing maps; the coverage, therefore, should be more complete than ever before.

There will be no "urban by special rule" towns in New England or townships in New Jersey or Pennsylvania, as there were in the 1960 census. Unincorporated places may, instead, be defined for the built-up areas in any of these towns and townships which fall outside the urbanized areas.

Within the urbanized areas, except in New England, only unincorporated places of 5,000 inhabitants or more are recognized, in contrast to the 10,000 cutoff level in 1960. Census tract committees have aided greatly in extending the identification of these unincorporated places.

One further change, made with the consent of the Department of Defense, is to recognize and delineate military installations outside incorporated places; the parts of the installations that are built-up, are recognized as unincorporated places.

15.3 Urban place-- This designates all incorporated and unincorporated places of 2,500 or more. In 1960, towns in New England and townships in New Jersey and Pennsylvania, which contained no incorporated municipalities, had 25,000 or more inhabitants, or had from 2,500 to 25,000 inhabitants with a population density of 1,500 or more persons per square mile, were regarded as urban places; also included were counties in other States which contained no incorporated municipalities and had a density of 1,500 inhabitants or more per square mile. These special rules will not be applied in the 1970 census. There were almost 5,500 urban places in 1960 and will be over 6,000 in 1970.

82

16. **Urban - rural areas** (population)--According to the definition adopted for use in the 1960 censuses, the urban population comprised all persons living in:

A. Places of 2,500 inhabitants or more incorporated as cities, boroughs, villages, and towns (except towns in New England, New York, and Wisconsin).

B. The densely settled urban fringe, whether incorporated or unincorporated, of urbanized areas.

C. Towns in New England and townships in New Jersey and Pennsylvania which contain no incorporated municipalities as subdivisions and have either 25,000 inhabitants or more, or a population of 2,500 to 25,000 and a density of 1,500 persons or more per square mile.

D. Counties in States other than the New England States, New Jersey, and Pennsylvania that have no incorporated municipalities within their boundaries and have a density of 1,500 persons or more per square mile.

E. Unincorporated places of 2,500 or more inhabitants.

NOTE: Rules (C) and (D) have been dropped for the 1970 census. Therefore, rural areas are those remaining areas not falling into one of the categories set forth by definition (A), (B), or (E).

The Bureau of the Census uses a one-digit numeric code on the summary tapes to classify enumeration districts as urban, rural, or a combination of these. The urban - rural code designations are as follows: 0=urban and 1=rural.

17. **Urbanized areas (UA)**-- An urbanized area contains a city (or twin cities) of 50,000 or more population (central city) plus the surrounding closely settled incorporated and unincorporated areas which meet certain criteria of population size or density. Beginning with the 1950 Censuses of Population and Housing, statistics have been presented for urbanized areas, which were established primarily to distinguish the urban from the rural population in the vicinity of large cities. They differed from SMSA's chiefly in excluding the rural portions of counties composing the SMSA's and excluding those places which were separated by rural territory from densely populated fringe around the central city. Also, urbanized areas are defined on the basis of the population distribution at the time of the

census, and therefore the boundaries are not permanent.

Contiguous urbanized areas with central cities in the same SMSA are combined. Urbanized areas with central cities in different SMSA's are not combined, except that a single urbanized area was established in each of the two Standard Consolidated Areas.

Essentially the same definition criteria are being, followed in 1970 as in 1960 with two exceptions:

A. The decision not to recognize selected towns in New England and townships in Pennsylvania and New Jersey as urban places under special rules will affect the definition of some areas in these States. Included in urbanized areas will be only the portions of towns and townships in these States that meet the rules followed in defining urbanized areas elsewhere in the United States. This also affects Arlington County, Virginia, which will be considered an urban unincorporated place rather than an urban by special rule county.

B. A change has been introduced with regard to the treatment of extended cities (previously called "overbounded") that contain large areas of very low density settlement. The decision to distinguish between urban and rural parts of extended cities in urbanized areas and to exclude the rural parts from the urbanized areas will help to present a more accurate representation of the population that is truly urban. Approximately sixty incorporated places are involved of which about twenty are central cities. An alphabetic code "A" appearing on the census summary tapes will identify these particular areas.

Pre-census planning indicated approximately fifty potential new urbanized areas. Those which prove to have a qualified central city or twin central cities in 1970 will appear in the published reports.

Maps in the Metropolitan Map Series essentially cover the urbanized areas of SMSA's and contain all recognized census boundaries down to the block level.

Two sets of four digit numeric codes for urbanized areas are contained in the 1970 census tabulations. The potential urbanized area code will identify each record (collection of related data items) in each urban fringe zone. This zone includes all of the area which has the potential of being part of an urbanized area after the 1970 census. The actual urbanized area code uniquely identifies all records in each urbanized area. The final extent of the urbanized area and, therefore, each of the specific records that will contain this code is not determined until after the 1970 census.

The components of UA's and their specific definitional criteria are as follows:

17.1 **Central city of an urbanized area**--An urbanized area contains at least one city which had 50,000 inhabitants in the census as well as the surrounding closely settled incorporated and unincorporated areas that meet the criteria for urban fringe areas. (There are a few urbanized areas where there are "twin central cities" that have combined population of at least 50,000.) All persons residing in an urbanized area are included in the urban population.

17.2 **Urban fringe**--In addition to its central city or cities, an urbanized area also contains the following types of contiguous areas, which together constitute its urban fringe:

A. Incorporated places with 2,500 inhabitants or more.

B. Incorporated places with less than 2,500 inhabitants, provided each has a closely settled area of 100 dwelling units or more.

C. Enumeration districts in unincorporated areas with a population density of 1,000 inhabitants or more per square mile. (The area of large nonresidential tracts devoted to such urban land uses as railroad yards, factories, and cemeteries is excluded in computing the population density.)

D. Other enumeration districts in unincorporated territory with lower population density provided that it serves one of the following purposes:

1. To eliminate enclaves.

2. To close indentations in the urbanized area of one mile or less across the open end.

3. To link outlying enumeration districts of qualifying density that were no more than 1-1/2 miles from the main body of the urbanized area.

A change in the definition since 1960 involves dropping the use of towns in the New England States, townships in New Jersey and Pennsylvania, and counties elsewhere which were classified as "urban by special rule." These areas or their parts, will qualify as part of the urbanized area only if they meet rule C above.

18. **Standard Metropolitan Statistical Areas (SMSA's)**--The concept of an SMSA has been developed in order to present general-purpose statistics. On the basis of the criteria listed below, the geographical boundaries of SMSA's are drawn by the Office of Statistical Policy in the Bureau of the Budget with the advice of representatives of the major Federal statistical agencies. A four-digit code identifies each SMSA. These codes are defined in Federal Information Processing Standard Publications, No. 8.

In 1960, there were 215 SMSA's in the United States and Puerto Rico; as of 1969, there are 233. Generally speaking an SMSA consists of a county or group of counties containing at least one city (or twin cities) having a population of 50,000 or more plus adjacent counties which are metropolitan in character and are economically and socially integrated with the central city. In New England, towns and cities rather than counties are the units used in defining SMSA's. The name of the central city or cities is used as the name of the SMSA. See Figure 2. There is no limit to the number of adjacent counties included in the SMSA as long as they are integrated with the central city nor is an SMSA limited to a single State; boundaries may cross State lines, as in the case of the Washington, D.C. - Maryland - Virginia SMSA.

Where the Current Population Reports series presents statistics for the metropolitan and nonmetropolitan populations, "metropolitan" refers to persons residing in SMSA's and "nonmetropolitan" refers to persons not residing in an SMSA even though they may live in a city.

Criteria for SMSA's:

A. Population size--each SMSA must include at least:

1. One city with 50,000 inhabitants or more, or

2. Two cities having contiguous boundaries and constituting, for general economic and social purposes, a single community with a combined population of at least 50,000, the smaller of which must have a population of at least, 15,000. If two or more adjacent counties each have a city of 50,000 inhabitants or more and the cities are within 20 miles of each other (city limits to city limits), they will be included in the same area unless there is definite evidence that the two cities are not economically and socially integrated.

84

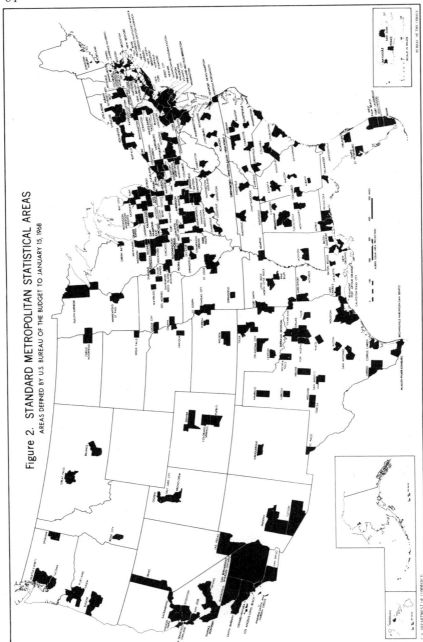

Figure 2. STANDARD METROPOLITAN STATISTICAL AREAS
AREAS DEFINED BY U.S. BUREAU OF THE BUDGET TO JANUARY 15, 1968

B. Metropolitan character of outlying counties--specifically, the following criteria must be met:

1. At least 75 percent of the labor force of the county must be in the nonagricultural labor force.

2. The county must meet at least one of the following conditions:

a. It must have 50 percent or more of its population living in contiguous minor civil divisions having a density of at least 150 persons per square mile, in an unbroken chain of minor civil divisions with such density radiating from a central city in the area.

b. The number of nonagricultural workers employed in the county must equal at least 10 percent of the number of nonagricultural workers employed in the county containing the largest city in the area, or the outlying county must be the place of employment of at least 10,000 nonagricultural workers.

c. The nonagricultural labor force living in the county must equal at least 10 percent of the nonagricultural labor force living in the county containing the largest city in the area, or the outlying county must be the place of residence of a nonagricultural labor force of at least 10,000.

C. Integration of central county and outlying counties--sufficient economic and social communication:

1. At least 15% of the workers living in the given outlying county must work in the county or counties containing the central city or cities of the area, or

2. At least 25 percent of those working in the given outlying county must live in the county or counties containing the central city or cities of the area.

D. In New England, where city and town are administratively more important than the county and data are compiled locally for those minor civil divisions, cities and towns are the units used in defining SMSA's. Here, a population density criterion of at least 100 persons per square mile is used as the measure of metropolitan character and the integration criteria for the towns and cities are similar to criterion C.

18.1 **Central city of an SMSA**--The largest city in an SMSA is always a central city. One or two additional cities may be secondary central cities in the SMSA on the basis and in the order of the following criteria:

A. The additional city or cities must have a population of one-third or more of that of the largest city and a minimum population of 25,000 except that both cities are central cities in those instances where cities quality under A, (2) of the criteria for SMSA's.

B. The additional city or cities must have at least 250,000 inhabitants.

18.2 **Ring of an SMSA**--The ring is all of the SMSA that is not part of the central city itself. This concept is used in the population census to provide information on commuting patterns of workers.

Reports from the 1970 census will include all existing SMSA's. Boundries will not be re-drawn until 1972.

19. **Standard Consolidated Areas (SCA's)**--In view of the special importance of the metropolitan complexes around two of the Nation's largest cities, New York and Chicago, several contiguous SMSA's and additional counties that do not meet the formal integration criteria but do have other strong interrelationships have been combined into SCA's known as the New York-Northeastern New Jersey SCA, and the Chicago-Northwestern Indiana SCA.

In census tabulations, a one-digit alphabetic code is assigned to these two SCA's. They are as follows: A ▪ New York SMSA, Newark (N.J.) SMSA, Jersey City (N.J.) SMSA, Paterson-Clifton-Passaic (N.J.) SMSA, and Middlesex and Somerset Counties in New Jersey; B ▪ Chicago SMSA and Gary-Hammond-East Chicago SMSA in Indiana.

20. **Universal Area Code (UAC)**-- All central cities of SMSA's, selected towns, and all counties and central business districts in the U.S. are assigned a five-digit numeric Universal Area Code. UAC's are assigned to each area requiring separate identification in the population census tabulations of mobility and place of work.

The level of a UAC is indicated in the tabulations by a one-digit numeric code. The levels are: County (1), Town (New England) (2), City (3), and Central Business District (4).

21. **Major Retail Center (MRC)**--This is a concentration of retail stores located in an SMSA

86

but not in the central business district of its chief city. (SMSA's may have more than one MRC.) To be considered an MRC, a shopping area must contain at least one major general merchandise store - usually a department store. MRC's include not only planned suburban shopping centers but also the older "string" street and neighborhood developments which meet the prerequisites. Frequently the boundaries of a single MRC include stores located within a planned shopping center as well as adjacent stores outside the planned portion. In general the boundaries of MRC's are established to include all the adjacent blocks containing at least one store in the general merchandise, apparel, or furniture-appliance groups of stores. In some cases MRC's are defined as census tracts.

The census of business is the only source of MRC statistics. In 1963, there were 972 MRC's identified in the 131 cities with CBD's; in 1967, there were 1,780 identified in almost all SMSA's.

22. **Central Business District (CBD)**-- The central business district is usually the downtown retail trade area of a city. As defined by the Census Bureau the CBD is an area of very high land valuation characterized by a high concentration of retail business offices, theaters, hotels, and service businesses, and with a high traffic flow. CBD's consist of one or more census tracts and have been defined only in cities with a population of 100,000 or more.

A one-digit numeric code, known as the Central Business District indication, denotes what tracted areas are to be tabulated as part of a CBD; 0 = yes and 1 = no.

23. **Census tract**-- Census tracts are small, relatively permanent areas into which large cities and adjacent areas are divided for the purpose of providing comparable small-area statistics. Tract boundaries are determined by a local committee and approved by the Census Bureau; they conform to county lines. Tracts are originally designed to be relatively homogeneous with respect to population characteristics, economic status and living conditions; the average tract has about 4,000 residents. From time to time, changes may be made in tract boundaries; they are not necessarily comparable from census to census.

Census tracts are often used by local agencies in tabulating their own statistics. In 1960, there were over 23,000 tracts identified in 180 areas of the U.S. and Puerto Rico.

All SMSA's presently recognized are completely tracted for the 1970 census. In addition, over 2,000 census tracts will be recognized in non-SMSA cities and counties. The 1970 total is about 34,600 tracts.

Each tract is assigned a six-digit numeric identification number. The first four digits are the "basic" code and the last two "suffix" code. The suffix is only used when necessary to identify two or more tracts formed from a former single tract. The six-digit code appears on the Metropolitan Map Series maps as a large printed number, i.e., 14 (representing 0014--with no suffix) or 14.01 (representing 0014 with the suffix .01). The maps are printed without leading zeroes on the left but when the code is used in publications, the summary tapes and the Master Enumeration District List (MEDList), then (1) no period occurs, and (2) leading zeroes are used, e.g., 0014 or 001401 for a split tract. Basic tract codes can range from 1 to 9999 and the suffix codes range from 00 (shown blank) to 95. Suffix code 99 is reserved for crews of ships. The number is always unique within county, usually unique within SMSA and, in a few instances, within State.

The Tracted Area Code, a four-digit number that uniquely identifies all records in each tracted area, appears on the summary tapes. Tracted areas are usually SMSA's and adjacent territory (non-SMSA cities and counties).

Maps defining census tract boundaries will be included in the published tract reports. These maps will be similar to those of earlier censuses, in that tract outlines and numbers will be shown; other streets and features will be omitted. The names of streets which form tract boundaries will be included except for very short street segments.

23.1 **Tract groups**-- In 1960, cities with a population of 1 million or more identified groups of tracts to form special areas for population and housing tabulations: Chicago - Community Areas; Detroit - Sub-community Areas; Los Angeles - Statistical Areas; New York - Health Areas; Philadelphia - Wards. No such areas will be identified in 1970 tabulations.

24. **Standard Location Area (SLA)**-- This is an area defined by the Office of Civil Defense. Special housing and population census tabulations are produced for use in OCD's damage assessment program. In 1960, SLA's consisted of tracts in tracted areas (tracts lying both inside and outside cities of 50,000 or more were regarded as split tracts and were treated as

two separate SLA's); wards in untracted cities of 25,000 or more where wards are identified; groups of enumeration districts (averaging 5,000 population) in cities of 25,000 or more with neither tracts nor wards; urban places of 2,500 to 25,000 outside tracted areas; MCD's or CCD's, grouped where necessary to attain a minimum population of 2,000 in remaining areas. Standard Location Areas for 1970 have not yet been established.

25. City block--A city block is normally a well-defined rectangular piece of land, bounded by streets and roads. However, it may be irregular in shape or bounded by railroad tracks, streams or other features. Blocks may not cross census tract boundaries, but may cross other boundaries such as city limits. A three-digit numeric identification number is assigned to each block; its first digit is always one or greater. Block numbers are unique within each census tract.

Block data will be tabulated and published for all cities with populations of 50,000 or more prior to 1970. There are about 350 such cities. Block data will also be tabulated and published for cities which may exceed 50,000 in 1970 and for approximately 900 cities and other areas that have contracted for block statistics. All block data discussed here will be available on census summary tapes.

   25.1 Block face-- The side of a city block; a segment of the periphery of a block or of a cul-de-sac into a block. Block faces can be identified using the Address Coding Guide and grouped to any specifications at request for a special tabulation.

25.2 Block group--This designation is new in 1970, and is used in census-by-mail areas where Address Coding Guides have been prepared. A block group is a combination of contiguous blocks having a combined average population of about 1,000. Block groups are approximately equal in area (discounting parks, cemeteries, railroads yards, industrial plants, rural areas, etc.); they are subdivisions of census tracts which simplify numbering and data control. Each block group is identified by the first digit of the three-digit block number. Block group "1" will contain any block in range 101-199, block group "2" in range 201-299, etc. However, normally only the first few numbers in a range are used. For purposes of providing small-area population and housing census data, they are the equivalent of enumeration districts within the mail-out/mail-back areas where Address Coding Guides have been prepared.

Block groups (and blocks) are typically defined without regard to the boundaries of political or administrative areas, such as cities, minor civil divisions, and congressional districts. When a block group straddles one or more of these boundaries, data for those parts in different areas will be tabulated separately. Where such a split occurs, the tapes contain two (or more) data records having the same block group number within the census tract but a different place, annexation, minor civil division, or congressional district code depending on the situation. The First Count tapes do not contain a "flag" to indicate that a block group has been split; however, the Master Enumeration District List (MEDList) can be used to identify split block groups.

In the Address Coding Guide areas, block groups are actually split into so-called "computer enumeration districts" by the Census Bureau to facilitate data processing and enumeration follow-up. No tabulations are produced for these so called enumeration districts since they cannot be mapped as coherent geographic entities.

25.3 Block numbering areas-- In untracted areas where city blocks will be tabulated on a contract basis, blocks have been numbered in block numbering areas which are identified by census tract-type numbers ranging from 9400.00 to 9999.00. Block numbering areas are also defined in parts of untracted counties that are within the 1970 potential urbanized area. Block numbering areas are unique within county boundaries and usually contain a population of about 4,000 people. Data are not tabulated for these areas.

Population and housing data from the decennial censuses have been published in a series of separate reports by census tract (primarily for SMSA's); a great many more statistics will be tabulated but not published and will be available at a nominal cost from the 1970 censuses.

Information about the census tract program is found in the Census Bureau's *Census Tract Manual* which tells how to get tracts established, outlines the responsibilities of local groups, tells about the large body of published and unpublished statistics which the Bureau has tabulated for census tracts, and reviews the ways tract statistics have been made more useful.

26. Enumeration Districts (ED's)--These small population areas average about 250 housing units and are defined by the Census Bureau. They are

88

used for the collection and tabulation of population and housing census data for the conventional enumeration areas and for portions of the mail-out/mail-back SMSA's not covered by the Address Coding Guide.

A four-digit numeric code (ED basic code) is assigned sequentially to each ED within a county, and in some instances within a District Office territory. A one-digit alphabetic suffix code is used to indicate splits of original ED's. Another one-digit code, commonly called ED type, identifies an ED as being in one of the following areas: Address Coding Guide (0), Prelist (1), and Conventional (non-mail) (2). ED's in Address Coding Guide areas are called block groups.

Two administrative factors play a part in determining the geographic definition of enumeration districts. These are: (1) the estimated population size of the ED should constitute an adequate enumerator workload; and (2) the enumeration district must fall within the boundaries of certain areas for which the results are to be tabulated, i.e., tracts, cities, minor civil divisions, etc.

City boundaries often subdivide a census tract or minor civil division into several separate parts. These separate parts are often combined into a single enumeration district on the basis of the workload and boundary considerations described above. Each part is identified on the census map by the same ED number followed by the word Part. However, the data on the summary tape for this split ED are aggregated for the different parts and presented as a single record.

26.1 **District office**--Temporary offices are set up in each of the Census Bureau's regional field office areas, the number for each area being determined primarily by the size and character of the population. Under the direction of the District Supervisor, crew leaders and their enumerators are selected and trained, and all follow-up work is carried out. Incomplete forms are completed and non-responses are eliminated if possible. For control purposes, each of the approximately 400 District Offices has been assigned a four-digit numeric code. The first two digits indicate the Census Field Region and the two remaining ones the specific District Office within the Region. The District Office boundaries contain about one-half million people and may cross State lines.

27. **Address Register**--This is a computer tape listing of all addresses for housing units (occupied or vacant) and other living quarters within addresses in selected areas receiving

city delivery postal service in the mail-out/mail-back census areas in 1970. It is used in preparing mailing labels and for drawing samples of housing units. This file is considered confidential and cannot be made available to anyone outside the Bureau of the Census.

28. **Address Coding Guide (ACG)**--New in 1970, this is an essential tool for the tabulation of census data in urbanized areas covered by the mail census. An ACG contains the actual or potential beginning and ending house numbers on every side of every block, (address range), the street name, the block and tract numbers and other geographic codes.

By referring to an ACG, persons requesting special tabulations from the Census Bureau will be able to define their own geographic units, to correspond to school districts, police precincts, etc. Copies of the guide are available to the public for the cost of reproduction.

28.1 **Address range**--In the ACG, the lowest and highest addresses of a range of addresses on a block face. Both are odd or even, never mixed except through error. In most cases the potential address range is shown. An address range of zero (0) to zero may appear when a block side contains no addresses or the potential range is unknown. High and low addresses may be the same if there is only one address on a block face.

28.2 **Area code**--During the preparation of ACG's, a three-digit numeric code, known as the area code, was devised solely to provide an identification combining MCD and place codes. The area code was assigned to MCD's or CCD's alphabetically within county, and to all places within each MCD or CCD. Numbers were assigned at intervals of five, beginning with 005, (except in Cook County, Illinois and Allegheny County, Pennsylvania, where numbers were assigned at intervals of four, beginning with 004) to provide for insertion of new places and changes in alphabetic listing of MCD/CCD's. The MCD-place combination would have required seven digits if the normal codes were used.

28.3 **Serial number**--Generally a five-digit identifier of a single record in the Address Coding Guide for an urbanized area. The serial number is unique within SMSA.

A suffix to the serial number, known as the check digit, is mathematically derived from the serial number and used to detect errors in transcribing or punching serial numbers. A typical use of the serial number is to identify records that are to be changed. When the serial number and its check digit are introduced into the computer (with other data), the check digit is recomputed and compared to that supplied. In the absence of a match, the entire record is rejected for correction of the error.

28.4 Street code--A five-digit numeric code for each street name in the Address Coding Guide. The street code is unique within postal finance areas. The postal finance area number (identification used by the Post Office Department) is the last 5-digits of the 7-digit Postal Data Code. The first two digits are a numeric State code. Postal Data Codes must be used with the street codes to distinguish among identical street codes in different postal finance areas.

29. Mail census areas (mail-out/mail-back areas)--For the 1970 census, in 145 SMSA's (and Appleton, Oshkosh, and Fond du Lac Wisconsin, areas which are not SMSA's) an Address Register compiled from a modified commercial mailing list was used to prepare individual address labels for the households in the city postal delivery area. In the balance of the metropolitan area, a special pre-listing was done and mailing pieces were addressed by hand; these addresses were not, however, put onto the computerized Address Register. About sixty percent of all households were part of this mail-out/mail-back system.

Questionnaires (either short forms or long forms) were left in mail boxes several days prior to April 1, 1970 and households were requested to place them in the return mail on census day. Follow-up-work was done by enumerator visit if the telephone did not suffice.

29.1 Pre-list areas--Some portions of SMSA's enumerated by the mail-out/mail-back procedure are not covered by the purchased mailing list. The Census Bureau made its own pre-listing of addresses in these areas. Mailing pieces for these areas were addressed by hand. There are no plans to add these addresses onto the computerized Address Register.

30. Non-mail areas (conventional enumeration areas)--Areas other than the 145 SMSA's in the mail-out/mail-back area were enumerated essentially as the census has previously been conducted. Letter carriers left unaddressed short-form questionnaires at each housing unit on April 1, and the enumerator visited each unit, bringing the sample long-form questionnaires at that time for particular households.

31. Census listing book--Printed address lists, usually containing between 300 and 600 addresses clustered within a given area are used to facilitate the control of receipts and work assignment for following enumerators. These books are confidential and are not available to the public. In 1970, these books exist only in pre-list and non-mail areas.

32. Master Reference File (MRF)--The numeric codes and associated place names (where relevant) for all areas recognized in regular or general census tabulations are carried on this computer tape. The smallest unit on the MRF is the city block in urbanized areas and the enumeration district in other areas. The codes are organized hierarchically (e.g., all the MCD codes for one county are grouped together, the county codes for the State, etc.). A selective printout of the MRF, called the Master Enumeration District List, is available for public use.

32.1 Master Enumeration District List (MEDList)--Contains relevant geographic codes and place names for the political and statistical subdivisions of States for which 1970 census data are tabulated. The MEDList is an expanded version of the *1960 Geographic Identification Code Scheme*. The smallest unit on the MEDList is the block group in areas where Address Coding Guides have been developed and the enumeration district in all other areas. The MEDList contains a total population and housing unit count for each block group and enumeration district. The MEDList is needed in conjunction with the census use summary tapes and microfilm.

33. Dual Independent Map Encoding (DIME)--This system of the 1970 census creates a geographic base file for computer mapping. DIME records contain address ranges and block numbers for both sides of each street segment. Block boundaries other than streets (e.g., rivers, shorelines, and city limits) are also included. DIME records can also carry grid coordinates for street intersections and other major map features.

The DIME system makes possible a computer display of data on maps, calculation of the area of blocks and tracts, calculation of the distance between two points, retrieval of data for areas specified as lying within a given distance of a particular point, and the accomplishment of other analyses. DIME also ensures greater accuracy of the Address Coding Guide through use of a topological edit.

The Bureau of the Census is working with other Federal and local agencies to permit the addition

90

of DIME features to already existing Address Coding Guides in most of the SMSA's included in the mail-out/mail-back census. Since the DIME system has several benefits over the ACG, it is also being used in nearly all non-mail SMSA's for creating a geographic base file. This geographic base file program was made possible through the cooperation and funding of local participants and Federal agencies.

34. ZIP Code areas-- These areas are established by the Post Office for directing and sorting mail. ZIP areas are identified by 5-digit codes. The first 3 digits indicate a major city or sectional distribution center; the last 2 digits signify a specific post office's delivery area within the center. Zip areas do not coincide with census areas and will change according to postal requirements. They are not mutually exclusive areas and their boundaries do not necessarily follow clearly identifiable physical features.

ZIP areas were not recognized in any previous census. Fifth Count summary tapes will be the only source for population and housing data by ZIP Code. The allocation of the data to the ZIP Code areas will be accomplished as accurately as possible by prorating ED sample counts to their respective ZIP Codes.

  34.1 City reference file-- This file lists all 35,000 U.S. post offices and the names of cities and towns within each (including variant names and spellings). The ZIP codes, 3-digit codes for multi-ZIP code post offices, and 5-digit codes for single ZIP code post offices, are included for each place name. The total file consists of over 100,000 records and is used for ZIP coding of incoming addresses which do not contain ZIP codes; it is available on request at cost.

  34.2 Street name reference file-- This computer tape or printed listing of street names within post office area includes street codes and number ranges within ZIP code. It can be used to assign 5-digit ZIP codes to addresses with missing ZIP codes and for assigning street codes. Since this file contains no confidential information, it can be made available to the public on a cost basis.

35. Summary and record-type codes-- The summary type code is a one-digit code designating the geographic level of summarization for each record on the First Count summary tape. Summary types are as follows: 0=State, 1=County, 2=MCD (CCD), 3-MCD (CCD) Place Segment, 4=Place, 5=Congressional District, 7=Enumeration District, 8=Block Group. Code 6 is not applicable.

The record type is a numeric code (one or two digits in length) designating a particular kind of tabulation on the Third and Fourth Count summary tapes. It is used for sorting the records within these files. For example, file A of the Fourth Count is sorted by 1970 county of tabulation, tract number, and record type.

  35.1 Sequencing keys-- Alphanumeric fields on the summary tapes which contain various geographic identifiers as determined by the designated summary area. These keys appear on various summary tapes for publication control purposes. Refer to the technical documentation of summary tapes for further details.

Part II. Population Census Concepts

(Concepts 50 through 149)

Introduction

This part of the Census Users' Dictionary defines the subject concepts recognized in 1960 and/or 1970 population census tabulations. Concepts and their categories and subcategories are included which appear in tabulations the Census Bureau makes available to users through printed publications, computer tapes, and microfilm or microfiche. Concepts are organized under broad headings such as Education, Financial Well-being, etc. Concept definitions indicate or are affected by:

Census questions from which the concept is derived. All concepts (tabulation categories) in this section are derived from responses to one or more census questions. In most cases the concepts are directly comparable to specific response categories. This is true for sex, type of school in which enrolled, year moved into present house, vocational training, etc. In other cases, concepts are derived by combining answers to two or more questions to obtain recodes, for instance, in the determination of labor force status and employment status. Where respondents write in answers, Census Bureau personnel determine a code for each handwritten entry according to specified rules. Occupation, industry, income, and mother tongue are among the concepts derived by coding.

Concept categories carried on basic records, but not on summary tapes. For reasons of cost, report size, usefulness, and reliability, fewer concept categories may be tabulated in a particular matrix than are included on the basic records. For instance, the basic records carry some 70 language codes for the concept mother tongue, but only 20 appear on any summary tape or in any printed report. Similarly, one hundred dollar intervals are used in coding income up to a certain maximum on basic records, but income tabulations employ broader income intervals.

Users may request special tabulations on a contract basis which recognize the full range of concept categories carried on the basic records. However, no information will be furnished which violates the confidentiality of the individual.

The universe to which the concept applies. Not all concepts are tabulated (or carried on basic records) for the entire population. Marital status, for instance, is tabulated for persons 14 years of age and over only, country of origin for the foreign stock only, occupation and industry for the experienced civilian labor force and labor reserve only. Quite a few tabulations are made for persons in households only, excluding groups quarters.

The census(es) to which the concept applies (year). Most concepts apply both to the 1960 and 1970 population censuses. A few are new in 1970; others have additional or different categories or different universes in 1970.

Whether related questions are complete-count or sample. A very few questions are are asked of the entire population--only those basic facts about people such as sex and age which are needed to make an accurate count of persons in each area. These are called complete-count or 100-percent items.

All other items about people are obtained from samples. Sampling permits the collection of data about an area which reflect the characteristics of all persons in the area even though only a small number of individuals were actually questioned. This process also allows the data to be obtained at a much lower cost. Sample cases are weighted to reflect the sampling percentages. In a tabulation based on the 20-percent sample, for example, all cases have weights which average 5; that is, all figures are multiplied by 5 so the final figures will be estimates for all the people in an area rather than just 20 percent of them. Control totals for the multiplication are obtained from the 100-percent items.

In 1960, there was a 25-percent sample; in 1970 there will be a 15-percent sample and a 5-percent sample (in order to reduce the length of the questionnaire for any one individual). Certain questions common to both samples will result in a 20-percent sample. Whether a question is asked of everyone or or of a sample of people depends in part on the size of the area for which statistics are to be tabulated and published. Basic population data, including that required for apportionment purposes, is collected on a 100-percent basis and published for city blocks. Data which is considered important for areas

91

447

92

as small as census tracts and minor civil divisions is to be collected on a 15- or 20-percent sample basis. The 5-percent sample includes items needed for larger cities, counties, standard metropolitan statistical areas, and States.

The sample percentages for population items included in the 1970 census schedules in comparison with items in the 1960 census are shown below.

Instructions for respondents in mail-census areas. The meaning of concepts and categories derived from replies on mailed-back questionnaires (except where Census editing procedures change the replies) depends on respondents' interpretation of the questions, which may or may not be as the Census intended. Some interpretive instructions were included with the questionnaire; these are reflected in the concept definitions included in this dictionary.

In the less densely populated areas of the country, enumeration procedures were the same as the single-stage procedure employed in 1960. In addition, special procedures were used to enumerate persons living in certain types of group quarters, such as college dormitories.

Population items	Complete-count or sample percentage	
	1960	1970
Relationship to head of household......................	100	100
Color or race...	100	100
Age (month and year of birth)...........................	100	100
Sex..	100	100
Marital Status...	100	100
State or country of birth.............................	25	20
Years of school completed.............................	25	20
Number of children ever born..........................	25	20
Activity 5 years ago..................................	–	20
Employment Status.....................................	25	20
Hours worked last week................................	25	20
Weeks worked last year................................	25	20
Last year in which worked.............................	25	20
Occupation, industry, and class of worker.............	25	20
Income last year:		
Wage and salary income.............................	25	20
Self-employment income.............................	25	[1]20
Other income.......................................	25	[2]20
Country of birth of parents...........................	25	15
Mother tongue...	25	15
Year moved into this house............................	25	15
Place of residence 5 years ago........................	25	[3]15
School or college enrollment (public or private)......	25	15
Veteran status..	25	15
Place of work...	25	[4]15
Means of transportation to work.......................	25	15
Mexican or Spanish origin or descent..................	–	5
Citizenship...	–	5
Year of immigration...................................	–	5
Marital history.......................................	25	[5]5
Vocational training completed.........................	–	5
Presence and duration of disability...................	–	5
Occupation-industry 5 years ago.......................	–	5

[1]Single item in 1960; two-way separation in 1970 by farm and nonfarm income.
[2]Single item in 1960; three-way separation in 1970 by social security, public assistance, and all other receipts..
[3]This item is also in the 5-percent sample but limited to State of residence 5 years ago.
[4]Street address included for 1970.
[5]In 1960, whether married more than once and date of first marriage; in 1970 also includes whether first marriage ended by death of spouse.

Editing and allocation procedures. Extensive efforts are made to ensure that data collected in the decennial population censuses are complete and accurate. Checking for completeness and consistency of replies began at the local district offices which received the mailed-back questionnaires. The questionnaires were then sent to a central processing center, microfilmed, and fed into an optical scanner (FOSDIC) which reads the information onto magnetic computer tapes. A computer edit program operates on these tapes to check further for completeness and consistency of the data. Certain entries are changed or "edited" according to fixed instructions. For instance, a person identified as the wife of a household head with a martial status of "single" is automatically changed to marital status of "married," if there is also a "head." Where single entries or whole questionnaires are missing, information is "allocated" for those persons. For example, if earnings were not reported for a male in a certain age group and occupation category who worked 40 or more weeks in 1969, the computer would supply to him the earnings of the last male processed living in the same area with the same age, occupation, and weeks worked characteristics.

Population and density

50. Total population-- The total population of a geographic area recognized in census tabulations comprises all persons enumerated whose usual place of residence at time of census was determined to be in that area.

Citizens of foreign countries temporarily visiting or traveling in the United States or living on the premises of an embassy, legation, etc. were not enumerated. Resident aliens were enumerated like other Americans.

51. Population density--Population density for a geographic area is calculated as the number of persons per square mile of land area (includes dry land; land temporarily or partially covered by water, such as swamps; streams, canals, etc. less than 1/8 mile in width; and lakes, reservoirs, etc. of less than 40 acres).

52. Place of residence at time of census--Each person enumerated was counted as an inhabitant of his usual place of abode, generally the place where he lived and slept. This place was not necessarily the same as his legal residence, voting residence, etc.

In the application of this rule, persons were not always counted as residents of the places where they happened to be found by the census enumerators or received a census questionnaire in the mail. Persons temporarily away from their usual place of residents--in a hospital, in a hotel, visiting another home, abroad on vacation--were allocated to their homes.

Certain groups in the population were allocated to a place of residence according to special rules. Persons in the Armed Forces quartered on military installations in the United States were enumerated as inhabitants of the places where their installations were located; college students as inhabitants of the places where they resided while attending college; crews of U.S. merchant vessels in harbor as inhabitants of the ports where their vessels were berthed; crews of U.S. naval vessels not deployed to an overseas fleet were enumerated as inhabitants of the home port of the vessel; inmates of institutions as inhabitants of the places where the institutions were located; persons without a usual place of residence and persons staying overnight at a mission, flophouse, jail, etc. as inhabitants of the places where they were enumerated.

American citizens abroad for an extended period (in the Armed Forces, working at civilian jobs, studying in foreign universities, etc.) are not included in the population of the United States or any subnational geographic area, but are tallied as the overseas population.

The place of residence of each individual is then defined in terms of the geographic areas--States, counties, etc.--recognized in census tabulations. The smallest area for which tabulations are generally prepared is the city block in areas with blocks and the enumeration district in other areas.

53. Urban-rural residence--This is one of the more important breakdowns of the population by geographic residence. The determination of urban-rural residence is made after census results have been tabulated. Geographic areas are classified as urban or rural on the basis of their population size or density at the time of the census.

> 53.1 Urban population-- Generally all persons residing in areas determined to be urbanized areas or in places of 2,500 or more outside urbanized areas. A common breakdown of the urban population is given below.
>
> 53.11 Population in central cities of urbanized areas.
>
> 53.12 Population in urban fringe of urbanized areas. Population in urbanized areas but not in central cities.

94

53.13 **Other urban population outside urbanized areas.** Population in places of 2,500 or more outside urbanized areas.

53.2 **Rural population**--Population not classified as urban constitutes the rural population.

53.21 **Rural farm population**--Rural population residing on farms, as ascertained from responses to a question on acreage and dollar sales of farm products.

Persons are classified as residing on farms if they indicate they live on places of 10 or more acres from which sales of crops, livestock, and other farm products amounted to $50 or more in the previous calendar year, or places of less than 10 acres from which sales of farm products amounted to $250 or more.

53.22 **Rural nonfarm population**--Population residing in rural territory but not on farms.

54. **Metropolitan residence**--This is another important breakdown of the population by geographic residence. It refers to residence in a standard metropolitan statistical area.

54.1 **Metropolitan population**--Population residing in standard metropolitan statistical areas.

54.2 **Nonmetropolitan population**--Population residing outside SMSA's.

55. **Place of residence five years ago**-- Ascertained for persons five years of age or over, who were asked to indicate if they lived in "this house" five years ago or a "different house," and, if the latter, to indicate the State (or foreign country, U.S. possession, etc.) county, and city or town where they lived. (Residence five years ago was to be indicated for the person's usual place of residence.)

In 1970, persons in the 15-percent sample only were asked the question on place of residence five years ago. Persons in the 5-percent sample were asked a less detailed question on State of residence five years ago. Persons fourteen years and over (in the 5-percent sample) were to indicate if they lived in "this State" five years ago and if not, to specify the State (or foreign country, U.S. possession, etc.) in which they lived.

56. **Mobility status**--Refers to the geographic mobility of the population aged five years and

older, comparing the place of residence at time of census with the place of residence five years ago.

56.1 **Nonmovers (in same house)**--Persons living in the same house at time of census as five years ago. Includes those who had moved but returned.

56.2 **Movers (mobile population)**--Persons living in a different house in the United States at time of census than five years ago. Includes only persons for whom sufficient information concerning place of residence five years ago is obtained. (Missing information is supplied where available from other members of the person's family.) A common breakdown of the mobile population is given below.

56.21 **Intracounty movers**-- Persons living in a different house but in the same county at time of census as five years ago. Includes those who had moved from the county but returned.

56.22 **Intercounty movers (migrants)**-- Persons living in a different county at time of census than five years ago. The migrant population is commonly broken down into intercounty migrants, same State and intercounty migrants, different State.

56.3 **Abroad**--Persons residing in a foreign country or an outlying area of the U.S. five years ago. (In 1960, persons living in Alaska or Hawaii in 1955 but in other States in 1960 were classified as living in a different State in 1955.

56.4 **Moved, place of prior residence not reported**--Includes persons living in a different house at time of census than five years ago, but did not provide sufficient or consistent information about their previous place of residence.

Also includes persons who gave no indication whether their place of residence at time of census was different from or the same as their place of residence five years ago, but who in response to the question on year moved into present house indicated that they moved into their present house within the five-year period before the census.

57. **Year moved into present house**--Persons were asked to indicate the most recent move they made by one of several time period categories. In 1970 the categories are: 1969-1970, 1968, 1967, 1965-1966, 1960-1964, 1950-

1959, 1949 or earlier, and "always lived in this house or apartment." The categories were comparable in 1960.

Persons who moved back into the same house or apartment where they lived previously were asked to give the year when they began the present occupancy. Persons who moved from one apartment to another in the same building were asked to give the year they moved into the present apartment.

Age and sex

58. Age--Age is usually determined in completed years as of the time of enumeration from replies to a question on month and year of birth. (Only year of birth by quarter is actually carried on census basic records.) Age is estimated from other information reported in the schedule if the respondent fails to indicate birth date.

Age is tabulated by single years from under 1 year, 1, 2, 3, ... to 98, 99, and 100 years or more; and by many different age groupings, such as five-year age groups.

Median age is calculated as the value which divides the age distribution into two equal parts, one-half the cases falling below this value, one-half above. Median age is generally computed from the age intervals or groupings shown in the particular tabulations, except that median age in tabulations of single years of age is based on five-year age groups.

59. Sex

   59.1 Males

   59.2 Females

Sex ratio is calculated as the number of males per 100 females.

Race

60. Race--refers to the division of the population into white, Negro, and several other racial categories. These racial categories do not correspond to strict scientific definitions of biological stock. Persons were asked to indicate their race by selecting one of the following: White; Negro or Black; Indian (American); Japanese; Chinese; Filipino; Hawaiian; Korean; Other (specify). (In Alaska, Hawaiian and Korean were omitted and Aleut and Eskimo were added.)

Written entries in the "other" category are checked against a list of possible written entries.

This list indicates whether the written entry should remain in the "other" category or be correctly classified in one of the printed categories. If the written entry does not appear on the list, the entry remains in the "other" category.

60.1 White population--Includes persons who indicated their race as white. Also includes persons who indicated the "other race" category and furnished written entries that should correctly be classified in the white category.

60.2 All other races population--Includes all persons who did not indicate their race as white or did not have their entry classified as white.

60.21 Negro and other races population--Includes persons who indicated their race as one of the following:

   60.211 Negro--Includes persons who indicated their race as "Negro or Black." Also includes persons who indicated the "other race" category and furnished a written entry that should be classified as "Negro or Black."

   60.212 American Indian--Includes persons who indicated their race as Indian (American) or reported an Indian tribe.

   In 1970 persons who indicated their race as American Indian were also asked to indicate their tribe.

   60.213 Japanese--Includes persons who indicated their race as Japanese and persons with written entries that should be classified as Japanese.

   60.214 Chinese--Includes persons who indicated their race as Chinese and persons with written entries that should be classified as Chinese.

   60.215 Filipino--Includes persons who indicated their race as Filipino and persons with written entries that should be classified as Filipino.

   60.216 Hawaiian and Korean--Includes persons in all the States (excluding Alaska) who indicated their race as Hawaiian or Korean. Also includes persons in the States who had written entries that should be classified as Hawaiian or Korean. In Alaska, persons who are Hawaiian and Korean are included in the "other races" category.

96

60.217 **Aleut and Eskimo**--Includes persons in the State of Alaska who indicated their race as Aleut or Eskimo. In the other 49 States persons who indicated Aleut and Eskimo are included in the "other race" category.

60.218 **Other races population**-- Includes persons who indicated the "other race" category and had a written entry that is not classified as another category.

During publication this is often considered as a residual category and includes statistics for all races not shown separately.

60.22 **Mixed parentage**-- Persons indicated racial mixture are classified according to the race of the father, if he was present in the household and his race was one of the races entered for the person. If the father's race cannot be determined, the first race listed is used.

Nativity, parentage, ethnic background

61. **Nativity**--Ascertained from a question on place of birth (State, foreign country or U.S. possession) or, in certain cases, parents' place of birth. The population is classified into two major groups: native and foreign born. Place of birth was to be reported for the mother's usual place of residence, rather than the location of the hospital, etc., where birth occurred.

61.1 **Native population**-- Includes persons born in the United States, Puerto Rico, or a possession of the United States. Also included are persons who, although they were born in a foreign country or at sea, have at least one native American parent.

The native population is classified by State of birth and related categories. Codes for each State and major U.S. possession are carried on census basic records. However, detailed tabulations of State of birth are not prepared. Rather, a more general categorization of State of birth related to State of residence, which is useful for migration analysis, is presented. The complete set of categories is as follows:

61.11 **Natives born in state of residence (persons living in state of birth)**--Persons born in the State in which they were residing at time of enumeration.

61.12 **Natives born in other states**--Persons born in a State other than one in which they were residing at time of enumeration. This category is further broken down into region of birth in some tabulations.

61.13 **Natives born in outlying area of U.S. (at Sea, etc.)**--Census basic records carry natives born in outlying areas of U.S. as a separate category from natives born at sea or abroad of American parents.

61.131 **Puerto Rican stock**--Includes persons known to have been born in Puerto Rico and other persons with one or both parents born in Puerto Rico. Also referred to as "natives of Puerto Rican origin" or as "persons of Puerto Rican birth or parentage."

61.14 **Natives State of birth not reported**--Persons whose place of birth was not reported are assumed to be native in the absence of contradictory information.

61.2 **Foreign born population**--Includes all persons not classified as native.

62. **Parentage**--Information obtained from a question on birthplace (country) of mother and father is used to classify the native population of the United States into two categories: native of native parentage and native of foreign or mixed parentage.

62.1 **Native of native parentage**-- Includes native persons, both of whose parents are also native of the United States.

62.2 **Native of foreign or mixed parentage**-- Includes native persons, one or both of whose parents are foreign born.

63. **Foreign stock**--Includes the native population of foreign or mixed parentage and the foreign born population. The foreign stock is classified by country of origin.

63.1 **Country of origin and country of birth**-- The foreign stock is classified by country of origin--either country of birth or country of birth of parents. Separate distributions are shown for the foreign born (based on country of birth) and for the native population of foreign or mixed parentage (based on country of birth of parents). Native persons of foreign parentage whose parents were born in different foreign countries are classified according to the father's country of birth.

Countries specified in the distributions comprise those officially recognized by the U.S. State Department at the time of the census. (Respondents were asked to report country of birth according to international boundaries recognized by the U.S. at the time of enumeration and to distinguish between Ireland and Northern Ireland.) Over 80 countries are separately shown in some country of origin tabulations.

64. Spanish-American population-- In the 1960 census, selected tabulations were prepared for the Puerto Rican population in areas outside the five Southwestern States where Spanish surname population was identified.

In the 1970 census, the Spanish-American population is defined differently according to the sample a person is enumerated in and his State of residence. All tabulations except those for 5-percent data are based upon a 15-percent sample, defined as follows:

a. In New York, New Jersey and Pennsylvania, persons of Puerto Rican stock, (See 61.131 above).

b. In the five southwestern States (Arizona, California, Colorado, New Mexico, and Texas), persons of Spanish language (see 67.1 below) or persons not of Spanish language but of Spanish surname identified by matching with a list of about 8,000 such names.

c. In the remaining States, persons of Spanish language. (See 67.1 below.)

Tabulations of 5-percent data are for persons who report Spanish origin or descent including Mexican, Puerto Rican, Cuban, Central or South American, and other Spanish. Spanish origin or decent is ascertained by means of a 5-percent sample question new with the 1970 census.

65. Citizenship--Not asked in 1960. In 1970 ascertained for persons born abroad, who were asked if they were naturalized citizens, aliens, or born abroad of American parents (native). The total population is then classified as native citizens, naturalized citizens, or aliens.

66. Year of immigration-- Not asked in 1960. In 1970 ascertained for the foreign born who were asked to indicate when they came to the United States to stay. The reply is categorized by several time periods: 1965-70, 1960-64, 1955-59, 1950-54, 1945-49, 1935-44, 1925-34, 1915-24 and before 1915.

67. Mother tongue--In 1960, only the foreign born were asked what language was spoken in the person's home before he came to the U.S. If a person reported more than one language, the code assigned was the mother tongue reported by the largest number of immigrants from his native country in the 1940 census.

In 1970, persons, regardless of place of birth, were asked what language, other than English, was usually spoken in the person's home when he was a child. If more than one foreign language were spoken, respondents were to indicate the principal one.

Tabulations are presented for over 20 common European languages, plus American Indian languages, Chinese, Japanese, and Arabic. Over 70 language categories are carried on census basic records.

67.1 Spanish language population-- Persons who report Spanish as their mother tongue, as well as persons in families in which the head or wife reports Spanish as his or her mother tongue.

Education

68. Enrollment status--In 1960, ascertained for persons 5 to 34 years of age, who were classified as enrolled in school if they attended regular school or college at any time since February 1, 1960. (Attendance at a nursery school, business or trade school, or adult education classes was not to be counted; "regular" schooling included kindergarten and schooling leading to an elementary school certificate, high school diploma, or college degree.) Persons enrolled in a regular school who did not actually attend because of illness, etc. were classified as enrolled in school. In 1970, ascertained for persons 3 years and older, who are classified as enrolled in school if they attended regular school or college at any time since February 1, 1970. ("Regular" schooling includes nursery school, kindergarten, and schooling leading to an elementary school certificate, high school diploma, or college degree.)

69. Level and year or grade of school in which enrolled--Persons enrolled in school were asked the year or grade in which enrolled up to 6 or more years of college. In 1960, enrollment was classified into four levels with separate years or grades identified within each level as indicated below. In 1970, enrollment is classified as in 1960 with the addition of nursery school.

98

69.1 **Nursery school**--Identified in 1970, but not in 1960.

69.2 **Kindergarten**

69.3 **Elementary school**--Includes grades 1 through 8, identified separately in some tabulations. (Persons enrolled in a junior high school are classified as enrolled in elementary school or high school according to year in which enrolled.)

69.4 **High school**--Includes grades 9 through 12, identified separately in some tabulations. (See elementary school, above, for treatment of junior high school enrollment.)

69.5 **College**--Includes 1 through 5 academic years and 6 years or more, identified separately in some tabulations. College enrollment is defined to include enrollment in junior or community colleges, regular 4-year colleges, and graduate or professional schools.

70. **Type of school in which enrolled**-- Persons enrolled in school are classified by type of school in terms of public or private, as indicated below.

70.1 **Public school enrollment**-- Includes persons attending schools controlled and supported primarily by local, State, or Federal governmental agencies.

70.2 **Private school enrollment**--Includes persons attending schools controlled and supported mainly by religious organizations (parochial schools) or private persons or organizations. In 1970, parochial school enrollment and other private school enrollment are identified as separate categories for each level of school except college.

71. **Years of school completed**--In 1960, ascertained for persons 5 years of age and over; in 1970, for persons 3 years of age and over, who were asked the highest grade or year of regular school they ever attended up to 6 or more years of college. Persons attending school were asked the year they were completing. Persons were also asked whether they finished the year specified as the highest grade attended (or were attending that year).

The number tabulated in each category of years of school completed includes persons who report completing that grade or year plus those who attended but did not complete the next higher grade. A common breakdown is no school

years completed; 1-4, 5-6, 7, 8 years elementary; 1-3, 4 years high school; 1-3, 4 academic years or more college. Single years of the highest grade attended are carried on census basic records. Tabulations are commonly produced for particular age groups such as persons 14 and over, persons 25 and over, persons 14 to 24 not enrolled in school.

Median school years completed is calculated as the value which divides the population in half. Years of school completed statistics are converted into a continuous series: the first year of high school becomes grade 9, the first year of college grade 13, etc. Persons who have completed a given year are assumed to be evenly distributed from .0 to .9 of the year. For example, persons who have completed the 12th grade are assumed to be evenly distributed between 12.0 and 12.9.

72. **Vocational training**-- Not asked in 1960. In 1970, ascertained for persons 14 to 64 years of age who were asked whether they ever completed a vocational training program; for example, in high school, as an apprentice, in a school of business, nursing, or trades, in a technical institute, or an Armed Forces school. Respondents were also asked to indicate the main field of such training as follows: business, office work; nursing, other health fields; trades and crafts; engineering or science technician, draftsman; agriculture or home economics; other field. Vocational training does not include courses received by correspondence, on-the-job training, or Armed Forces training not useful in a civilian job.

Marital status and history

73. **Marital status**-- Persons were asked whether they were "now married," "widowed," "divorced," "separated," or "never married."

73.1 **Single (never married)**--Includes persons whose only marriage was annulled.

73.2 **Ever married**--Includes persons married at time of enumeration including separated, plus widowed and divorced.

73.21 **Now married**--Includes persons married only once plus persons who remarried after being widowed or divorced. Enumerators were instructed to report persons in common-law marriages as married.

73.211 **Married, spouse present**--Persons whose spouse was enumerated as a member of the same household, even though he or she may have been temporarily absent on vacation, visiting, in hospital, etc. This category is recorded as a sample item only.

The number of married males, wife present by definition equals the number of married females, husband present, but may not do so in tabulations of the sample because of the method used to weight information on persons enumerated in the sample portion of the census.

73.212 **Married, spouse absent**

73.2121 **Separated**--Persons who reported they were separated. (Includes persons deserted or living apart because of marital discord, as well as legally separated persons.)

73.2122 **Married, spouse absent, other**--Married persons whose spouse was not enumerated as a member of the same household, excluding separated. Includes those whose spouse was employed and living away from home, whose spouse was absent in the Armed Forces, or was an inmate of an institution, all married persons living in group quarters, and all other married persons whose place of residence was not the same as that of their spouse. This category is recorded as a sample item only.

73.22 **Widowed**

73.23 **Divorced**--Persons legally divorced.

74. **Times married**-- Ascertained for persons ever married, who were asked if they had been married once or more than once.

75. **Age at first marriage**--Shown in completed years for persons ever married. Ascertained from a question on month and year of marriage if married once, and month and year of first marriage if married more than once.

76. **Termination of first marriage**--Not asked in 1960. In 1970, persons ever married who reported they had been married more than once were asked if their first marriage ended because of death of spouse. This information is used in conjunction with current marital status to classify the entire ever married population by marital history as follows.

76.1 **Widowed only**-- Persons married only once who were widowed at the time of enumeration, plus persons married more than once whose first marriage ended by the death of the spouse and who were not divorced. (In printed reports, this group is combined with 76.3 to represent known to have been widowed.

76.2 **Divorced only**-- Persons married only once who were divorced; plus persons married more than once whose first marriage did not end by the death of the spouse and who were not widowed. (In printed reports, this group is combined with 76.3 to represent known to have been divorced.)

76.3 **Widowed and divorced**-- Persons married more than once whose marital status at the time of enumeration was widowed and whose first marriage did not end in death of spouse, or whose marital status was divorced and whose first marriage ended in death of spouse.

76.4 **Neither widowed nor divorced**-- All other married persons married only once.

Fertility

77. **Children ever born**--In 1960 total live births of women age 14 or over (in some tabulations 15 or over) who reported they were ever married. In 1970, total live births are ascertained (and carried on census basic records) for all women age 14 or over, regardless of marital status. (Tabulations generally are still for married women). Respondents were asked to indicate number of children ever born as none, 1, 2, 3, ...up to 12 or more. (For purposes of computing total children ever born, the terminal category is given a mean value of 13.)

The questionnaire instructed respondents to exclude stepchildren or adopted children. Enumerators were instructed to include children born to the woman before her present marriage, children no longer living, and children away from home, as well as children still at home.

This information is used in fertility analysis. The number of children ever born per 1,000 women of several age groups is calculated for all women and for ever married women.

100

78. **Fertility ratio**-- This is calculated as the number of children under 5 years of age per 1,000 women 15 to 49 years old. (The base includes single women as well as women ever married.)

Living arrangements

79. **Household/group quarters membership**-- All persons enumerated are classified as living in households or group quarters.

79.1 **Household membership**-- All persons occupying a single housing unit (see Part III, Housing Concepts) are referred to as a household. Average population per household is calculated as the population in households divided by the number of households. (See also persons per unit in Part III.)

79.2 **Group quarters membership**--All persons who are not members of households are regarded as living in group quarters. (See Part III, Housing Concepts.)

Quarters occupied by 5 or more persons unrelated to the head of the household are called group quarters. Quarters with no designated head but with 6 or more unrelated persons are also group quarters.

Some quarters occupied by only one or two persons may also be group quarters. For example, one to five persons occupying a surgical ward of a general hospital, who have no usual residence elsewhere, are in group quarters, as are students living in dormitories. Institutional quarters occupied by one or more patients or inmates are institutional group quarters.

All members of group quarters are classified as either secondary individuals or as inmates of institutions. Group quarters members are classified by type of group quarters as shown below.

79.21 **Inmates of institutions**-- Persons for whom care or custody is being provided in institutions. Includes inmates of mental hospitals, inmates of homes for the aged, and inmates of other institutions. Census sample basic records include type of institution categories.

79.22 **Other persons in group quarters (Noninmates)**--Further classified as shown below. (See also secondary individual.)

79.221 **Persons in rooming houses**-- In addition to rooming and boarding houses, this category includes group quarters in ordinary homes, tourist homes, residential clubs, and Y's. (Not all persons living in these types of quarters are classified as living in group quarters; some are classified as living in housing units.) (See Concept No. 151.1, housing units.)

79.222 **Persons in military barracks**-- Quarters for military personnel which are not divided into separate housing units. In 1960, data on persons in such quarters were shown only for men. In 1970, they will include both men and women as well as being shown separately for men.

79.223 **Persons in college dormitories**--Includes dormitories and fraternity and sorority houses.

79.224 **Persons in other group quarters**--Includes general hospitals (including quarters for staff), missions or flophouses, ships, religious group quarters such as convents, dormitories for workers (such as logging camps or quarters for migratory workers). In 1960, women in military barracks were also classified as in other group quarters in tabulations. In 1970, resident staff members of institutions (persons occupying group quarters on institutional grounds who provide care or custody for inmates) are classed as in other group quarters in tabulations (but carried separately on census basic records); in 1960, such persons were shown as a separate category.

80. **Household relationship**-- Ascertained from replies to a question on relationship to household head. Respondents were asked if they were the "head of household," "wife of head," "son or daughter of head," "other relative of head" (and to specify exact relationship), "roomer, boarder, lodger," "patient or inmate," "other not related to head" (and to specify exact relationship).

80.1 **Head of household**--One person in each household was designated as the "head," that is, the person who was reported as the head by the members of the household. However, if a married woman living with her husband was reported as the head, her husband is

101

considered as the head for the purpose of simplifying the tabulations.

Two types of household head are distinguished--head of a family and primary individual. A family head is a household head living with one or more persons related to him by blood, marriage, or adoption. A primary individual is a household head living alone or with non-relatives only.

80.2 **Wife of head**--A woman married to and living with a household head. This category includes women in common-law marriages as well as women in formal marriages. In complete-count tabulations, the number of wives of head is the same as the number of husband-wife households and the number of husband-wife families. The number does not equal the number of married women, husband present, since it excludes those married women whose husbands are not household heads (as in subfamilies, Concept No. 81.111).

80.3 **Child of head**--A son, daughter, stepchild, or adopted child of the head of the household of which he is a member, regardless of the child's age or marital status. The category excludes sons-in-law and daughters-in-law. (Also see own children Concept No. 84.1.)

80.4 **Other relative of household head**--Household member related to head by blood, marriage, or adoption, but not included specifically in another relationship category. In the sample they are classified as grandchild of head, parent of head or son- or daughter-in-law of head, brother or sister of head, parent-in-law of heads or brother- or sister-in-law of head and other relative of head, and are identified as separate categories in some tabulations.

80.5 **Nonrelative of household head**-- Any household member not related to the head; further classified as lodger, resident employee, and friend or partner. These categories are recorded as sample items only.

80.51 **Lodger**--Persons identified as "roomer, boarder, lodger." In the sample it includes foster children not already identified as "roomer, boarder or lodger."

80.52 **Resident employee**--An employee of the household (such as maid, cook, hired farm hand, companion, nurse), who

usually resides in the housing unit. Also includes the employee's relatives living in the housing unit.

80.53 **Friend or partner**--This is a residual category, including all persons not identified as "roomer, boarder, or lodger" or "resident employee." In tabulations, it is often combined with "roomer, boarder, lodger."

Family structure

81. **Family/unrelated individual status**-- All persons enumerated are classified as family members, unrelated individuals, or inmates of institutions.

81.1 **Family**--Two or more persons living in same household who are related by blood, marriage, or adoption. (No families are recognized in group quarters.) All persons living in a household related to each other are regarded as one family. For instance, a son of the head and his wife living in the household are treated as part of the head's family.

The number of families does not necessarily equal the number of households, since not all households include families. Families are classified in the complete-count basic records by family size or number of persons in a family from 2 persons to 35 persons. Average number of persons per family is calculated.

81.11 **Family (primary)**--Family whose head is also the household head. In 1970, primary families are simply termed families.

81.111 **Subfamily**--Married couple with or without own children, or one parent with one or more own children (parent-child group), living in a housing unit and related to the household head, but excluding the head (for example, a son, his wife and children, living with the household head). Since subfamily members are counted as part of the head's (primary) family, too, the number of subfamilies is not included in the count of families per se or in any tabulations for families. Census basic records include categories of sub-families by family type.

81.12 **Secondary family**--In 1960, a family in a household whose head was not related

102

to the household head. In 1970, secondary families are not recognized (since there are so few); persons formerly classed as secondary family members are classed as secondary individuals.

81.2 **Unrelated individual**--Persons not living with relatives, but living in a household entirely alone or with one or more persons not related to him, or living in group quarters (excepting inmates of institutions).

81.21 **Primary** individual--Household head living alone or with nonrelatives only. The number of primary individuals living alone equals the number of one-person households.

81.22 **Secondary** individual--Unrelated individual who is not a household head or who lives in group quarters (excepting inmates of institutions).

82. **Family Type (family head)**--Families (primary) and subfamilies are classified by type according to sex and marital status of the family head as indicated below.

82.1 **Husband-wife families**--The head and his wife were enumerated as members of the same household.

82.2 **Other families with male head**--Family with male head, but no spouse of head present.

82.3 **Families with female head**--Family where the head is female and there is no spouse of head present.

83. **Married couples**--Husband and his wife were enumerated as members of the same household. (No married couples were recognized in group quarters.) This category is recorded as a sample item only. The number of married couples equals the number of married males, wife present. By definition it also equals the number of married females, husband present, but may not do so in tabulations because of the method used to weight information on persons enumerated in the sample portion of the census. The number of married couples bears no necessary relationship to the number of families, since some married couples may constitute subfamilies of household heads' families, while some families may be headed by single individuals.

83.1 **Married couples with own household**--In 1960, the same as husband-wife primary families. In 1970, the same as husband-wife families.

83.2 **Married couples without own household**--In 1960, two subcategories were recognized: married couples without own household living with nonrelatives, i.e., husband-wife secondary families; and married couples without own household living with relatives, i.e., subfamilies with both spouses present.

In 1970, only the second category of married couples without own household living with relatives is recognized.

84. **Children**

84.1 **Own children**--Never-married persons under 18 who are son, daughter, stepchild, or adopted child of the family head.

84.2 **Related children**--Own children under 18 plus all other family members under 18 (regardless of marital status) related to the family head.

Military status and history

85. **Military status**--Ascertained as of time of enumeration for all persons 14 years of age and over. This information is used in connection with labor force concepts.

85.1 **Civilians**--Persons 14 and over not in the Armed Forces at the time of enumeration.

85.2 **In the Armed Forces**--Persons 14 and over on active duty with the U.S. Army, Navy, Air Force, Marine Corps, or Coast Guard.

86. **Veteran status and history**--Veterans are civilian males (persons on active duty at the time of enumeration are excluded), 14 years of age and over, who have served in the Armed Forces of the United States, regardless of whether their service was in war or peace-time. Veterans in 1960 were asked whether they served in World War I (April 1917 to Nov. 1918), World War II (Sept. 1940 to July 1947), the Korean War (June 1950 to Jan. 1955), and "any other time, including present service." Persons who reported serving in both the Korean War and World War II were tabulated as a separate group. All others who reported more than one period of service were classified according to the most recent wartime period of service reported.

In 1970, veterans were asked whether they served in World War I, World War II, the Korean War, the Vietnam Conflict (August 1964 to present), and any other time.

Work patterns: labor force and employment concepts

87. Labor force status--Ascertained for persons 14 years of age and over as of the calendar week prior to data of enumeration (reference week). In 1970, most labor force tabulations will be presented for persons 16 years and over.

87.1 Labor force--Includes persons classified as employed or unemployed plus members of the Armed Forces.

87.11 Civilian labor force-- All persons employed or unemployed, excluding members of the Armed Forces.

87.111 Experienced civilian labor force--Employed plus experienced unemployed.

87.2 Not in labor force--All persons 14 and over not classified as members of the labor force, including persons doing only incidental unpaid work on a family farm or business (less than 15 hours during the reference week). Most of the persons in this category are students, housewives, retired workers, seasonal workers enumerated in an "off" season who are not looking for work, inmates of institutions, or persons who cannot work because of long-term physical or mental illness or disability.

87.21 Labor reserve--Persons classified as not in the labor force during the reference week, but who indicated in reply to the question on year last worked that they did work within the ten-year period preceding the census.

88. Employment status--Ascertained for persons 14 years of age and over from replies to several questions relating to work activity and status during the reference week. These questions were: Did this person work at any time last week (include part-time work such as Saturday job or helping without pay in family business or farm and active duty in the Armed Forces; exclude housework, school work, or volunteer work)? How many hours did he work last week (at all jobs)? Does this person have a job or business from which he was temporarily absent either because of illness, vacation, labor dispute, etc., or because he was on layoff last week? Has he been looking for work during the past four weeks, and if so, was there any reason why he could not take a job last week?

88.1 Employed--Civilians 14 years and over who during the reference week were either "at work"--who did any work for pay or profit or worked without pay for 15 hours or more on a family farm or business; or "with a job but not at work"--were temporarily absent because of reasons such as illness, vacation, etc. The two categories, at work and with a job but not at work, are shown separately in some tabulations.

88.2 Unemployed--In 1960, civilians 14 years and over who were neither "at work" nor "with a job but not at work" during the reference week but were "looking for work" within the past 60 days. (Examples of looking for work include registering at an employment office, writing letters of application, etc.) Persons waiting to be called back to a job from which they were laid off or furloughed were also counted among the unemployed.

In 1970, civilians 14 years and over who were neither "at work" nor "with a job but not at work" within the past 4 weeks and were "available for work" during the reference week. Persons waiting to be called back to a job from which they had been laid off or who were waiting to report to a new wage or salary job within 30 days were counted among the unemployed.

(Availability for work is indicated by replies to a question--new in 1970--whether there was any reason why the respondent could not take a job last week.)

88.21 Experienced unemployed--Those unemployed who indicate in reply to the year last worked question that they have worked at some time in the past.

89. Unemployment rate-- Represents the number of unemployed as a percent of the civilian labor force. Unemployment rates shown for occupation and industry groups are based on the experienced civilian labor force, since occupation and industry cannot be ascertained for those unemployed who have never worked.

90. Hours worked--Ascertained for persons 14 years of age and over who indicate they were "at work" during the reference week. Respondents were asked how many hours they worked last week at all jobs, excluding time off and including overtime or extra hours. The information was collected for the following categories: 1 to 14 hours, 15 to 29 hours, 30 to 34 hours, 35 to 39 hours, 40 hours, 41 to 48 hours, 49 to 59 hours, 60 hours or more.

104

Tabulations are shown for hours worked by several categories. The information is also used to classify employed persons "at work" into full-time employed (persons working 35 hours or more during the reference week) and part-time employed (persons working less than 35 hours during the reference week).

91. Weeks worked--Ascertained for persons 14 years of age and over who worked at all during the calendar year preceding the census. Two questions on this subject were asked: "Last year (1969), did this person work at all, even for a few days?" If yes, then How many weeks did he work in 1969, either full-time or part-time?" Paid vacations, paid sick leave, and military service are counted as weeks worked. The following time categories were presented: 13 weeks or less, 14 to 26 weeks, 27 to 39 weeks, 40 to 47 weeks, 48 to 49 weeks, and 50 to 52 weeks.

It should be noted that the determination of weeks worked during the previous year was essentially independent of the determination of the current employment status of the respondent.

92. Year last worked--Ascertained for persons 14 years of age and over who were not classified in one of the following categories: "at work," "Armed Forces," "with a job but not at work." Respondents were asked when they last worked at all, even for a few days (including any work for pay or profit, unpaid work on a family farm or business, and active service in the Armed Forces), for several time period categories. For 1960, these were 1960, 1959, 1955 to 1958, 1950 to 1954, 1949 or earlier, and never worked; for 1970 categories were 1970, 1969, 1968, 1964 to 1967, 1960 to 1963, 1959 or earlier, and never worked.

Year last worked was tabulated for persons classified as not in the labor force or unemployed.

93. Disability status-- Not asked in 1960. In 1970, ascertained for persons age 14 to 64. Respondents were asked if they had a health condition or disability which limited the kind or amount of work they could do at a job and whether their health prevented them from doing any work at all. Persons who answered "yes" to either or both questions are classified as disabled; persons who responded that they had a disability but were not prevented from doing any work at all as disabled, able to work; persons who responded their health prevented them from doing any work at all as disabled, cannot work.

94. Duration of disability--Not asked in 1960. In 1970, persons who indicated they had a disability affecting the kind or amount of work they could do on a job were asked how long they had been disabled: less than 6 months, 6 to 11 months, 1 to 2 years, 3 or 4 years, 5 to 9 years, 10 years or more.

Work patterns: occupation, industry, and related concepts

95. Occupation--Ascertained for persons 14 years of age and over in the experienced civilian labor force or in the labor reserve. Employed persons were to report the occupation at which they worked the most hours during the reference week. The experienced unemployed and persons in the labor reserve were to report their last occupation. (Excludes the small number of experienced unemployed persons who last worked more than 10 years ago).

In 1960, respondents were asked to describe what kind of work they were doing, for example, 8th grade English teacher, farmer, grocery checker, etc. In 1970, respondents were asked to give this information and, in addition, to specify their most important activities on duties on the job, such as types, cleans building, sells cars, etc., and to indicate their job title. This additional information was requested so that occupation can be coded more accurately.

Information supplied by respondents is assigned an occupation code by clerks. In 1960, there were 11 major occupation groups and an occupation not reported category (listed below). The major occupation groups were divided into 494 items: 297 specific occupations and 197 subcategories which were mainly industry distributions of 13 specific occupations. Tabulations which present the complete range of specific occupations and subcategories are referred to as detailed occupation tabulations. Other tabulations present intermediate levels of classification, combining specific occupations and subcategories into broader groupings.

The occupation classification scheme employed in 1960 is fully described in Bureau of the Census, 1960 Census of Population, *Classified Index of Occupations and Industries*, available from the Superintendent of Documents.

1960 Major Occupation Groups

95.1 Professional, technical and kindred workers

95.2 Farmers and farm managers

95.3 **Managers, officials, and proprietors, except farm**

95.4 **Clerical and kindred workers**

95.5 **Sales workers**

95.6 **Craftsmen, foremen, and kindred workers**

95.7 **Operatives and kindred workers**

95.8 **Private household workers**

95.9 **Service workers, except private household**

95.10 **Farm laborers and foremen**

95.11 **Laborers, except farm and mining**

95.12 **Occupation not reported**

For 1970, there are 12 major groups instead of 11 as in 1960. The new major group, entitled "Transport Equipment Operatives," is made up of bus drivers, parking attendants, truck drivers, and others similarly employed. The categories comprising the new major group were moved from the 1960 major group "Operatives and Kindred Workers," affording a basis for comparability.

A second revision (shown below) was the recasting of the arrangement of the major groups to reflect four broad occupational areas. The major groups and the occupational areas to which they relate are as follows:

1970 Major Occupational Groups	Occupational Areas
Professional, technical, and kindred workers	
Managers and administrators, except farm	White collar workers
Sales workers	
Clerical and kindred workers	
Craftsmen and kindred workers	
Operatives, except transport	Blue collar
Transport equipment operatives	workers
Laborers, except farm	
Farmers and farm managers	
Farm laborers and farm foremen	Farm workers
Service workers, except private household	Service workers
Private household workers	

A third revision to the major groups relates to the processing of the data. Individuals who did not report an occupation were allocated to a major group through an allocation matrix based on selected demographic and economic characteristics. Thus, major group totals in 1970 include persons allocated to the major groups.

Fourth, instead of having the categories within each major group listed alphabetically, subgroupings, or occupation "families," have been established in several major groups. For example, the service workers group, to clarify its content, has been reclassified into 5 subcategories--cleaning service, food service, health service, personal service, and protective service.

The *Classified Index of Occupations* to be used in the 1970 census is scheduled to be published this year. Copies will be available from the Superintendent of Documents of the Government Printing Office. Also, see the *Statistical Reporter*, December 1969, for a relevant article and occupations listing.

96. **Industry**--Ascertained for persons 14 years of age and over in the experienced civilian labor force or in the labor reserve. Employed persons were to report the job at which they worked the most hours during the reference week. The experienced unemployed and persons in the labor reserve were to report the job they last held. Respondents were asked the name of their employer (company or organization); what kind of business or industry this was (describe activity at location where employed, for example, county junior high school, auto assembly plant, retail supermarket, farm, etc.); and to indicate whether this was primarily manufacturing, wholesale trade, retail trade, or other. The name of employer is a basic tool in coding industry, since coders refer to lists of establishments showing their industrial classification from the quinquennial Economic Censuses.

Information supplied by respondents is assigned an industry code. In 1960, there were 12 major industry groups and an industry not reported category (listed below). These groups were further classified into 150 specific categories. Intermediate levels of industry classification scheme were presented in some tables. The 1960 industry classification scheme is described fully in Bureau of the Census, 1960 Census of Population, *Classified Index of Occupations and Industries*, available from the Superintendent of Documents.

106

96.1 Agriculture, forestry, and fisheries

96.2 Mining

96.3 Construction

96.4 Manufacturing

96.5 Transportation, communication, and other utilities

96.6 Wholesale and retail trade

96.7 Finance, insurance, and real estate

96.8 Business and repair services

96.9 Personal services

96.10 Entertainment and recreation services

96.11 Professional and Related services

96.12 Public administration

96.13 Industry not reported

For the 1970 census, the Industry Classification System, like that for occupation, has been revised. This system is designed for use in classifying the industry returns for the 1970 population census and demographic surveys to be conducted by the Bureau of the Census during the decade of the seventies. The system is patterned after the classification outlined in the 1967 edition of the *Standard Industrial Classification Manual (S.I.C.).*

For 1970, there are 226 uniquely identified groups in the classification in contrast to the 150 groups in the 1960 classification. These 76 additional codes stemmed from revisions to 24 specific 1960 industry categories. For the most part, the changes represent establishment of smaller, more homogeneous groups. The 1960 "Industry Not Reported" category has been eliminated. Cases where codes are not reported will be allocated to the major groups.

The 1970 *Classified Index of Industries* will be published some time this year. Copies may be obtained from the Superintendent of Documents. Also, see the *Statistical Reporter,* April 1969, for a relevant article and industries listing.

97. Class of worker--Ascertained for persons 14 years of age and over in the experienced civilian labor force or in the labor reserve. Employed persons were to report class of worker for the

job at which they worked the most hours during the reference week. The experienced unemployed and persons in the labor reserve were to report class of worker for the job last held. Respondents were asked to indicate class of worker by one of several categories shown below. Note that the determination of class of worker was independent of occupation and industry classification but refers to the same job.

97.1 Private wage and salary workers--Includes persons who indicated they were employees of a private company, business, or individual, working for wages, salary, or commissions, and those noted in 97.3 below.

97.2 Government workers--Includes persons who indicated they worked for a governmental unit (Federal, State, or local). In 1970, employees of the Federal government, State governments, and local governments were ascertained as separate subcategories.

97.3 Self-employed workers--Persons who indicated they were self-employed in own business, professional practice, or farm. In 1970, respondents were asked to specify whether their own business was incorporated or unincorporated. Those who said the business was incorporated are classified as private wage and salary workers rather than as self-employed.

97.4 Unpaid family worker-- Persons who indicate they worked without pay in a family business or farm (the business or farm was operated by a family relative).

98. Activity five years ago-- Asked in 1970 for the first time and ascertained for persons 14 years of age and over who were asked if, in April, 1965, they were "working at a job or business (full or part-time)," if they were "in the Armed Forces," or if they were "attending college."

99. Occupation five years ago-- Asked in 1970 for the first time and ascertained for persons 14 years of age and over who indicated they were working at a job or business five years ago. Respondents were asked to specify their occupation or "kind of work" in 1965. The questions on major activities and job titles were not included. Occupation five years ago is then coded as for current occupation.

100. Industry five years ago--Asked in 1970 for the first time and ascertained for persons 14 years of age and over who reported they were

working at a job or business five years ago. Respondents were asked to specify the industry for which they worked five years ago. The supplementary questions on name of employer and manufacturer, wholesaler, etc., were not included. Industry five years ago is then coded as for current industry.

101. Class of worker five years ago--Asked in 1970 for the first time and ascertained for persons who reported they were working at a job or business five years ago. The information was obtained from a question which asked if they were "an employee of a private company or government agency" or "self-employed or an unpaid family worker."

Work patterns: place of work and means of transportation to work

102. Place of work--Ascertained for persons 14 years of age and over who reported working at some time during the reference week, (except those on leave, sick, etc.). They were asked where they worked "last week." (Persons who worked at more than one job are to report place of work for the job at which they worked the greatest number of hours; persons who traveled in their work or worked in more than one place are to report where they began work if they reported to a central headquarters, or where they worked the most hours.)

In 1960, respondents were to specify city or town, county and State where they worked. Place of work replies were tabulated in various ways by the worker's place of residence; for example, working in same county or different county as worker's place of residence; working in same State, contiguous State, or noncontiguous State as place of residence. Place of work by place of residence was also tabulated for universal area code areas.

In 1970, respondents were to specify State, zip code, county, city or town, and exact street address where they worked. Tabulations similar to 1960 will be produced. In addition, since street address was ascertained, place of work may be coded to small geographic areas such as tracts or enumeration districts and made available on that basis if requested as a special tabulation.

103. Means of transportation to work--Ascertained for persons 14 years of age and over who reported working during the reference week, including Armed Forces personnel. Respondents

were asked that principal mode of travel or type of conveyance used to get to their place of work on the last day they worked.

In 1960, the categories were private auto or car pool; railroad, subway or elevated (the latter two categories were combined in tabulations); bus or streetcar; taxicab, other means (taxicab was included in other means in tabulations); walked only; worked at home.

In 1970, the categories are driver, private auto; passenger, private auto; bus or streetcar; subway or elevated; railroad; taxicab, walked only; worked at home; other means.

Financial well being: income and poverty concepts

104. Total income--Ascertained for all persons 14 years of age and over for the preceding calendar year, even if they had no income. Total income is the sum of the dollar amounts of money respondents reported receiving (best estimate if exact amount not known) as wages or salary income, net nonfarm and farm self-employment income, and other income, as specified below. In statistics on family income or household income, the combined incomes of all members of each family or household are treated as a single amount. For unrelated individual income and for income statistics of persons 14 years and over, the classification is by the amount of their own (individual) total income.

Income is tabulated by several intervals. For example, under $1,000, $1,000 - $1,999 . . . $9,000 - $9,999, $10,000 - $14,999, $15,000 - $24,999, $25,000 and over. The 1960 census basic records included dollar amounts for each type of income in intervals of $10 from $1 - $9 to $9,990 - $9,999 and in intervals of $1,000 from $10,000 - $10,999 to $24,000 - $24,999. Two separate categories were provided for each of the following items: (1) no income and (2) incomes of $25,000 and over. (Net loss from self-employment and all other sources was included in intervals of $100 from $1 - $99 to $9,800 - $9,899. Net losses of $9,900 and over were tabulated separately.)

In the 1970 census basic records, for dollar amounts of each type of income, questionnaire dollar entries within $100 intervals from $1 - $99, $100 - $199, to $99,900 - $99,999 are shown as one-tenth of the midpoint value for that interval. For example, any questionnaire entry between $100 and $199 is represented as "15" on the basic record; any questionnaire entry between

108

$99,900 and $99,999 is represented as "9995" on the basic record. Similarly, dollar amounts within $10,000 intervals from $100,000 - $109,999, $110,000 - $119,999 to $980,000 - $989,999 are shown as one-tenth of the midpoint value for that interval. For example, any questionnaire entry between $100,000 and $109,999 is represented as "10500" on the basic record; any questionnaire entry between $980,000 and $989,999 is represented as "98500" on the basic record. Separate categories are provided for no income and incomes of $990,000 or more. Net losses from self-employment income (section 104.12 below) and income from all other sources (section 104.23 below) are included in intervals of $100 from $1 - $99 to $9,800 - $9,899. Net losses of $9,900 or more are carried as one category.

Median and mean incomes are calculated for families, unrelated individuals, and persons 14 years and over for total income and for each type of income. (In the 1960 derivation of aggregate amounts for calculating mean income, persons in the open-ended interval "$25,000 and over" were assigned an estimated mean of $50,000 for each income type. In the 1970 derivation of aggregate amounts for each income type, persons in the open-ended interval "$990,000 and over" are assigned an estimated mean of $995,000.)

104.1 **Earnings**--The sum of wage or salary income and net self-employment income.

104.11 **Wage or salary income**--Money respondents reported receiving as wages, salary, commissions, bonuses, or tips from all jobs (before deductions for taxes, bonds, dues, etc.). Respondents were to include sick leave pay, but exclude military bonuses, reimbursements for business expenses, and pay "in kind."

104.12 **Self-employment income**--Money respondents reported receiving as profits or fees (net income after business expenses) from their own business, professional practice, partnership, or farm. (If the enterprise lost money, respondents were to report the amount of loss.) In 1970, self-employment income from a farm (including earnings as a tenant farmer or sharecropper, excluding payment "in kind") was reported separately from other self-employment income.

104.2 **Income other than earnings (other income)**--Money respondents reported receiving from sources other than wages or salary and self-employment. In 1960, res-

pondents were asked to report other income as a single amount. In 1970, respondents were asked to specify other income as follows.

104.21 **Income from social security or railroad retirement**--Includes U.S. Government payments to retired persons, to dependents of deceased insured workers, or to disabled workers; excludes Medicare reimbursements.

104.22 **Income from public assistance or welfare**--Includes amounts received from Federal, State, and local public programs such as aid for dependent children, old-age assistance, general assistance, and aid to the blind or totally disabled. Excludes separate payments for hospital or other medical care.

104.23 **Income from all other sources**--Includes interest; dividends; veterans payments of all kinds; retirement pensions from private employers, unions, and governmental agencies; and other regular payments such as net rental income, unemployment insurance benefits, workmen's compensation, private welfare payments, alimony or child support, Armed Forces allotments and regular contributions from persons not members of the household. Excludes receipts from sale of personal property, capital gains, lump-sum insurance or inheritance payments, or payments "in kind."

105. **Poverty level**--Not ascertained in 1960. In 1970, families and unrelated individuals (excluding college students in dormitories and Armed Forces personnel in barracks) are classified as being above or below the poverty level, using the poverty index adopted by a Federal Interagency Committee in 1969. This index takes into account such factors as family size, number of children, and farm-nonfarm residence, as well as the amount of money income. The poverty level is based on an "economy" food plan designed by the Department of Agriculture for "emergency or temporary use when funds are low." The definition assumes that a family is classified as poor if its total money income amounts to less than approximately three times the cost of the "economy" food plan. These cutoff levels are updated every year to reflect changes in the Consumer Price Index.

In 1970, percent below poverty level is calculated as the proportion of the total universe which reports income below the poverty level: for example, below poverty level families as a percent of all families.

106. **Income deficit**-- Not ascertained in 1960. In 1970, the income deficit is calculated as the difference between the total income of families and unrelated individuals and their respective poverty levels. Families and unrelated individuals can then be classified both by the absolute amount of their income deficit and by the ratio of their income to the poverty level. 1970 census tabulations express the income deficit in both absolute and relative terms.

107. **Poverty areas**-- All census tracts and MCD's outside tracted areas will be classified as poverty or non-poverty areas on the basis of population census data. Poverty areas will be those tracts and MCD's with an incidence of poverty at least one and one-quarter times the national average. 1970 is the first census for which such statistics will be a part of the regular printed reports.

Part III. Housing Census Concepts

(Concepts 150 through 250)

Introduction

This part of the Census Users' Dictionary defines the subject concepts recognized in 1960 and/or 1970 housing census tabulations. Concepts and their categories and subcategories are included which appear in tabulations the Census Bureau makes generally available to users through printed publications, summary tapes, and microfilm or microfiche. Housing Census Concepts (Part III) is subdivided into three sections: Basic Housing Concepts, Components of Inventory Change (CINCH) Survey, and Survey of Residential Finance. The last two sections include introductory material specifically relevant to the programs involved (pp. 124-129).

Concepts in the Basic Housing Concepts section are organized under broad headings such as Financial Characteristics, Household Equipment, etc. Concept definitions indicate or are affected by:

Census questions from which the concept is derived. All concepts (tabulation categories) in this section are derived from responses to one or more census questions. In most cases, the concepts are directly comparable to specific response categories. This is true for year structure built, value of unit, rooms in unit, etc. In other cases, concepts are derived by combining answers to two or more questions to obtain recodes, for instance in the determination of plumbing facilities or gross rent. Two questions (H-12 and H-14) include an "other" category where respondents can write in replies, which Census Bureau personnel then code into one of the specified categories.

Concept categories carried on basic records, but not on summary tapes. For reasons of cost, report size, usefulness, and reliability, fewer categories may be tabulated than are included in the basic records. For instance, sample basic records carry specific dollar amounts of contract rent, but contract rent tabulations are commonly by ten-dollar intervals. Users may request special tabulations on a contract basis which recognize the full range of concept categories carried on the

basic records. However, no information will be furnished which violates the confidentiality of the individual.

The universe to which the concept applies. Not every concept is tabulated (or carried on basic records) for every housing unit. Tenure, for instance, is only tabulated for occupied housing units, and value is only tabulated for "single-family, owner-occupied" and "vacant for sale" units which are on lots of less than 10 acres and with no business on the property.

The census(es) to which the concept applies (year). Most concepts apply to both the 1960 and 1970 housing censuses. A few 1960 concepts have been dropped for 1970; a few new concepts have been added.

Whether related questions are complete-count or sample. Some of the questions are asked about each unit and are called complete-count or 100-percent items. These are necessary because of the need for housing data on a city block basis where a sample would not be reliable because of the small number of cases.

All other items about housing units are obtained from samples. Sampling permits collection of data about an area which reflect the characteristics of the housing inventory at a much lower cost than complete enumeration. Sample cases are weighted to reflect the sampling percentages. In a tabulation based on the 20-percent sample, for example, all cases have weights which average 5; that is, all figures are multiplied by approximately 5 so the final figures will be estimates for the total housing inventory in an area rather than just 20 percent of it. Control totals for the multiplication are obtained from the 100-percent items.

In 1960, there was a 25-percent sample (although some items asked of the entire 25-percent sample were processed onto basic record tapes for only a 20- or 5-percent sample); in 1970, there will be a 15-percent sample and a 5-percent sample (in order to

111

112

reduce the length of the questionnaire for any one household). Certain questions common to both samples will result in a 20-percent sample. Whether a question is asked for every housing unit or only a sample depends primarily on the size of the area for which statistics are to be tabulated and published. Information needed for city blocks is collected on a 100-percent basis; that which is considered important for areas as small as census tracts and minor civil divisions is to be on a 15- or 20-percent sample basis. The 5-percent sample includes items needed for larger cities, counties, standard metropolitan statistical areas, and States.

The sample percentages for housing items included in the 1970 census schedules in comparison with items in the 1960 census are shown below.

Housing items	Complete-count or sample percentage	
	1960	1970
Number of units at this address........................	-	[1]100
Telephone available....................................	25	[2]100
Access to unit..	100	100
Kitchen or cooking facilities.........................	100	-
Complete kitchen facilities...........................	-	100
Condition of housing unit.............................	100	-
Rooms...	100	100
Water supply..	100	100
Flush toilet..	100	100
Bathtub or shower.....................................	100	100
Basement..	20	100
Tenure..	100	100
Commercial establishment on property..................	[3]100	100
Value...	[3]100	100
Contract rent...	[3]100	100
Vacancy status..	100	100
Months vacant...	25	100
Heating equipment.....................................	25	20
Components of gross rent..............................	25	20
Year structure built.................................	25	20
Number of units in structure and whether a trailer.....	20	20
Farm residence (acreage and sales of farm products)....	[4]25	20
Land used for farming.................................	[5]25	-
Source of water......................................	[4]20	15
Sewage disposal......................................	[4]20	15
Bathrooms...	20	15
Air conditioning.....................................	5	15
Automobiles...	[6]20	15
Stories, elevator in structure.......................	[7]20	5
Fuel--heating, cooking, water heating.................	5	5
Bedrooms..	5	5
Clothes washing machine...............................	5	5
Clothes dryer..	5	5
Dishwasher..	-	5
Home food freezer.....................................	5	5
Television..	5	5
Radio...	5	5
Second home...	-	5

[1]Collected primarily for coverage check purposes.
[2]Required on 100-percent basis for field follow-up purposes in mail areas.
[3]100-percent in places of 50,000 or more inhabitants, 25-percent elsewhere.
[4]Omitted in places of 50,000 or more inhabitants..
[5]For renter-occupied and vacant-for-rent units outside places of 50,000 or more inhabitants.
[6]20-percent in places of 50,000 or more inhabitants, 5-percent elsewhere.
[7]Collected only in places of 50,000 or more inhabitants.

Instructions for respondents in mail census areas. The meaning of concepts and categories derived from replies on mailed-back questionnaires (except where Census editing procedures change the replies) depends on respondents' interpretation of the questions, which may or may not be as Census intended. Some interpretive instructions are included with the questionnaire; these are reflected in the concept definitions included in this dictionary.

In the housing census there is the special problem that people are being asked to supply information not about themselves but about the housing unit they occupy. In some cases, the questions apply to the entire structure in which the housing unit is located; for instance, heating equipment, year structure built, or fuels. Where respondents live in a large apartment building for example, they may be less than familiar with these items for their building. There is also the problem that vacant units as well as occupied units are included in the housing inventory. Enumerators must obtain information about these units from landlords, owners, neighbors, etc.

Editing and allocation procedures. Extensive efforts are made to ensure that data collected in the decennial housing censuses are complete and accurate. Checking for completeness and consistency of replies begins at the local district offices which receive the mailed-back questionnaires. The questionnaires are then microfilmed and fed into a FOSDIC machine which reads the information onto magnetic computer tapes. A computer edit program operates on these tapes to check further for completeness and consistency of the data. Certain entries are changed or "edited" according to fixed instructions. For example, if a housing unit is enumerated as having "no piped water" but having bathing and toilet facilities, the computer changes water supply to "hot and cold piped water." Where single entries or whole questionnaires are missing, information is "allocated" for those units from other information reported on the questionnaire or from information reported for a similar unit in the immediate neighborhood.

Housing inventory

150. **Total housing units** (housing inventory)-- Total housing units in a geographic area recognized in census tabulations (see Part I, Geographic Areas) comprise all living quarters located in that area which are determined to be housing units, including occupied and vacant units.

Living arrangements: Definition of a housing unit

151. Living quarters--All structures occupied or intended for occupancy as living quarters are classified as housing units or group quarters. Group quarters are not included in the housing inventory; no housing information is collected about them.

151.1 Housing units--Housing units comprise houses, apartments, groups of rooms, or single rooms, which are occupied, or vacant but intended for occupancy, as separate living quarters. Specifically, there is a housing unit when the occupants live and eat separately from any other persons in the structure and there is either (1) direct access to the unit from the outside or through a common hall, or (2) in 1960, a kitchen or cooking equipment for the occupants' exclusive use; in 1970, complete kitchen facilities for the occupants' exclusive use.

Structures intended primarily for business or other non-residential use may contain housing units; for example, the living quarters of a merchant in back of his shop. Separate living quarters occupied by staff personnel (but not inmates) in institutions which meet the definitional criteria constitute housing units; as do separate living quarters of supervisory staff in dormitories, nursing homes, etc. Any separate living quarters, which meet the above criteria, in rooming or boarding houses are classified as housing units, as are entire rooming or boarding houses where there are four or fewer roomers unrelated to the person in charge. Trailers, tents, boats, railroad cars, hotel and motels occupied by usual residents which meet the definitional criteria constitute housing units; as do vacant rooms or suites in hotels where 75 percent or more of the accommodations are occupied by usual residents.

151.2 Group quarters--Living arrangements for other than ordinary household life. Includes institutions such as mental hospitals, homes for the aged, prisons, etc., plus other quarters containing 6 or more persons where five or more are unrelated to the head. Such quarters are most commonly found in dormitories, military barracks, etc.; but may also

114

be in a house or apartment used as a rooming house or occupied on a partnership basis, if five or more of the occupants are unrelated to the head. Group quarters are not included in the housing inventory.

152. **Access (entrance to unit)**--Living quarters are classified as having direct access or access through other living quarters as indicated below.

> 152.1 **Direct access**--Living quarters have direct access if the entrance to the unit is direct from the outside of the structure or through a common or public hall, lobby, or vestibule used by occupants of more than one housing unit. (The common hall must not be part of any unit, but clearly separate from all units in the structure.)

> 152.2 **Access through other living quarters**--Living quarters have access through another housing unit when the only entrance is through a hall or room which is part of the other unit.

153. **Kitchen facilities**--The 1960 concept of kitchen facilities was a kitchen or cooking equipment. A kitchen was a room used primarily for cooking and meal preparation; cooking equipment was defined as a range or stove, whether or not regularly used, or other equipment such as a hotplate regularly used to prepare meals.

The 1970 concept of kitchen facilities is complete kitchen facilities, defined as including a sink with piped water, a range or cook stove (excluding portable cooking equipment), and a refrigerator (excluding ice boxes). These facilities must be located in the same building as the living quarters but need not be all in the same room.

Kitchen facilities are further classified as indicated below.

> 153.1 **This unit only**--Kitchen facilities used or, in the case of vacant units, intended for use only by the occupants of the unit.

> 153.2 **Also used by another household**--The kitchen facilities also used or intended for use by someone else in the building not a member of the respondent's household.

> 153.3 **None**--In 1970, means that one or more of the specified equipment items is lacking.

Location of housing units

154. **Urban-rural location**--This is one of the more important breakdowns of the housing inventory by geographic location. The determination of urban-rural location is made after census results have been tabulated. Geographic areas are classified as urban or rural on the basis of certain criteria as to their population size or density at the time of the census. (See Concept No. 12, urban-rural areas.)

> 154.1 **Urban housing units**--Generally units located in areas determined to be urbanized areas or urban places outside urbanized areas.

> 154.2 **Rural housing units**--Units not classified as urban comprise rural units.

> > 154.21 **Rural farm housing units**--Rural occupied units located on farms, as ascertained from questions on acreage of place where located and gross dollar sales of farm products. In 1960, occupied units located on farms where the occupants reported paying cash rent for the house and yard only were not classified as rural farm units. (In 1970, there was no question on whether rent paid includes any land used for farming.) Vacant rural units are classified as nonfarm.

> > Farms are places of 10 acres or more from which sales of crops, livestock, and other farm products amounted to $50 or more in the previous calendar year, or places of less than 10 acres (other than city or suburban lots) from which sales of farm products amounted to $250 or more.

> > In 1970, rural farm housing units are classified by five classes of dollar sales of farm products: $50-249 (places of 10 or more acres only), $250-2,499, $2,500-4,999, $5,000-9,999, $10,000 or more.

> > 154.22 **Rural nonfarm housing units**--All other rural units, including occupied units located in rural territory but not on farms, and all vacant rural units.

155. **Metropolitan location**--This is another important breakdown of the housing inventory by geographic location.

> 155.1 **Metropolitan housing units**--Units located in standard metropolitan statistical areas (SMSA's).

> 155.2 **Nonmetropolitan housing units**--Units located outside SMSA's.

Occupancy status

156. Occupancy status--All housing units are classified as occupied or vacant.

156.1 Occupied housing units (Households)-- A unit is considered occupied if it was the usual place of residence of the person(s) living in it at the time of enumeration. (See Concept No. 52, place of residence.) Included are units occupied by persons only temporarily absent (on vacation, etc.) and units occupied by persons with no usual place of residence (for example, migratory workers).

156.2 Vacant housing units--Generally a unit is considered vacant if no persons were living in it at the time of enumeration. However, units temporarily occupied by persons having a usual place of residence elsewhere are classified as vacant; whereas units where the usual residents were only temporarily absent are not classified as vacant.

Newly constructed vacant units are included in the housing inventory if all exterior doors and windows and final usable floors were in place. Vacant units under construction, units being used for nonresidential purposes, units unfit for human habitation, condemned, or scheduled for demolition, and vacant trailers excluded from the housing inventory.

Occupancy characteristics of occupied housing units

157. Population in units--Total number of persons living in quarters, located in a specific geographic area, which are classified as housing units, excluding persons living in group quarters.

Persons per unit is also calculated. Occupied housing units are classified by the number of persons in the unit from 1 to 8 persons or more in 1960 and 1 to 9 persons or more in 1970. 1960 census basic records carried number of persons in the unit up to 29; 1970 records up to 35.

Median number of persons per occupied unit is calculated as the value which divides the distribution in half. In computing the median, a continuous distribution is assumed, with the whole number of persons as the midpoint of the class interval (for example, 3 as the midpoint of the interval 2.5 to 3.5 persons per unit).

Average number of persons per occupied unit is also calculated.

158. Persons per room--Occupied housing units are classified by the number of persons per room, calculated by dividing the number of persons by the number of rooms in each unit. In 1960, categories of persons per room tabulated were: 0.50 or less, 0.51 to 0.75, 0.76 to 1.00, 1.01 to 1.50, 1.51 or more; in 1970, the terminal category is broken down into 1.51 to 2.00, 2.01 or more. Persons per room data are used to determine overcrowding in housing units.

In 1960, the highest category for number of rooms was 10 or more; this was given an assumed mean value of 11. In 1970, the highest category is 9 rooms or more, given an assumed value of 10 in calculating averages.

159. Characteristics of persons in occupied housing units--It is possible on a special tabulation basis to associate any and all characteristics of persons in households with characteristics of the housing units they occupy. Characteristics of occupied housing units are cross-tabulated in standard data products by a limited number of person characteristics, most commonly for the household head, as indicated below. (See Part II, "Population Census Concepts," for detailed definitions.)

159.1 Age--Age group of household head, and (primary) family head is shown in some housing census tabulations.

Number of own children under 18, under 6, and 6 to 17 is shown in some housing census tabulations. In 1960, extensive tabulations of senior citizen housing (persons 60 and over and 65 and over) were made.

159.2 Race of household head--The race of household head is reflected in some census tabulations. In 1960 counts were presented for units with white and nonwhite household heads; in 1970 for units with white and Negro household heads. Separate tabulations were presented for nonwhite-occupied units in 1960; in 1970, for Negro-occupied units. Selected tabulations are also prepared for the Spanish-American population (see Concept No. 64).

159.3 Household (head) type and household relationship--Type of household head is shown in some housing tabulations as (primary) family head, further broken down by family type (husband-wife, other male head, female head); and as primary individual (male, female).

116

Household relationship is shown in some tabulations as households or families with or without nonrelatives.

Number of persons per family and per household is also shown in some housing census tabulations.

159.4 **Income**--Total income by several income intervals and median income of (primary) families, primary individuals, and household heads, are shown in some housing census tabulations.

160. **Year moved into (occupied) unit**--Determined from information reported for the household head's most recent move by one of several time period categories. (The question was the same as used to determine year moved into present house for the total population in population census tabulations, concept No. 57.) In 1960, these categories were: 1959 to 1960, 1958, 1957, 1954 to 1956, 1950 to 1953, 1940 to 1949, 1939 or earlier, and "always lived here." In 1970, the questionnaire categories are 1969 or 1970, 1968, 1967, 1965 or 1966, 1960 to 1964, 1950 to 1959, 1949 or earlier, and "always lived in this house or apartment."

Respondents who moved back into the same house or apartment where they lived previously were asked to give the year when they began the present occupancy. Respondents who moved from one apartment to another in the same building were asked to give the year they moved into the present apartment.

161. **Tenure**--For occupied housing units.

161.1 **Owner-occupied housing units**--A housing unit is owner-occupied if respondent living in the unit reported that it was "owned or being bought" (i.e., owned outright, mortgaged, or being bought on land contract) by someone in the household. (The owner need not be the head of the household and may be either the sole owner or co-owner.)

161.11 **Cooperatives or condominiums**--In 1960, cooperative apartments or houses owned or being bought by someone in the household were classed as part of the owner-occupied category. In 1970, cooperatives or condominiums constitute a separate category from other owner-occupied units.

161.2 **Renter-occupied housing units**--All occupied housing units which were not owner-occupied are classified as renter-occupied.

161.21 **Occupied units rented for cash**--Includes units where respondents reported that money rent was paid or contracted for.

The rent may have been paid by persons who were not members of the household; for example friends, relatives, a welfare agency, etc.

161.22 **Occupied units rented without payment of cash**--Includes units where respondents reported the unit was occupied without payment of cash rent and was not being owned or bought; for example, houses or apartments provided free of rent by friends or relatives who owned the property but lived elsewhere, parsonages or houses or apartments occupied by janitors or caretakers in full or partial payment for services, units occupied by tenant farmers or share-croppers who paid no cash rent.

162. **Second Homes**--There was a 1970 question on whether any member of the household owned a second home which he occupied sometime during the year. Second homes included single family homes, vacation cottages, hunting cabins, etc. Respondents were to exclude vacant trailers, tents, or boats, and second homes used only for investment purposes. Note that this question obtained information about the number of households with second homes and not the number of second homes itself.

Vacancy characteristics of vacant units

163. **Vacancy status**--Vacant housing units were classified by vacancy status as of the time of enumeration. Vacancy status classification was based on whether the unit was for year-round or seasonal occupancy, and if year-round the purpose for which the unit was being held (sale, rent, etc.). Vacancy status, as other characteristics of vacant units, was determined by enumerator questioning of landlords, owners, neighbors, rental agents, etc.

163.1 **Vacant year-round units**--Vacant units which were intended for occupancy at any time of the year, even if used only occasionally throughout the year.

163.11 **Vacant year-round units**--Vacant units intended for year-round occupancy which were offered for sale or rent. In 1960, the concept of "available" vacant units was used. A unit for rent or for sale was classified as available if it was in sound or deteriorating condition, but not if in dilapidated condition. In 1970, the item on housing condition was not included in the census, so the concept of "available" unit was not utilized.

163.111 **Vacant units for sale only**--Vacant year-round units offered for sale only, usually one-family houses, but also including vacant units in a cooperatively owned apartment building which were for sale only, and vacant units in a multi-unit structure which was for sale as an entire structure, if the particular unit was intended to be occupied by the new owner and was not also for rent.

163.112 **Vacant units for rent**--Vacant year-round units offered for rent or for rent or sale at the same time; including vacant units in a multiunit structure which was for sale as an entire structure if the particular unit is intended for rent.

163.12 **Vacant year-round units rented or sold awaiting occupancy**--Vacant units for year-round occupancy which were rented or sold, but the new occupants had not moved in as of the time of enumeration. (In 1960, included only sound or deteriorating vacant units.)

163.13 **Vacant year-round units held for occasional use**--Vacant units for year-round occupancy which were held for weekend or other occasional use. (In 1960, included only sound or deteriorating vacant units). In 1960, the intent of this category was to identify homes reserved by their owners as second homes.

Because of the difficulty of distinguishing between this category and seasonal vacancies, it is possible that some units which should be included in the occasional use category are classified as seasonal.

163.14 **Vacant year-round units held for other reasons**--Vacant units for year-round occupancy which were held off the market for reasons not specified above; for example, units held for a janitor or caretaker, settlement of an estate, pending repairs or modernization, or personal reasons of the owner. (In 1960, included only sound or deteriorating units.)

163.2 **Vacant seasonal units**--Vacant units intended for occupancy during only a season

of the year; for example, units for summer or winter recreational use, units for herders or loggers.

In 1970, complete-count and sample housing characteristics are tabulated only for year-round units; i.e., occupied units plus vacant year-round units, excluding vacant seasonal and migratory units. This is because "not reported" rates for sample housing items are extremely high for seasonal and migratory vacancies.

163.3 **Vacant migratory units**--Units for migratory workers employed in farm work during the crop season. In 1970, vacant migratory units are identified as a category separate from vacant seasonal units, and counts of each are included in the tabulations. (1960 census basic records also carried such units as a separate category.)

In 1970, complete-count and sample housing characteristics are tabulated only for year-round units; i.e., occupied units plus vacant year-round units, excluding vacant seasonal and migratory units. This is because "not reported" rates for sample housing items are extremely high for seasonal and migratory vacancies.

164. **Vacancy rates**--Vacancy rates for the homeowner housing inventory and the rental housing inventory are calculated as indicated below.

164.1 **Homeowner vacancy rate**--Calculated as the number of vacant units for sale as a percentage of the total homeowner inventory (owner-occupied units plus vacant units for sale).

164.2 **Rental vacancy rate**--Calculated as the number of vacant units for rent as a percentage of the total rental inventory (renter-occupied units plus vacant units for rent).

165. **Duration of vacancy**--The length of time from the date the last occupants moved away from the unit to the date of enumeration. For newly constructed units which have never been occupied, duration of vacancy was the time period from the date construction was completed to the date of enumeration.

In 1960, the basic record for duration of vacancy was categorized as: less than 1 month, 1 up to 2 months, 2 to 4 months, 4 to 6, 6 or more. In

118

1970, the categories are: less than one month, 1 up to 2 months, 2 up to 6 months, 6 months up to 1 year, 1 up to 2 years, 2 years or more.

Financial characteristics of housing units

166. Value of unit--The respondent's estimate of how much the property would sell for on the current market or (for vacant units) the asking price at the time of enumeration. Value was collected only for one-family houses (one-unit structures), detached and attached, which were owner-occupied or vacant for sale, and which were not on places of 10 or more acres, or on properties which also had a business establishment (a retail store, gasoline station, etc.) or a medical or dental office. Cooperatives, condominiums, mobile homes, and trailers were excluded from the value tabulations.

One-family houses on places (lots) of 10 acres or more, or with a business establishment or medical office on the property, were identified by a separate question on the schedule. No estimate of the value of such units was obtained.

A property is defined as the house and land on which it stands. Respondents were to estimate the value of the entire property even if the occupant owned the house but not the land or owned the property jointly with another owner.

Respondents were to indicate estimated value by several categories. In 1960, these were: less than $5,000, $5,000 - 7,400, $7,500 - 9,900, 10,000 - $12,400, $12,500 - 14,900, $15,000 - 17,400, $17,500 - 19,900, $20,000 - 24,900, $25,000 - 34,900, $35,000 or more. In 1970 the categories ended in "99," e.g., $5,000 - 7,499, and the following categories were added: $35,000 - 49,999 and $50,000 or more.

Total value, median value, and average value of housing units are calculated. (Midpoints of intervals are used in calculating average values. In 1960, values under $5,000 were assigned a mean of $3,500, and values of $35,000 or more a mean of $42,000; in 1970 values of $50,000 or more are assigned a mean of $60,000.)

167. Rent--Rent was asked only for renter-occupied housing units rented for cash rent and vacant units, for rent, excluding one-family houses on places of 10 or more acres. Respondents were to indicate rent only for the housing unit being enumerated and to exclude any rent paid for additional units or for business premises.

167.1 Contract rent--The monthly dollar rent agreed upon or (for vacant units) the monthly dollar rent asked at the time of enumeration, regardless of any furnishings, utilities, or services that were included. Respondents were to indicate monthly contract rent to the nearest dollar. (If rent was paid by the week or some other time period, respondents were to indicate the amount and the time period so that their monthly contract rent can be entered by census employees.)

Contract rent is tabulated by several distributions; for example, less than $30, $30 - 39 $90 - 99, $100 - 119, $120 - 149, $150 - 199, $200 - 249, $250 - 299, $300 or more. The category "no cash rent" is also included in tabulations of contract rent for all renter-occupied units. (Census samples basic records carry dollar amounts on contract rent from $1 to $999.)

Total, median, and average contract rents are calculated for rental units.

Vacant units for rent are also classified as with all utilities included in rent and with some or no utilities included in rent.

167.2 Gross rent--Gross rent is calculated for renter-occupied units rented for cash rent (with the exclusions noted above for rent). It represents the contract rent plus the average monthly cost of utilities (water, electricity, gas,) and fuels, to the extent that these are paid for by the renter (or paid for by a relative, welfare agency, or friend) in addition to the rent. Gross rent thus eliminates differentials which result from varying practices with respect to the inclusion of utilities and fuel in contract rent.

In 1960, respondents were to indicate if they paid for electricity, gas, water or fuels (oil, coal wood, kerosene) in addition to rent; and if yes, to indicate the estimated average monthly dollar cost for electricity, gas, water, and the total yearly cost for fuel. In 1970, respondents were to answer similarly but further specify if they did not use particular utilities or fuels.

Gross rent is calculated from this information. Gross rent is tabulated by several distributions; for example, less than $30, $30 - 39 . . . $90 - 99, $100 - 119, $120 - 149, $150 - 199, $200 - 249, $250 - 299, $300 or more. (Census basic records carry dollar amounts of gross rent up to $999.)

Total, median, and average gross rent are calculated.

168. Ratios of income to value and rent

168.1 Value-income ratio--Calculated for owner-occupied units (with the exclusion noted in the discussion of value of unit. Concept No. 166). Value-income ratio is the value of the unit in relation to the total income reported by the (primary) family or primary individual for the preceding year.

Value-income ratio is tabulated as follows: value as less than 1.5 times income, 1.5 to 1.9, 2.0 to 2.4, 2.5 to 2.9, 3.0 to 3.9, 4.0 or more, and not computed. (The category not computed includes units occupied by families or primary individuals who report no income or a net loss.)

168.2 Rent-income ratio (Gross Rent as Percentage of Income)--Calculated for renter-occupied units for which gross rent is tabulated. Rent - income ratio is the yearly gross rent expressed as a percentage of the total income reported by the (primary) family or primary individual for the preceding year.

Rent-income ratio is commonly tabulated as follows: yearly gross rent as less than 10 percent of total income, 10 to 14 percent 15 to 19, 20 to 24, 25 to 34, 35 or more, and not computed. (The category not computed includes renter-occupied units rented without payment of cash rent and units occupied by families or primary individuals who report no income or a net loss.)

Structural characteristics

169. Units in structure (type of structure)-- Housing units are classified by the number of units (including occupied and vacant, excluding business units or group quarters) in the structure in which they are located, as indicated below. Data are tabulated only in terms of housing units. Except for one-family houses (detached and attached), there is no information regarding number of structures. In 1960, determination of units in structure was by enumerator observation or, by inquiring of the landlord, the janitor, etc.; in 1970, by respondent replies to a question on whether this is a building for one family, 2 families, etc. (Categories which respondents could specify are indicated below.)

A structure is defined as a separate building that either has open space on all four sides (detached),

or is separated by dividing walls that extend from ground to roof (attached). Tabulations of this and other structural characteristics are in terms of number of housing units rather than number of structures.

169.1 1-Unit--Structures containing only one housing unit, further classified as indicated below. (1-unit structures may contain business units.)

169.11 1-Unit Detached--1-unit structures detached from any other house, i.e., with open space on all four sides. Such structures are considered detached even if they have an adjoining private garage or contain a business unit.

169.111 Trailers--Occupied trailers or mobile homes are shown separately from other 1-unit detached structures in some tabulations. In 1960, trailers were further classified as mobile (resting on wheels or on a temporary foundation, such as blocks or posts), or on permanent foundation (mounted on a regular foundation of brick or concrete, etc.). In 1970, this breakdown was omitted.

169.12 1-Unit attached--1-unit structures which have one or more walls extending from ground to roof separating th m from adjoining structures; for example, a row house.

169.2 2 or More Units--Structures containing 2 or more housing units; further broken down as 2-units, 3 or 4-units, 5 to 9-units, 10 to 19, 20 to 49, 50 or more units.

In 1970, to reflect the wording on the questionnaire, tabulations of units in structure are sometimes in terms of buildings for two families, 3 - 4 families, etc.

170. Type of foundation--Housing units are classified by the type of foundation of the structure or building in which they are located, as indicated below.

170.1 With a basement--Structures are considered to have basements if they have an enclosed space beneath all or part of the structure, are accessible to the occupants, and are of sufficient depth so that an adult can walk upright. The basement floor must be below ground level or all or part of its perimeter.

120

170.2 **On a concrete slab**--Structures built on a concrete slab have no basement and no crawl space or air space below the first floor.

170.3 **Other**--Structures built with other types of foundations include structures supported on piers or posts, built on a continuous masonry foundation (without a basement), built directly on the ground, or built in unconventional ways, such as with a central supporting mast.

171. **Number of stories in structure**--Housing units are classified by the number of stories in the structure in which they are located. In 1960, the categories were 1 to 3 stories or floors and 4 or more; in 1970, 1 to 3 stories, 4 to 6, 7 to 12, 13 or more. Respondents were not to count basements as stories.

In 1960, number of stories was collected only for housing units located in places of 50,000 or more inhabitants; in 1970 for all units.

172. **Elevator in structure**--Only for housing units in structures with four stories or more. In 1960, elevator in structure was obtained only for such housing units located in places of 50,000 or more inhabitants; in 1970, for all such units.

172.1 **With passenger elevator**--4 or more story structures have elevators if there is an elevator which passengers may use.

172.2 **Walkup**--4 or more story structures where there is no passenger elevator.

173. **Year structure built**--Housing units are classified by the year the structure in which they are located was built, i.e., the date the original construction was completed (not the date of any later remodeling, addition, or conversion).

In 1960, the categories were: 1959 to March 1960, 1955 to 1958, 1950 to 1954, 1940 to 1949, 1930 to 1939, 1929 or earlier. The 1970 categories were updated by ten years.

Tabulations on the number of units built during a given period relate to the number of units in existence at the time of enumeration, which may not be the same as the original number. Year built data are particularly susceptible to response errors and nonreporting, since respondents must rely on their memory or estimates of persons who have lived in the neighborhood a long time, etc.

174. **Rooms in unit**--The categories were from 1 room to 10 rooms or more in 1960 and from 1 to 9 or more in 1970. Respondents were to count only whole rooms used for living purposes, such as living rooms, dining rooms, kitchens, bedrooms, finished recreation rooms, etc.; and to exclude kitchenettes, strip or pullman kitchens, bathrooms, porches, balconies, foyers, halls, half-rooms, utility rooms, unfinished attics or basements, or other space used for storage.

Total, median, and average number of rooms for all (or certain kinds of) housing units are calculated. Persons per room is also calculated.

174.1 **Bedrooms in unit**--Number of bedrooms, from 1 bedroom to 4 bedrooms or more in 1960 and from 1 to 5 or more in 1970. Respondents were to count rooms used mainly for sleeping even if used also for other purposes; for example, dens, enclosed porches, and rooms reserved for sleeping, such as guest rooms, even though used infrequently. They were not to count rooms used incidentally for sleeping, such as a living room with a hideaway bed.

Substandard housing

175. **Substandard housing**--Statistics on substandard housing are tentatively scheduled for publication in 1973 in **Housing,** Volume VI, "Estimates of Substandard Housing." The estimates of "substandard" housing are based on the number of units lacking some or all plumbing facilities in 1970, plus updated 1960 proportions of dilapidated units with all plumbing facilities applied to units with all plumbing facilities in 1970.

Plumbing characteristics

176. **Plumbing facilities**--Plumbing facilities include toilet facilities, bathing facilities, and water supply. Tabulations of plumbing facilities are considered a measure of housing quality. Housing units are classified by plumbing facilities as follows.

176.1 **With all plumbing facilities**--Housing units which have piped hot and cold water inside the structure, flush toilet and bathtub or shower inside the structure for use only by the occupants of the unit (including roomers, boarders, and other non-relatives) are considered to have all plumbing facilities.

176.2 **Lacking some or all plumbing facilities**--Housing units which lack one or more plumbing facilities; i.e., which lack piped hot and/or cold water, lack toilet or bathtub or have a toilet or bathing facilities used also by occupants of another unit.

176.21 Lacking hot water only--Units which have all facilities except hot water.

176.22 Lacking other plumbing facilities--Units which lack one or more of the following: piped water, a flush toilet used only by the occupant household, or a bathtub or shower used only by the occupant household.

176.3 Use of plumbing facilities

176.32 For this household only--Describes plumbing (flush toilet, or bathtub or shower) used only by the occupants of one housing unit or, in the case of vacant units, plumbing intended only for the use of future occupants.

176.32 Also used by other household--Describes plumbing used by the occupants of more than one housing unit or, in the case of vacant units, plumbing intended for use by more than one unit.

177. Toilet facilities--Housing units are classified by toilet facilities as follows.

177.1 Flush toilet--Housing units have flush toilets (supplied with piped water) if they are inside the structure and available for the use of the occupants. Flush toilets are classified according to whether they are used only by the occupant household or are used also by occupants of another unit.

177.2 No flush toilet facilities--Includes privies, chemical toilets, outside flush toilets, as well as no toilet on the property.

178. Bathing facilities--Housing units are classified by bathing facilities as follows.

178.1 Bathtub or shower--Housing units have a bathtub or shower if either facility is supplied with piped water (not necessarily hot), is located inside the structure and available for the use of the occupants of the unit. Bathing facilities are classified according to whether they are used by the occupant household only or are also used by occupants of another unit.

178.2 No bathtub or shower--Includes units with only portable facilities as well as units

with no bathing facilities inside the structure and available for the use of the occupants.

179. Water supply--Housing units are classified by water supply in terms of piped hot or cold versus no piped water as indicated below.

179.1 Hot and cold piped water inside structure--Water must be available to the occupants of the unit. The hot water need not be supplied continuously.

179.2 Only cold piped water inside structure--Water must be available to the occupants of the unit.

179.3 No piped water--In 1960, units with no piped water inside structure but piped water outside structure available on the same property (either outdoors or in another structure) constituted a separate category from units with no piped water available at all (i.e., the only source of water was a hand pump, open well, spring, cistern, etc., or the occupants obtained water from a source not on the same property).

In 1970, all units with no piped water available inside the building are treated as a single category, regardless of whether piped water is available outside the building on the same property.

180. Bathrooms--Housing units are classified by the number of complete and partial or half bathrooms in the unit. In 1960, the categories were 1 (complete) bathroom, 1 plus partial, 2 or more, and none or only a partial bathroom (including shared). In 1970, the categories were 1 complete bathroom, 1 plus half, 2 complete, 2 plus half, 3 or more complete, and none or only a half bathroom (including a bathroom also used by another household).

180.1 Complete bathroom--A bathroom with all plumbing facilities, including hot piped water, a flush toilet, bathtub or shower, and wash basin for the use of only the occupant household. The facilities must be located inside the structure and located in one room.

180.2 Partial or half bathroom--A partial (1960 terminology) or half (1970 terminology) bathroom has toilet facilities (flush toilet) or bathing facilities for exclusive use but not both. Units with partial or half bathrooms are included under units with "more than 1 bathroom," "1 plus partial," etc., if the unit also has a complete bathroom. Units with only a partial or half bathroom are included under units with none or only a half bathroom.

122

180.3 **None** or only a half bathroom--Includes units with no bathroom, units with only a partial or half bathroom, and units with bathroom facilities also used by occupants of another unit.

181. **Source of water**--Housing units are classified by the source of their water supply as indicated below. In 1960, source of water was not collected for housing units in places of 50,000 or more. In 1970, source of water was obtained for all units.

181.1 **Public system or private company**-- A common source supplying running water to more than five units. This source may be a city or county water department, a water district, a private water company, cooperative or partnership group, or a well which supplies 6 or more houses or apartments.

181.2 **Individual well**--A source serving five or fewer units from a well on the property of the unit being enumerated or on a neighboring property.

181.3 **Other**--Water coming directly from springs, creeks, rivers, etc., and all other sources.

182. **Sewage disposal**--Housing units are classified by the sewage disposal system for the structure in which the unit is located as indicated below. In 1960, sewage disposal was not collected for housing units in places of 50,000 or more, but in 1970 this item was collected and tabulated for all units.

182.1 **Public sewer**--Includes units connected to a city, county, sanitary district, neighborhood, or subdivision sewer system.

182.2 **Septic tank or cesspool**--An underground tank or pit into which sewage flows from the plumbing fixtures in the building.

182.3 **Other or none**--Includes on individual sewer line running to a creek, lake, swamp, etc., units with a privy, and other arrangements.

Heating equipment and fuels

183. **Heating equipment**--All housing units are classified by type of heating equipment used. Vacant units are classified by the type of heating equipment available for use or used by the previous occupants (if the unit is without heating equipment). Respondents were to report only the principal kind of equipment. Respondents

indicated heating equipment by one of the following categories; or they described the means, in which case their response is coded into an appropriate category.

183.1 **Steam or hot water system**--A central heating system which supplies steam or hot water to conventional radiators, baseboard radiators, heating pipes embeded in walls or ceilings, heating coils or equipment which are part of a combined heating ventilating or heating-air conditioning system.

183.2 **Central warm air furnace with ducts or central heat pump**--A central warm air furnace is a system which provides warm air through ducts (passageways for air movement) leading to the various rooms. In 1970, central heat pumps or reverse cycle systems were specified as part of this category.

183.3 **Built-in electric units**--Electric heating units permanently installed in floors, walls, or ceilings. Does not include electric heaters plugged into an electric outlet.

183.4 **Floor, wall, or pipeless furnaces**-- Floor and pipeless furnaces deliver heated air to the room in which the furnace is located, or, in some types of floor furnaces, to two adjoining rooms on either side of a partition. Wall furnaces, installed in walls or partitions, deliver heated air to the room(s) on one or both sides. None of the three types of furnaces have ducts leading to other rooms.

183.5 **Other means with flue**--In 1960, included circulating heaters, radiant and other gas room heaters, freestanding room heaters, parlor stoves, ranges or cook stoves used for heating, and fireplaces, regardless of fuel, if equipped with a flue, vent or chimney for removal of smoke, fumes, and combustion gases.

In 1970, this category is termed **room heaters** with flue or vent, burning gas, oil, or kerosene. It consists of circulating heaters, convectors, radiant gas heaters that burn gas, oil, kerosene or other liquid fuel, and which are connected to a flue, vent, or chimney. The category excludes fireplaces or stoves burning coal or wood.

183.6 **Other means without flue**--In 1960, included any of the following, if used as the principal source of heat: room heaters that burn gas, kerosene, or any other fuel, but do not have a flue or chimney; electric heaters that get current through a cord plugged into an ordinary electrical outlet; portable heaters.

In 1970, the category is termed **room heaters without flue or vent, burning gas, oil, or kerosene.** It consists of unvented room heaters (circulating and radiant) burning gas or liquid fuel. The category excludes portable heaters.

183.7 **Fireplaces, stoves, or portable room heaters of any kind**--This category is new in 1970, and consists of heating devices transferred from the two other means categories of 1960. Fireplaces as the principal source of heat is self-explanatory; in 1960 they were included in other means with flue. Stoves means room heaters that burn coal or wood--parlor stoves, circulating heaters, cookstoves also used for heating, etc. These must be vented if the rooms in which they are located are to be usable when they are burning; they also were included in other means with flue in 1960. Portable heaters (classified in 1960 as other means without fuel) can be picked up and moved around at will, either without limitation (kerosene, oil, gasoline heaters) or within the radius allowed by a flexible gas hose or an electric cord (gas, electric heaters). This classification includes all electric heaters that get current through a cord plugged into a convenience outlet.

183.8 **Not heated**--Consists of units without heating equipment-most common among units in warmest part of the country (Hawaii, southern Florida, etc.) and seasonal units not intended for winter occupancy.

184. **Heating fuel**--For occupied housing units only. Respondents are to indicate the fuel most used for heating the unit by one of the following categories.

184.1 **Coal or coke**

184.2 **Wood**--May be salvage wood as well as trees felled by users and purchased wood. In some tabulations "wood" is not shown separately but included in the category other fuel.

184.3 **Utility gas**--Gas from underground pipes serving the neighborhood supplied by a public utility company, municipal government, etc.

184.4 **Bottled, tank, or LP gas**--Bottled, tank, or liquefied petroleum gas stored in tanks which are replaced or refilled as necessary.

184.5 **Electricity**

184.6 **Fuel oil, kerosene, etc.**--Fuel oil, distillate, residual oil, kerosene, gasoline, alcohol, and other combustible liquids and semi-fluids.

184.7 **Other fuel**--All other fuels not specified elsewhere, including purchased steam, fuel briquettes, waste materials such as corncobs, etc.

184.8 **No fuel used or none**

185. **Cooking fuel**--For occupied housing units only; the fuel most used for cooking. The same categories as for heating fuel; respondents who eat all meals elsewhere were to report "no fuel."

186. **Water heating fuel**--For occupied housing units only; the fuel most used for heating water. The same categories as for heating and cooking fuel; units which reported no hot piped water are classified as using no fuel for heating water

Household equipment*

187. **Clothes washing machine**--In 1960, respondents were to report only washing machines owned by a member of the household (whether located in the unit or elsewhere on the property). In 1970, respondents could also report machines provided as part of the equipment in their living quarters, but not coin-operated machines or machines in storage.

187.1 **Wringer or spinner**--A power-operated washing machine which requires handling of the laundry between washing and rinsing.

187.2 **Automatic or semi-automatic**--In 1960, a machine which washes, rinses, and damp dries in the same tub, without intermediate handling. Housing units with both automatic and wringer washing machines were included in the automatic washing machine category. In 1970, washer-dryer combinations were also included.

---

*Household equipment items are only collected for occupied housing units.

124

187.3 **Washer-Dryer Combination**--In 1960, a single machine which washes and fully dries the laundry in the same tub. Combined with automatic or semi-automatic in 1970.

187.4 **None**--The housing unit has no washing machine.

188. **Clothes dryer**--Basis for inclusion the same as for clothes washing machines. Occupied housing units are classified as having gas heated clothes dryer, electrically heated clothes dryer, or no clothes dryer (none).

In 1960, units with **washer-dryer combination** were not included in the clothes dryer category; in 1970, such units are tabulated as having a clothes dryer.

189. **Home food freezer**--Basis for inclusion the same as for clothes washing machines. Occupied housing units are classified as having 1 or more home food freezers (separate from the refrigerator) or as having none.

190. **Dishwasher**--Not collected in earlier censuses. Basis for inclusion the same as for clothes washing machines. Occupied housing units are classified as having a dishwasher (built-in or portable) or as having no dishwasher.

191. **Telephone available**--Occupied housing units are classified as having a telephone available, if there is a telephone on which the occupants can receive calls. The telephone may be located in the housing unit or elsewhere, as in the hall of an apartment building. in another apartment, or in another building entirely.

192. **Automobiles available**--Occupied housing units are classified by the number of passenger automobiles owned or regularly used by any member of the household (including nonrelatives, such as lodgers) as follows: 1 automobile available, 2, 3 or more, none.

Respondents were to include company cars kept at home and to exclude taxicabs, pickups, larger trucks, cars being junked or permanently out of working order.

193. **Air conditioning**--For occupied housing units by the following categories. Respondents were to include only equipment with a refrigeration unit to cool air and to exclude evaporative coolers and fans or blowers not connected to a refrigerating apparatus.

193.1 **1 room unit**--An individual window or through-the-wall air conditioner unit designed to deliver cooled air to the room in which it is located.

193.2 **2 or more room units**

193.3 **Central system**--An installation designed to deliver cooled air to each principal room of a house or apartment.

193.4 **No air conditioning or none**

194. **Television sets**--Categories the same as in 1960: 1, 2 or more, none. Respondents were to include sets of all kinds that were located in the unit and were in working order--floor models, built-in, table, portable, combination with radio or phonograph.

In 1970, a further question asked whether the household had a television set equipped to receive UHF broadcasts (i.e., channels 14 to 83). Respondents were to include sets which could be tuned directly to channels 14 to 83, sets which could receive UHF broadcasts by means of a converter, or through a community (CATV) or master antenna which receives incoming UHF signals and transfers them to a vacant VHF channel (2 to 13).

195. **Radio sets**--In 1960, occupied housing units were classified by the number of radio sets located in the unit: 1, 2 or more, and none. Respondents were to include floor models, built-in, table, combination with TV or phonograph or clock, and to include sets being repaired as well as sets in working order. Respondents were to exclude "ham radio" sets automobile radios, and sets not in working order which were not being repaired.

In 1970, occupied housing units are classified only by the number of battery operated radios owned by any member of the household as follows: 1 or more, and none. Respondents were to include only sets in working order and sets needing only new batteries; specifically included were car radios, transistor sets, and battery-operated sets which can also operate on house current.

Components of Inventory Change (CINCH) Survey

During the early 1950's, pressures developed from many sources for more sophisticated and useful data on the dynamics of the housing inventory than that provided by the decennial census of housing. There was a need to analyze the changes in the national housing inventory by type of addition, loss, new construction, demolition, conversion, and merger. This need was first met by the Census Bureau through data collection in the 1956 National Housing Inventory. In 1959 a Components of Inventory Change survey was conducted as a part of the 1960 Census of Housing; this study provided information on the changes in the housing inventory for the decennial period April 1950 to December 1959 as well as on changes which had occurred since the December 1956 National Housing Inventory. Requests for data on changes in the housing inventory during the 1960's have resulted in the inclusion of the Components of Inventory Change Survey as part of the 1970 Census of Housing.

The components of change. The basic components of housing inventory change are additions, losses, conversions, mergers, and "sames." With reference to the 1970 survey, they are defined as follows.

Additions are those units which did not exist in 1960 and have been newly constructed, moved to site, created from group quarters, or created from space previously used for nonresidential purposes.

Losses comprise those units which existed in 1960 but do not exist now. These include units now used as group quarters, changed to nonresidential use, demolished, moved from site, destroyed by fire or flood, etc.

Conversions cover those units created by division of a 1960 unit into two or more 1970 units. Mergers are those 1970 units resulting from merger of two or more 1960 units.

Sames are the great bulk of the 1960 units which were not affected by these changes. That is, they are living quarters enumerated as one housing unit in 1960 and as one housing unit in 1970.

Note: For operational definitions see concepts 200-203 below.

The 1970 CINCH program. The CINCH program gathers and reports information on the demographic and housing characteristics pertaining to the various components. Among such characteristics are the following: number of rooms, year built, condition and plumbing facilities, tenure, value, rent, household composition, and family income. In addition, selected 1970 and 1960 characteristics such as tenure, value, rent, and family income will be cross-tabulated for units found to be the "same." Such information will reflect changes occurring to these

units during the decade. Information on recent movers (i.e., heads of households who moved after January 1, 1969) will also be provided. This will be done in cross-tabulations relating characteristics of the present and previous units occupied by recent movers.

The 1970 survey will be conducted in the fall and winter of 1970-71. It will be based on a sample of housing units located in the 357 PSU (primary sampling unit) areas designated for Bureau of the Census current survey programs plus non-PSU counties in the 15 selected standard metropolitan statistical areas. The data will be summarized for the U.S., for the four geographic regions, and for 15 SMSA's. As in 1960, the survey's finding will be published as Volume IV in the series of 1970 publications from the census of housing.

Use of the data. CINCH data is widely used by analysts in business and government. The major value of the information is the basis it provides for projecting future housing requirements for the various sectors of our population. The home building industry needs the information to estimate the extent to which the overall demand for new houses is being met, while bankers and mortgage lenders use it as a guide to lending practices. Producers of building materials and home equipment are aided by information on changes in the housing inventory in their planning for volume and type of production; public utility companies use the information to plan their rate of expansion.

The Federal government, as well as State and local governments, is concerned with changes in both the housing supply and its characteristics because housing construction constitutes an important part of the Nation's economy and credit structure. Various departments and agencies use information on housing trends to establish policies for regulating the flow of mortgage credit, maintaining high levels of employment, planning, and developing housing programs and goals.

Description of the 1970 Survey

Unit of Enumeration. The unit of enumeration, or measurement, will be the housing unit. The definition of a housing unit is the same as the one used in the 1970 census and is essentially the same as the 1960 census definition.

The sample. The sample for measuring counts of the components will consist of about 320,000 units. A subsample of about 120,000 units will

126

be used for compiling the detailed demographic and housing characteristics.

One part of the sample will be selected from 1960 census addresses of those housing units in the 25 percent sample. The 1960 sample will provide counts and characteristics for all components except new construction and whole structure "additions" such as changes from nonresidential to residential use, moved to site, etc.

The count and most of the characteristics for new construction are obtained from the 1970 census tabulations for units reported built in the 1960-70 period. However, several items which will not be obtained in the census will come from the 1970 CINCH sample. These include "condition of unit" and detailed information on "recent movers." The 1970 CINCH sample will also provide information on the counts and characteristics for whole structure additions.

Procedures. The procedures for interviewing will largely be determined by the adequacy or inadequacy of the addresses for the sample units. After the sample units have been selected, the areas in which the units are located will be screened according to the criteria used for current surveys in order to determine the adequacy of the addresses. The results of the screening operation will then determine the procedures.

1. Areas with adequate addresses (urban areas).

In areas where the addresses are sufficiently specific, interviewers will visit the specific addresses of the 1960 sample units and will determine the components for all units in the building. If the building is no longer standing, interviewers will determine the disposition (e.g., demolished or moved from site) from neighbors or other reliable respondents. For buildings still containing living quarters, interviewers will make a unit-by-unit comparison based on information from the 1960 census. This procedure will provide the counts for the following components: sames, conversions, mergers, part structure additions and losses, and whole structure losses such as demolitions, moved from site, etc.

For the 1970 sample the addresses of units will be matched with the 1960 census enumerators' listing books. All non-matching units will be checked for additional character-

istics of new construction and for the count and characteristics of whole structure additions.

2. Areas with inadequate addresses (rural areas).

In these areas, a sample of units in small land area segments will be used. Interviewers will list selected characteristics for all 1970 units located on the segments and make a unit-by-unit comparison with information from the 1960 census to determine the following components: sames, conversions, mergers, part structure additions and losses and whole structure additions. A sub-sample of the existing units will be interviewed for more detailed characteristics. The 1970 sample units in these areas will be matched against the 1970 census schedules. All units built in 1960 or later will be interviewed for the additional items which are not collected in the 1970 census.

Timing. The enumeration is tentatively scheduled for the fall and winter of 1970-71. The information for the first SMSA is scheduled to be published in late fall 1972 with the publications for the remaining SMSA's and the United States following very closely.

Scope. For the 1970 program, separate data will be published for the United States, four geographic regions, and fifteen individual standard metropolitan statistical areas. In response to requests from users of the data, reports for each SMSA will present data separately for the total SMSA, the central city, and possibly the area outside the central city.

The SMSA's, by region, are: **Northeast Region**--Boston, Buffalo, New York, Philadelphia; **North Central Region**--Chicago, Cleveland, Detroit, St. Louis; **South**--Atlanta, Houston, Miami, Washington; **West**--Los Angeles, San Francisco, Seattle.

All the above SMSA's except Houston and Miami were included in the 1959 Components of Change Survey. The SMSA's not to be included in 1970 which were included in 1959 are: Minneapolis-St. Paul, Pittsburgh, Baltimore, and Dallas. Of these, only Dallas was in both the 1956 National Housing Inventory and the 1959 survey.

Unpublished tabulations. It is currently planned to tabulate selected information for analytical purposes. These analytical tabulations will be available to the users at nominal cost. Some examples of these tabulations are: current use of site for 1960 buildings which are no longer in the inventory; a more detailed breakdown of other additions and losses; more detailed tables of the 1960 and 1970 characteristics for "same" units; and previous and present residences of "recent movers;" etc. Special tabulations can be obtained by users on a contract basis. It is not planned, however, to provide users with the computer tapes.

Components of Inventory Change Concepts

A majority of concepts in the Components of Inventory Change (CINCH) Survey are identical to those in the 1970 Census of Population and Housing. Some of the demographic characteristics will not, however, be collected or reported in as much detail as in the decennial census. There are no changes in definition of components of change from the 1960 program.

The following concepts (in alphabetical order) employed in CINCH are used in the same manner as in the census.

Age of persons in household
Bathtub or shower
Color or race of head of housheold
Gross rent
Contract rent
Duration of vacancy
Flush toilet
Group quarters
Heating equipment
Highest year of school completed by head
Hot and cold piped water
Housing unit
Income of primary families and individuals
Number of bathrooms
Number of bedrooms
Number of housing units in structure
Number of persons in unit
Number of rooms in unit
Relationship to head of unit
Sex
Tenure
Type of living quarters
Vacancy status
Value of property
Year head moved into unit
Year structure built

Concepts Unique to CINCH

The components of change

200. **Same unit**--A unit which existed in 1960 in the same form as it does in 1970. Living quarters enumerated in 1970 as one housing unit are "same" if they were classified as one and only one housing unit in 1960. A "same unit" may have a different number of rooms, changes in architecture, changes in plumbing equipment, or other changes in characteristics.

201. **Different unit**--A unit which has been altered in some way since 1960 to create either more or less units in 1970. The housing unit was created by dividing (converting) one 1960 unit into two or more or by combining (merging) two or more 1960 units into one.

A 1970 unit may have more or less space than in 1960 because of remodeling or alterations, but this does not necessarily make it a "different unit." Only if the alteration or remodeling changes the number of units is it a "different unit."

201.1 **Conversion**--Converison is the creation of two or more 1970 housing units from fewer 1960 units through structural alteration or change in use. Structural alteration includes such changes as adding a room or installing partitions to form another housing unit. "Change in use" is a simple rearrangement in the use of space without structural alteration, such as locking a door which closes off one or more rooms to form a separate housing unit. Each unit involved in the change is a converted unit.

201.2 **Merger**--Merger is the opposite of conversion. It is the combining of two or more 1960 housing units into fewer 1970 units through structural alteration or change in use. Structural alteration includes such changes as the removal of partitions or the dismantling of kitchen facilities. Change in use may be a simple rearrangement in the use of space without structural alteration, such as unlocking a door which formerly separated two housing units. In other instances, a household on the first floor may occupy both the first and second floors which formerly constituted separate housing units.

202. **Lost unit**--A unit which existed in 1960 but does not exist as a housing unit in 1970.

202.1 **To group quarters**-- A 1960 housing unit may have become a "group quarters" by 1970. For example, a large single housing unit structure may have become a lodging house.

202.2 **To nonresidential**-- A 1960 housing unit may now be used for nonresidential purposes, such as for a store, office space, permanent storage, etc.

128

202.3 **Unfit**-- A housing unit which existed in 1960 and is now both vacant and unfit for human habitation is unfit. Unfit for human habitation is defined as a building intended for residential use where the roof, walls, windows, or doors no longer protect the interior from the elements.

202.4 **Condemned**--A housing unit which existed in 1960 and is now designated as condemned by a sign, notice, or mark on the building or in the neighborhood is in this category. The sign may show that the unit is condemned for reasons of health or safety.

202.5 **Demolished**--A housing unit in a building which has been torn down since 1960 is considered demolished. This category does not include units lost by fire, flood, etc.

202.6 **Moved from site**--A housing unit which has been moved from its 1960 site would be in this category. Included are mobile homes and trailers which have been moved from their 1960 sites.

202.7 **Other (burned, etc.)**--This includes all other recorded 1960 units which are no longer in existence. Examples are units which have been lost by fire, flood, wind, or hail, and vacant units which are scheduled for demolition.

203. **Added unit**--An added 1970 unit is a unit which did not exist in 1960 and has been newly constructed, moved to the site, created from group quarters or created from space previously used for nonresidential purposes.

203.1 **From group quarters**--Group quarters (rooming houses, dormitories, transient hotels, etc.) which have been changed to housing units are in this category.

203.1 **From nonresidential**--These are housing units which have been created from nonresidential space such as a store, garage, barn, and the like.

203.3 **Moved to site**--A housing unit which has been moved to its present site since 1960 would be in this category. This includes mobile homes and trailers which have been moved to their present site since 1960, if they were occupied in 1960 and are occupied in 1970.

203.4 **New construction**--Any unit in a building which has been built since 1960 is new construction.

Condition

204. **Condition of housing unit**--In the 1970 CINCH Survey, as in the 1960 Census of Housing, housing units will be classified by condition (categories indicated below) on the basis of enumerator observation. Enumerators are to look for specified visible defects relating to weather tightness, extent of disrepair, hazards to the physical safety of the occupants, and inadequate or makeshift construction.

204.1 **Sound condition**--Housing units are to be classified as sound if they have no visible defects or only slight defects that are normally corrected during the course of regular maintenance. Examples of slight defects are lack of paint, small cracks in the plaster, cracked windows, etc.

204.2 **Deteriorating condition**-- Units are to be classified as deteriorating if they need more repair than would be provided in the course of regular maintenance. Deteriorating units have one or more defects such as several broken or missing window panes, a shaky or unsafe porch, holes or open cracks over a small area of a wall, etc.

204.3 **Dilapidated conditions**--Units are to be classified as dilapidated if they do not provide adequate shelter and in their present condition endanger the health and safety of their inhabitants. They might have such defects as holes, open cracks, etc., over a large area of the foundation or walls, substantial sagging of floors and roof, or extensive damage by storm, fire, or flood.

Recent movers

205. **Recent mover**--A household head who moved into his residence after January 1, 1969 is considered a recent mover.

206. **Previous residence**--The previous residence is the last housing unit in which the present head lived before moving into the present unit. If the household is in the recent mover category, a series of questions about the previous unit is asked: location, whether he was head there at the time of moving, the number of rooms, the year originally built, the number of units in the structure, tenure, value, disposition, contract monthly rent, main reason for moving, and number of times the head moved since January 1, 1969.

1970 Residential Finance Program

The Residential Finance Survey is designed to provide data about the financing of nonfarm, privately-owned, residential properties. Similar surveys were conducted in 1950 and 1960, and, in a more limited fashion, in 1956.

The Residential Finance program makes a distinction between two types of properties-- homeowner properties, which have from one to four housing units, one of which is occupied by the owner, and all other properties.

Data are collected from the owner of the property, and, if the property is mortgaged, from the holder of the mortgage. The property owner is asked to provide information about the property itself, e.g., number of units, when it was built, when it was acquired, purchase price, current market value. Homeowners are also asked about their housing expenses and about themselves, e.g., age, income, color. Owners of rental property are also asked about some of their expenses and about their rental receipts. All owners are asked if their property is mortgaged, and if so, to whom they make their mortgage payments. The lender is then asked to provide information about the mortgage, e.g., interest rate, face amount, term, current outstanding debt.

The 1970 survey will be based on a national sample of approximately 65,000 properties, of which about half will be home-owner properties. Data will be available for the total United States and for the four Census regions.

In addition, the U.S. data will be presented by size of place and by type of area (e.g., inside central cities of SMSA's).

Tabulation plans are not final, but it is expected that the basic tables will provide all of the information collected by type of mortgage (FHA-insured, VA-guaranteed, or converntional) and by type of mortgage holder (commercial banks, life insurance companies, etc.). In addition, some analytical tables will be provided. Publication is planned for the middle of 1972.

The data compiled in the Residential Finance Survey are particularly useful to economists and financial analysts who guide and counsel home and apartment builders, financial institutions and institutional investors (pension funds, endowments, etc.), producers of building materials, real estate companies, community planners, and governmental agencies at the Federal, State, and local level.

In essence, the 1970 publications will be comparable to those from 1960. Persons desiring detailed information on the 1970 program are advised to address inquiries to the Housing Division, Bureau of the Census, Washington, D.C. 20233.

Residential finance concepts

In general the 1970 concepts and their definitions will be those used in 1960. Anyone needing detailed information should address inquiries to the Housing Division, Bureau of the Census, Washington, D.C. 20233.

INDEX

131

132

134

Bibliography and Index

BIBLIOGRAPHY

        This bibliography contains all documents referred to in
the body of the glossary. Some of the documents do not provide
complete information as to its place of publication, publisher,
or date of publication. Therefore, the designation (no date)
denotes the absence of such information.

<u>Air</u>

    1. Boston, Air Pollution Control Commission. <u>Regulations
            for the Control of Atmospheric Pollution</u>. (no date).

    2. California, Air Resources Board. <u>Air Resources Board
            Glossary of Air Pollution Terms</u>. (no date).

    3. Connecticut, Air Conservation Committee. <u>A Glossary of
            Air Pollution Terms</u>. (no date).

    4. Puget Sound Air Pollution Control Agency. <u>Air Quality
            Impact Analysis Interim Guidelines for Indirect
            Carbon Monoxide Sources</u>. January 1978.

<u>Earth</u>

    1. Pacific Northwest River Basin Commission. "Land and
            Mineral Resources" in <u>Columbia-North Pacific Region
            Framework Study</u>. 1970-1972.

    2. Soil Conservation Society of America. <u>Resource Conserva-
            tion Glossary</u>. Ankeny, Iowa: Soil Conservation
            Society of America, 1976.

    3. Virginia, Soil and Water Conservation Commission. <u>Virgin-
            ia Erosion and Sediment Control Handbook Glossary</u>.
            (no date).

4. Soil Conservation Society of America. <u>Resource Conser-
    vation Glossary</u>. Ankeny, Iowa: Soil Conservation
    Society of America, 1976. (Same as Earth 2).

Energy/Utility

1. American Gas Association. <u>Glossary for the Gas Industry</u>.
    Arlington, Virginia. (no date).

2. California, The Resources Agency, State Water Resources
    Control Board. <u>Glossary of Terms for Water Resources
    Management and Water Quality Control</u>. June 1976.

3. Edison Electric Institute, Statistical Committee. <u>Glos-
    sary of Electric Utility Terms</u>. EEI Publication
    Number 70-40. (no date).

4. Michigan, Department of Agriculture. Letter from Emmanuel
    T. Van Nierop, Environmental Advisor. August 17,
    1978.

5. National Center for Resource Recovery. <u>Glossary of
    Solid Waste Management and Resource Recovery</u>.
    Washington, D.C. (no date).

6. Pacific Northwest River Basins Commission. "Electric
    Power" in <u>Columbia-North Pacific Region Framework
    Study</u>. 1970-1972.

7. Pacific Northwest River Basins Commission. "Municipal
    and Industrial Water Supply" in <u>Columbia-North
    Pacific Region Framework Study</u>. 1970-1972.

8. Pacific Northwest River Basins Commission. "Water Qual-
    ity and Pollution Control" in <u>Columbia-North Pacific
    Region Framework Study</u>. 1970-1972.

9. San Francisco, Department of City Planning, Office of
    Environmental Review. <u>Standard Definitions--Water
    and Wastewater Terms</u>. July 26, 1977.

General

1. U.S. Environmental Protection Agency. <u>Common Environmen-
    tal Terms</u>. Stock Number 1977-0-245-772. Washington,
    D.C.: U.S. Government Printing Office, September
    1977.

2. U.S. General Services Administration, Public Building
   Service, Prepared by Resource Planning Associates,
   Inc., Cambridge, Massachusetts. A Guide for Assess-
   ing Environmental Impacts. March 1977.

3. U.S. Department of Interior, Bureau of Reclamation. En-
   vironmental Glossary with Metric Conversion Tables.
   Stock Number 024-003-00111-2. Washington, D.C.:
   U.S. Government Printing Office, 1977.

4. U.S. Department of the Interior, Bonneville Power Admin-
   istration. The Alumax Environmental Statement:
   Draft Environmental Impact Statement. DES-77-22.
   July 25, 1977.

5. City of Dallas, Department of Urban Planning. The Escarpment Report: Environmental Assessment and Development Guidelines for the White Rock Escarpment. June
   1977.

6. U.S. Department of Commerce, National Oceanic and Atmospheric Administration, Office of Coastal Zone Management, and State of New Jersey, Department of Environmental Protection, Division of Marine Services,
   Office of Coastal Zone Management. New Jersey Coastal Zone Management Program, Bay and Ocean Shore Segment and Draft Environmental Impact Statement. May
   1978.

7. Pacific Northwest Laboratory, prepared for ERDA. Waste-
   water Management Operations, Hanford Reservation,
   Richland, Washington: Final Environmental Impact
   Statement. 1974.

8. City of San Francisco, Department of City Planning, Office
   of Environmental Review. Standard Definitions--Gen-
   eral Definitions. July 18, 1977.

9. U.S. Department of the Army, Office of the Chief of Engineers, prepared by the Institute of Ecology. An
   Ecological Glossary for Engineers and Resource Managers. Pamphlet Number 1105-2-2. November 30, 1973.

10. Appalachian Regional Commission. Guidebook for Regional
    and State Environmental Analysis of Projects in Ap-
    palachia, Volume I: Guidebook and ARC Role, Final
    Report. Contract Number 75-180. August 1977.

11. U.S. Department of Housing and Urban Development, Office
     of Policy Development and Research, prepared by Abt
     Associates, Inc., Cambridge, Massachusetts. <u>Environ-
     mental Considerations in the Comprehensive Planning
     and Management Process</u>. Contract Number H-217R.
     August 1977.

12. Montana, Department of Natural Resources. <u>Yellowstone
     River Water Reservation: Environmental Impact
     Statement</u>. 1976.

13. Pacific Northwest River Basin Commission. <u>Columbia-North
     Pacific Comprehensive Framework Study</u>. 1970-1972.

14. U.S. Department of Commerce, National Oceanic and Atmos-
     pheric Administration, Office of Coastal Zone Man-
     agement, and State of New Jersey, Department of En-
     vironmental Protection, Division of Marine Services,
     Office of Coastal Zone Management. "Definitions"
     in <u>State of New Jersey Coastal Management Program</u>,
     <u>Bay and Ocean Segment: Draft Environmental Impact
     Statement</u>. May 1978.

15. U.S. Department of Agriculture, Forest Service, Pacific
     Southwest Forest and Range Experiment Station. <u>Wild-
     land Planning Glossary</u>. Stock Number 001-001-00413-
     8. Washington, D.C.: U.S. Government Printing
     Office, 1976.

Health

1. U.S. Department of Health, Education and Welfare, Public
    Health Service, National Institutes of Health. <u>Pre-
    venting Environmentally-Related Diseases</u>. DHEW Pub-
    lication Number NH 76-1071.

Legal Jargon

1. Virginia, Council on the Environment. <u>Procedural Manual
    and Guidelines for the EIS Program in the Common-
    wealth of Virginia</u>. June 1978.

2. City of Los Angeles. <u>Guidelines for the Implementation
    of the California Environmental Quality Act of 1970</u>.
    Revised August 1, 1978.

3. State of Hawaii, Environmental Quality Commission. <u>En-
    vironmental Impact Statement Regulations</u>, Sub-Part A.

4. Delaware River Basin Commission. Code of Federal Reg-
    ulations, Part 401, Chapter III. 1977.

5. South Dakota Codified Laws. Chapter 34A-9-1.

6. Commonwealth of Massachusetts, Office of Environmental
    Affairs. Environmental Policy Act. May 19, 1978.

7. Texas, The Interagency Council on Natural Resources and
    the Environment. The Environment, Policy--Guide-
    lines and Procedures for Processing Environmental
    Impact Statements. November 1975.

8. South Dakota, Board of Environmental Protection. Amend-
    ment of Article 34:10 of the Administrative Rules
    of South Dakota. November 17, 1975.

9. Indiana, Environmental Management Board. Regulation
    EMB-5, Environmental Quality Review. 1971.

10. Montana, Commission on Environmental Quality. Uniform
    Rules Implementing the Montana Environmental Policy
    Act. January 15, 1976.

11. U.S. Department of Housing and Urban Development. Hand-
    book of Departmental Policies, Responsibilities and
    Procedures for Protection and Enhancement of Environ-
    mental Quality. Federal Register, Volume 38, Number
    137, July 18, 1973.

12. U.S. Department of Transportation, Federal Aviation Ad-
    ministration. Policies and Procedures for Consid-
    ering Environmental Impacts. Federal Register,
    Volume 42, Number 123, June 27, 1977.

13. U.S. Department of Transportation, Federal Highway Ad-
    ministration. Design Approval and Environmental
    Impact. Federal Register, Volume 39, Number 232,
    December 2, 1974.

14. New Jersey, Department of Environmental Protection.
    Rules and Regulations Under Coastal Area Facility
    Review Act. Docket Number DEP 005-76-03. April 1,
    1977.

15. California. State EIR Guidelines, Guidelines for Imple-
    mentation of the California Environmental Quality
    Act. California Administrative Code, Title 14,
    Natural Resources, Division 6, Resources Agency,
    Chapter 3. March 4, 1978.

16. U.S. General Services Administration. GSA Order, Sub-
    ject: Environmental Considerations in Decision-
    making. ADM 1095.1A. April 27, 1977.

17. U.S. Department of the Interior, Bureau of Land Manage-
    ment. Manual Transmission Sheet, Subject: 1792--
    Environmental Statements. Release 1-1033. March
    15, 1976.

18. U.S. General Services Administration. GSA Order, Sub-
    ject: HB, PBS Preparation of Environmental Impact
    Assessments and Statements. PBS P 1095.4. March
    17, 1978.

19. U.S. Environmental Protection Agency. Preparation of
    Environmental Impact Statements, New Source NPDES
    Permits. Federal Register, Volume 42, Number 7,
    January 11, 1977.

20. Wisconsin, Environmental Impact Program Definitions.
    (no date).

21. Wisconsin, Proposed Repeal and Recreation of Chapter
    NR 150, Wisconsin Administrative Code, Regarding
    Environmental Impact Statement Procedures and
    Preparation Fees. (no date).

22. North Carolina, Council on State Goals and Policy. En-
    vironmental Policy Act, and Environmental Impact
    Statement System. (no date).

23. City of New York. Executive Order Number 91, City En-
    vironmental Quality Review. August 24, 1977.

24. State of New York. Part 617, State Environmental Quality
    Review, Statutory Authority: Environmental Conser-
    vation Law. Section 8-80113. January 24, 1978.

25. Nebraska, Department of Environmental Control. Nebraska
    Environmental Protection Act of 1971.

26. Minnesota, Environmental Quality Council. Environmental
    Impact Statements. State Register, January 24, 1977.

27. Michigan. Guidelines for the Preparation and Review of
    Environmental Impact Statements Under Executive
    Order 1974-4. November 20, 1975.

28. Massachusetts. Regulations Under the Wetlands Protection
    Act. General Laws, Chapter 131, Section 40.

29. U.S. Department of Commerce. Statement on Proposed Federal Actions Affecting the Environment. Department Administrative Order 216-6. November 27, 1974.

30. U.S. Council on Environmental Quality. National Environmental Policy Act, Proposed Regulations for Implementing Procedural Provisions. Federal Register, Volume 43, Number 112, June 9, 1978.

31. U.S. Department of Defense, Corps of Engineers. Environmental Considerations, Proposed Policies and Procedures. Federal Register, Volume 42, Number 36, February 23, 1977.

32. U.S. Department of Energy. Compliance with the National Environmental Policy Act, Proposed Rulemaking: Public Hearing. Federal Register, Volume 43, Number 35, February 21, 1978.

33. U.S. Environmental Protection Agency. Preparation of Environmental Impact Statements, Final Regulations. Federal Register, Volume 40, Number 72, April 14, 1975.

34. U.S. Advisory Council on Historic Preservation. Procedures for the Protection of Historic and Cultural Properties, Establishment of New Chapter and Part. Federal Register, Volume 39, Number 18, January 25, 1974.

35. U.S. Department of Transportation, Urban Mass Transportation Authority. UMTA Environmental Guidelines. Circular UMTA C 5620. (no date).

36. U.S. Water Resources Council. The Utilization of Comprehensive Regional Water Resource Management Plans, Policy Statement Number 4. (no date).

37. State of Washington, Department of Ecology. Guidelines in Interpreting the State Environmental Policy Act. Chapter 197-10 WAC. Revised January 21, 1978.

38. Executive Order 11987. Exotic Organisms. May 24, 1977.

39. Executive Order 11990. Protection of Wetlands. May 24, 1977.

Noise

  1. Boston, Air Pollution Control Commission. <u>Regulations</u>
      <u>for the Control of Noise in the City of Boston</u>.
      (no date).

  2. U.S. National Academy of Sciences, Committee on Hearing,
      Bioacoustics, and Biomechanics, Assembly of Behav-
      orial and Social Sciences, National Research Council.
      <u>Guidelines for Preparing Environmental Impact State-</u>
      <u>ments on Noise, Report of Working Group 69 on Eval-</u>
      <u>uation of Environmental Impact on Noise</u>. 1977.

Plant/Animal

  1. Pacific Northwest River Basins Commission. "Fish and
      Wildlife" in <u>Columbia-North Pacific Region Frame-</u>
      <u>work Study</u>. 1970-1972.

  2. U.S. Department of Commerce, National Oceanic and Atmos-
      pheric Administration, National Marine Fisheries
      Service. <u>Fisheries of the United States</u>. 1977.

Recreation

  1. Pacific Northwest River Basin Commission. "Recreation"
      in <u>Columbia-North Pacific Comprehensive Framework</u>
      <u>Study</u>. 1970-1972.

  2. U.S. Department of the Army, Office of the Chief of
      Engineers. <u>Recreation and Natural Resources: In-</u>
      <u>vestigation and Reporting</u>. Circular Number 1105-
      2-87. July 14, 1978.

  3. U.S. Department of the Interior, Bureau of Outdoor Rec-
      reation. <u>A Glossary of Terms Used by the Bureau</u>
      <u>of Outdoor Recreation</u>. Washington, D.C.: U.S.
      Government Printing Office, September 1975.

Transportation

  1. Georgia, Department of Transportation. <u>Environmental</u>
      <u>Training Course, Terms or the Jargon of EIS's</u>.
      (no date).

2. U.S. Department of Transportation, Federal Highway Administration. <u>Federal-Aid Highway Program Manual,
    Volume 7, Right-of-Way and Environment: Chapter 7,
    Environment, Section 2, Environmental Impact and
    Related Statements</u>. Transmittal 177, HEV-11.
    January 2, 1976.

3. U.S. Department of Transportation, prepared by Skidmore,
    Owings and Merrill. <u>Environmental Assessment Reference Book, A Reference Manual of Environmental
    Assessment Materials Related to Highway Facility
    Improvements</u>. Contract DOT-OS-40175. Stock Number
    050-000-00109-1. Washington, D.C.: U.S. Government
    Printing Office, 1975.

<u>Water</u>

1. American Society of Agricultural Engineers, Nomenclature
    Committee. <u>Glossary, Soil and Water Terms</u>. January
    1967.

2. Delaware River Basin Commission. <u>Water Management of
    the Delaware River Basin, Part One, Chapter 1</u>.
    April 1975.

3. Pacific Northwest River Basins Commission. "Flood Control" in <u>Columbia-North Pacific Region Framework
    Study</u>. 1970-1972.

4. Pacific Northwest River Basins Commission. "Irrigation"
    in <u>Columbia-North Pacific Region Framework Study</u>.
    1970-1972.

5. Pacific Northwest River Basins Commission. "Water Resources" in <u>Columbia-North Pacific Region Framework
    Study</u>. 1970-1972.

6. U.S. Water Resources Council. <u>Floodplain Management
    Guidelines for Implementing E.O. 11988</u>. Federal
    Register, Volume 43, Number 29, February 10, 1978.

INDEX (All Terms Listed Alphabetically)

A